ANSYS Workbench 2023 R2: A Tutorial Approach (6th Edition)

CADCIM Technologies
525 St. Andrews Drive
Schererville, IN 46375, USA
(www.cadcim.com)

Contributing Author
Sham Tickoo
Professor
Department of Mechanical Engineering Technology
Purdue University Northwest
Hammond, Indiana, USA

CADCIM Technologies

ANSYS Workbench 2023 R2: A Tutorial Approach
Sham Tickoo

CADCIM Technologies
525 St Andrews Drive
Schererville, Indiana 46375, USA
www.cadcim.com

ISBN 978-1-64057-232-4

www.cadcim.com

DEDICATION

*To teachers, who make it possible to disseminate knowledge
to enlighten the young and curious minds
of our future generations*

*To students, who are dedicated to learning new technologies
and making the world a better place to live in*

THANKS

*To the faculty and students of the MET department of
Purdue University Northwest for their cooperation*

To employees of CADCIM Technologies for their valuable help

Online Training Program Offered by CADCIM Technologies

CADCIM Technologies provides effective and affordable virtual online training on various software packages related Computer Aided Design, Manufacturing and Engineering (CAD/CAM/CAE), computer programming languages, animation, architecture, and GIS. The training is delivered 'live' via Internet at any time, any place, and at any pace to individuals and the students of colleges, universities, and CAD/CAM training centers. The main features of this program are:

Training for Students and Companies in a Classroom Setting

Highly experienced instructors and qualified engineers at CADCIM Technologies conduct the classes under the guidance of Prof. Sham Tickoo of Purdue University Northwest, USA. This team has authored several textbooks that are rated "one of the best" in their categories and are used in various colleges, universities, and training centers in North America, Europe, and in other parts of the world.

Training for Individuals

CADCIM Technologies with its cost effective and time saving initiative strives to deliver the training in the comfort of your home or work place, thereby relieving you from the hassles of traveling to training centers.

Training Offered on Software Packages

We provide basic and advanced training on the following software packages:

CAD/CAM/CAE: ANSYS Workbench, CATIA, SOLIDWORKS, Autodesk Inventor, Solid Edge, Siemens NX, Creo Parametric, Creo Direct, Autodesk Fusion 360, SOLIDWORKSSimulation, AutoCAD, AutoCAD LT, Customizing AutoCAD, EdgeCAM, and AutoCAD Plant 3D

Computer Programming: C++, VB.NET, Oracle, AJAX, and Java

Animation and Styling: Autodesk 3ds Max, 3ds Max Design, Maya, and Autodesk Alias

Architecture and GIS: Autodesk Revit (Architecture, Structure, MEP), AutoCAD Civil 3D, and AutoCAD, Map 3D

For more information, please visit the following link: https://www.cadcim.com

---

Note
If you are a faculty member, you can register by clicking on the following link to access the teaching resources: ***https://www.cadcim.com/Registration.aspx***. The student resources are available at ***https://www.cadcim.com***. We also provide **Live Virtual Online Training** on various software packages. For more information, write us at ***sales@cadcim.com***.

Table of Contents

Chapter 2: Introduction to ANSYS Workbench

Chapter 3: Part Modeling - 1

Chapter 4: Part Modeling- II

Chapter 5: Part Modeling- III

Chapter 6: Defining Material Properties

Chapter 7: Generating Mesh- I

Chapter 8: Generating Mesh - II

Chapter 11: Thermal Analysis

Preface

ANSYS Workbench 2023 R2

ANSYS, a product of ANSYS Inc., is a world's leading, widely distributed, and popular commercial CAE package. It is widely used by designers/analysts in industries such as aerospace, automotive, manufacturing, nuclear, electronics, biomedical, and many more. ANSYS provides simulation solution that enables designers to simulate design performance directly on the desktop. In this way, it provides fast, efficient, and cost-effective product developement from design concept stage to performance validation stage of the product developement cycle. It helps accelerate and streamline the product developement process by helping designers toresolve issues related to structural , thermal, fluid, flow, electrimagnetic effects, a combination of these phenomena acting together, and so on.

ANSYS Workbench 2023 R2: A Tutorial Approach textbook has been written with the intention to assist engineering and practicing designers who are new to the field of FEM. The textbook covers the basis of FEA concepts, modeling, and the analysis of engineering problems using ANSYS Workbench. In addition, the description of the latest tools introduced, the enhancement, and new tutorials based on new and enhanced tools are provided so that the users learn and understand their usage properly and effectively. This textbook covers the following simulation streams of ANSYS:

1. Structural Analysis
   Static Structural Analysis
   Vibration Analysis
2. Thermal Analysis
   Steady State Thermal Analysis
   Transient Thermal Analysis
   Thermal Stress Analysis

The main features of the textbook are as follows:

- **Tutorial Approach**

  The author has adopted the tutorial point-of-view and learn-by-doing approach throughout the textbook. This approach helps the users learn the concepts faster and apply them effectively and efficiently. Sufficient theoretical explanation has been provided during the tutorial whenever required.

- **Real-World Projects as Tutorials**

  The author has used about 30 real-world mechanical engineering projects as tutorials in this book. This will enable the readers to relate the tutorials to the real-world models in the mechanical engineering industry. In addition, there are about 15 exercises based on the real-world mechanical engineering projects.

- **Tips and Notes**
  The additional information related to various topics is provided to the users in the form of tips and notes.

- **Learning Objectives**
  The first page of every chapter summarizes the topics that are covered in that chapter.

- **Self-Evaluation Test, Review Questions, and Exercises**
  Every chapter ends with Self-Evaluation Test so that the users can assess their knowledge of the chapter. The answers to Self-Evaluation Test are given at the end of the chapter. Also, Review Questions and Exercises are given at the end of the chapters and they can be used by instructors as test questions and exercises.

Symbols Used in the Textbook

Note
The author has provided additional information to the users about the topic being discussed in the form of notes.

Tip
Special information and techniques are provided in the form of tips that will increase the efficiency of the users.

Formatting Conventions Used in the Textbook
Refer to the following list for the formatting conventions used in this textbook.

- Names of tools, buttons, options, tabs, toolbars, and windows are written in boldface.

  Example: The **Extrude** tool, the **Save** button, the **Toolbox** window, the **Graph** tab, and so on.

- Names of Details windows, drop-downs, drop-down lists, edit boxes, selection boxes, areas, check boxes, dialog boxes and radio buttons are written in boldface.

  Example: The Details of "Revolve" window, the **Geometry** selection box, the **Blend** drop-down of the **Features** toolbar; the **OK** button of the **ANSYS Workbench** dialog box, the **millimeter** radio button of the **ANSYS Workbench** dialog box, and so on.

- Values entered in edit boxes are written in boldface.

  Example: Enter **5** in the **Max Element Size** edit box.

- Names and paths of the files are written in italics.

  Example: C:\ANSYS_WB\c03\Tut01\ c03_ansWB_tut02, and so on

Naming Conventions Used in the Textbook

Tool

If a command is invoked on clicking an item, then that item is termed as tool.

For example:

To Create: Line tool, **General** tool, **Extrude** tool, **Pattern** tool, and so on.

To Generate: **General** tool, **Horizontal** tool, **Vertical** tool, and so on.

To Edit: **Fillet** tool, **Extend** tool, **Replicate** tool, and so on.

Action: **Rotate** tool, **Pan** tool, **Box Zoom** tool.

If on clicking an item, corresponding **Details View** window is displayed just below the Tree Outline, wherein you can set the parameters to create/edit an object, then that item is also termed as tool, refer to Figure 1.

For example:

To Create: **Revolve** tool, **Skin/Loft** tool

To Edit: **Slice** tool, **Chamfer** tool

Figure 1 Partial view of the toolbars having different tools

Button

The item in a dialog box that has a 3d shape like a button is termed as Button. For example, **OK** button, **Cancel** button, **Apply** button, and so on.

Drop-down

A drop-down is the one in which a set of common tools are grouped together. You can identify a drop-down with a down arrow on it. These drop-downs are given a name based on the tools grouped in them. For example, **Blend** drop-down, **Mesh** drop-down, **Mesh Control** drop-down, **Support** drop-down, and so on; refer to Figure 2.

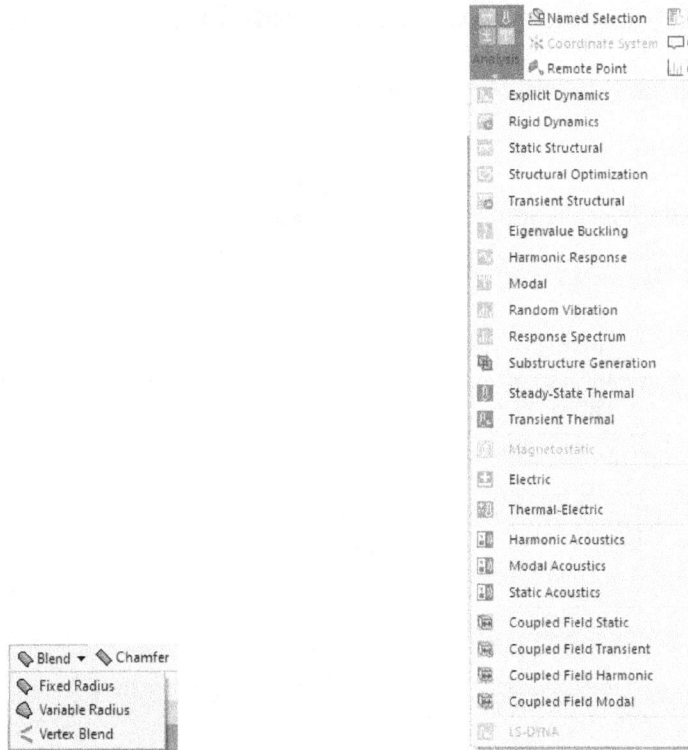

*Figure 2 The **Blend** and **Analysis** drop-downs*

Drop-down List

A drop-down list is the one in which a set of options are grouped together. You can set values for various parameters using these options. You can identify a drop-down list with a down arrow on it. For example, **Extents** drop-down list, **Color Override** drop-down list, and so on; refer to Figure 3.

*Figure 3 The **Physics Preference** and **Mesh Metric** drop-down lists*

Options

Options are the items that are available in shortcut menu, Marking Menu, drop-down list, dialog boxes, and so on. For example, choose the **Select All** option from the shortcut menu displayed on right-clicking in the Graphics screen; choose the Concrete option from the **Assignment** flyout; choose the **Front** option from the **Orientation** area, refer to Figure 4.

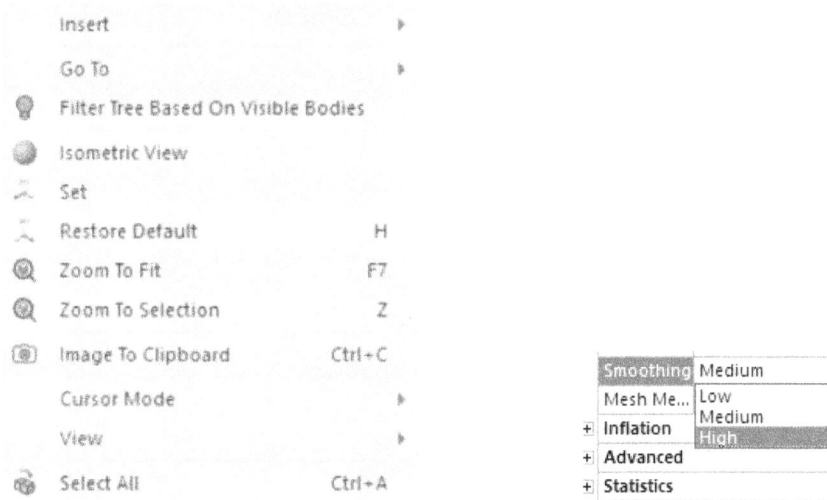

Figure 4 *Options in the shortcut menu and the* **Smoothing** *drop-down list*

Selection Box

Many operations in ANSYS Workbench require you to select entities in the graphics screen or from the **Outline** pane. After you select the entities/features, you need to confirm the selection in the selection box. For example, if you want to extrude a sketch, you need to select the sketch and then confirm the selection in the selection box. A typical **Geometry** selection box is shown in Figure 5.

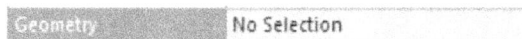

Figure 5 *The* **Geometry** *selection box*

Free Companion Website

It has been our constant endeavor to provide you the best textbooks and services at affordable prices. The free Companion website provides access to all the teaching and learning resources that are required during the course of this textbook. If you purchase this textbook, you can access the resources on the Companion website.

The following resources are available for the faculty and students in this website:

Faculty Resources

• **Technical Support**
   You can get online technical support by contacting **techsupport@cadcim.com.**

• **Instructor Guide**
   Solutions to all review questions and exercises in the textbook are provided in this guide to help the faculty members test the skills of the students.

- **Input Files**

  The input hies used in exercises are available for free download.

Student Resources

- **Technical Support**

  You can get online technical support by contacting ***techsupport@cadcim.com***.

- **Part Files**

  The part files used in illustrations and examples are available for free download.

Note that you can access the faculty resources only if you are registered as faculty at ***www.cadcim.com/Registration.aspx***

If you face any problem in accessing these files, please contact the publisher at ***sales@cadcim.com*** or the author at ***stickoo@pnw.edu*** or ***tickoo525@gmail.com***.

Video Courses

CADCIM offers video courses in CAD, CAE Simulation, BIM, Civil/GIS, and Animation domains on various e-Learning/Video platforms. To enroll for the video courses, please visit the CADCIM website using the link **https://www.cadcim.com/video-courses.**

Stay Connected

You can now stay connected with us through Facebook and Twitter to get the latest information about our text books, videos, and teaching/learning resources. To stay informed of such updates, follow us on Facebook *(www.facebook.com/cadcim)* and Twitter *(@cadcimtech)*. You can also subscribe to our You Tube channel *(www.youtube.com/cadcimtech)* to get the information about our latest video tutorials.

Chapter *1*

Introduction to FEA

Learning Objectives

After completing this chapter, you will be able to:
- *Understand the different design validation techniques*
- *Understand the basic concepts and general working of FEA*
- *Understand the types of elements*
- *Understand the advantages, limitations, and applications of FEA*
- *Understand the types of analysis*
- *Understand important terms and definitions in FEA*

DESIGN VALIDATION TECHNIQUES

Design validation of a component refers to the sustainability of that component under various loading conditions. In other words, design validation is a process to find the results such as stress, displacement, strain, fatigue life, eigenvalues, heat flux, and so on which leads failure of a component.

There are three methods to validate any design:
1. Analytical method
2. Numerical method
3. Experimental method

All these methods are discussed next:

1. Analytical Method

An analytical method is a classical approach that involves solution techniques based on formulas and theorems such as bending theory, torsion theory, failure theories, and so on. This is a widely used method in curriculum research. It is a closed-form solution method that gives 100 % accurate results. Most of the time, solutions have been obtained for very trivial problems such as cantilever and simply supported beams. This method is mainly applicable only for isotropic materials such as glass and metals.

2. Numerical Method

The numerical method is used to determine a numerical solution by satisfying the governing equations and boundary conditions for the most complex engineering problems. In this method, real-life complex problems can be modeled mathematically. However, several assumptions need to be considered while simulating the problems. Therefore, this method always provides the approximate results to the given problem. This method does not require physical prototypes or models to calculate the response of the models. The numerical method can handle the problem with anisotropic materials such as woods and composites. There are numerous numerical methods available such as Finite Element Method (FEM), Finite Volume Method (FVM), Boundary Element Method (BEM), Finite Difference Method (FDM), and so on. Now a days, these numerical methods can be referred to as computational methods where mathematical models can be written in software codes.

3. Experimental Method

In the experimental method, the model is tested physically and actual measurements are carried out in real-time operating conditions. This method is highly reliable and hence used in product prototype testing in the industry. It is possible to accommodate all types of physical and manufacturing errors such as surface finish, surface heat treatment, alloying elements, decarburizing, actual welding pattern, and so on in the testing. In order to carry out physical real-time testing, a prototype of the product must be available. For reliable outcomes, 3 to 5 prototypes must be required to test. This iterative process of testing makes the experimental method more time-consuming and requires an expensive experimental setup. There are various types of equipments are available such as strain gauges, photoelasticity measuring setup, vibrometers, fatigue test sensors for temperature & pressure measurements, and so on.

The fundamental concepts of Finite Element Method (FEM) also referred as Finite Element Analysis (FEA), which is from numerical method category, have been described next.

INTRODUCTION TO FEA

The Finite element analysis (FEA) is a computing technique that is used to obtain approximate solutions to maximum engineering boundary value problems. It uses a numerical method called finite element method (FEM). FEA involves the computer model of a design that is loaded and analyzed for specific results, such as stress, deformation, deflection, natural frequencies, mode shapes, temperature distributions, and so on.

The concept of FEA can be explained through a basic example involving measurement of the perimeter of a circle. To measure the perimeter of a circle without using the conventional formula, divide the circle into equal segments, as shown in Figure 1-1. Next, join the start point and the endpoint of each of these segments by a straight line. Now, you can measure the length of straight line very easily, and thus, the perimeter of the circle by adding the length of these straight lines.

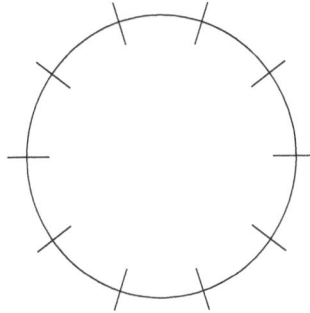

Figure 1-1 *The circle divided into small equal segments*

If you divide the circle into four segments only, you will not get accurate results. For accuracy, divide the circle into more number of segments. However, with more segments, the time required for getting the accurate result will also increase. The same concept can be applied to FEA also, and therefore, there is always a compromise between accuracy and speed while using this method. This compromise between accuracy and speed makes it an approximate method.

The FEA was first developed to be used in the aerospace and nuclear industries, where the safety of structures is critical. Today, even the simplest of products rely on FEA for design evaluation.

The term Finite Element Method has been described next.

Finite

All real-life objects have infinite degrees of freedom and thus solving infinite degrees of freedom problems is very difficult. The Finite Element Method (FEM) reduces the infinite degrees of freedom of a continuous domain into the finite discrete domain with the help of the meshing technique.

Element

In the finite element analysis, all unknowns are calculated on their values at the limited number of points. These points are called nodes. The entity connecting nodes and forming a particular shape such as quadrilateral, triangular, tetrahedral, hexahedral, and so on is known as an element. This particular shape of an element is used to define the unknown field variables that predict the individual element response to applied loads. The assembly of the responses of all elements in a domain determines the total response of the complete domain. To calculate the value of a field variable (say displacement) at any point other than nodes, an interpolation function is used. This interpolation function is called shape function and the value of this interpolation function may vary with the predefined shapes of the element.

Method

There are three methods to solve any engineering problem. Finite element analysis belongs to the numerical method category.

General Working of FEA

A better knowledge of FEA helps in building more accurate models. Also, it helps in understanding the back-end working of ANSYS. Here, a simple model is discussed to give you a brief overview of the working of FEA.

Figure 1-2 shows a spring assembly that represents a simple two-spring element model. In this model, two springs are connected in series and one of the springs is fixed at the left most endpoint, refer to Figure 1-2. In this figure, the stiffness of the springs has been represented by the spring constants K_1 and K_2. The movement of endpoints of each spring is restricted to the X direction only. The change in position from the undeformed state of each endpoint can be defined by the variables X_1 and X_2. The two forces acting on the end points of the springs are represented by F_1 and F_2.

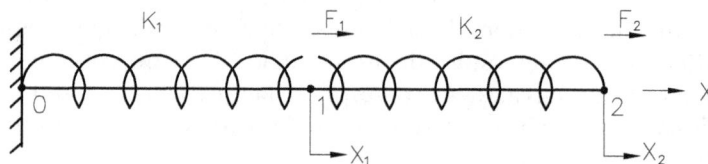

Figure 1-2 Representation of a two-spring assembly

To develop a model that can predict the state of this spring assembly, you can use the linear spring equation given below:

$$F = KX$$

where,
F = force applied,
X = displacement, and
K = spring constant

If you use the spring parameters defined above and assume a state of equilibrium, the following equations can be written for the state of each endpoint:

$$F_1 - X_1 K_1 + (X_2 - X_1)K_2 = 0$$
$$F_2 - (X_2 - X_1)K_2 = 0$$

Therefore,

$$F_1 = (K_1 + K_2)X_1 + (-K_2)X_2$$
$$F_2 = (-K_2)X_1 + K_2 X_2$$

If the set of equation is written in matrix form, it will be represented as follows:

$$\begin{bmatrix} K_1 + K_2 & -K_2 \\ -K_2 & K_2 \end{bmatrix} \begin{bmatrix} X_1 \\ X_2 \end{bmatrix} = \begin{bmatrix} F_1 \\ F_2 \end{bmatrix}$$

In the above mathematical model, if the spring constants (K_1 and K_2) are known and the deformed shapes (X_1 and X_2) are defined, then the resulting forces (F_1 and F_2) can be determined. Alternatively, if the spring constants (K_1 and K_2) are known and the forces (F_1 and F_2) are defined, then the resulting deformed shape (X_1 and X_2) can be determined.

Various terminologies that are used in the previous example are discussed next.

Stiffness Matrix

In the previous equation, the following part represents the stiffness matrix (K):

$$\begin{bmatrix} K_1 + K_2 & -K_2 \\ -K_2 & K_2 \end{bmatrix}$$

This matrix is relatively simple because it comprises only one pair of springs, but it turns complex when the number of springs increases.

Degree of Freedom

Degree of freedom is defined as the least number of independent coordinates required to define the configuration of a system in space. In the previous example, you are only concerned with the displacement and forces. By making one endpoint fixed, you will restrict all degrees of freedom for that particular node. Which means that, there will be no translational or rotational degrees of freedom for that node. But, there are two nodes still have some degrees of freedom. As these two nodes are allowed to translate along the X axis only, they have 1 degree of freedom each considering that no rotational degree of freedom exist in them. The number of the degrees of freedom on free nodes in a model determines the number of equations required to solve a mathematical model.

Boundary Conditions

The boundary conditions are used to eliminate the unknowns in the system. A set of equations that is solvable is meaningless without the input. In the previous example, the boundary condition

$X_0 = 0$, and the input forces are F1 and F2. In either ways, the displacements could have been specified in place of forces as boundary conditions and the mathematical model could have been solved for the forces. In other words, the boundary conditions help you reduce or eliminate the unknowns in the system.

Note

The solutions generated by using FEA are always approximate.

Types of Element

Before proceeding further, you must be familiar with the concepts of element shapes which are the building blocks of FEA. These concepts are discussed next.

Element is an entity into which the system under study is divided. An element shape is specified by nodes. The shape (area, length, and volume) of an element depends on the nodes with which it is made. Based on the shapes elements can be classified as below.

Line/(1D) Element

A line element, also called 1D element, has the shape of a line or a curve. Therefore, a minimum of two nodes are required to define it. There can be higher order elements that have additional nodes (at the middle of the edge of an element). An element that does not have a node in between its edges is called a linear element. The elements that have nodes in between edges are called quadratic or second order elements. Figure 1-3 shows some line elements. There are some practical applications such as beams, columns, long shafts, trusses, connection elements, and so on that can be modeled in a 1-D element. The change in material properties along the cross-section of a model is assumed to be negligible.

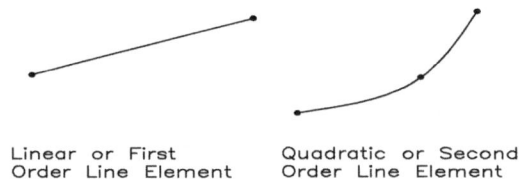

Linear or First
Order Line Element

Quadratic or Second
Order Line Element

Figure 1-3 Line elements

Surface/(2D) Element

A surface or 2D element has the shape of a triangle or a quadrilateral; therefore, it requires a minimum of three or four nodes to define it. These surface elements are also called as shell elements. Some surface elements are shown in Figure 1-4. There are some practical applications such as sheet metal parts, sheet metal cabinets, engine pallets, and so on that can be modeled in 2-D elements. The change in material properties along the thickness is assumed to be negligible.

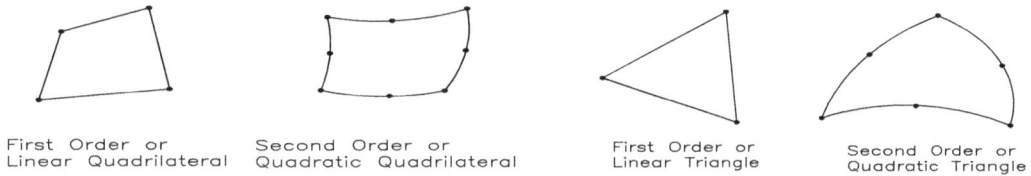

Figure 1-4 The surface elements

Volume/(3D) Element

A volume element has the shape of a hexahedron (8 nodes), a wedge (6 nodes), a tetrahedron (4 nodes), or a pyramid (5 nodes). Some of the volume elements are shown in Figure 1-5. These volume elements are also called solid elements. There are some applications such as gearbox, engine cylinder block, crankshaft, and so on that can be modeled in 3-D elements.

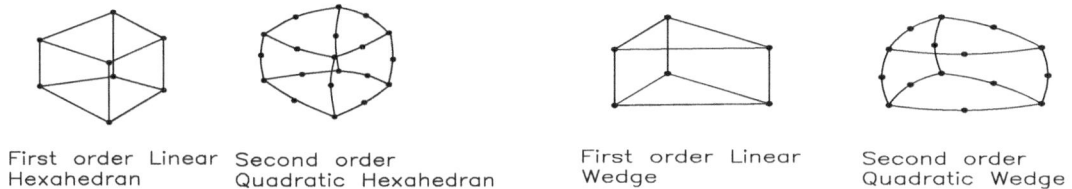

First order Linear Hexahedran   Second order Quadratic Hexahedran   First order Linear Wedge   Second order Quadratic Wedge

Figure 1-5 The volume elements

General Procedure to Conduct Finite Element Analysis

To conduct the finite element analysis, you need to follow certain steps that are given next.
1. Set the type of analysis to be used.
2. Create model.
3. Define the element type.
4. Divide the given geometry into nodes and elements (mesh the model).
5. Apply material properties and boundary conditions.
6. Derive element matrices and equations.
7. Assemble element equations.
8. Solve the unknown parameters at nodes.
9. Interpret the results.

The general process of FEA by using software is divided into three main phases: preprocessing, solution, and postprocessing, refer to Figure 1-6.

Preprocessor

The preprocessor is a phase that processes input data to produce output, which is used as input in the subsequent phase (solution). Following are the input data that need to be given to the preprocessor:

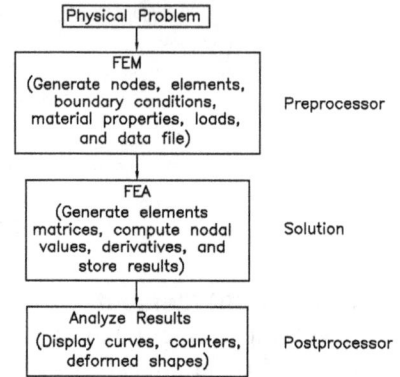

Figure 1-6 Flow diagram of FEA process through software

1. Type of analysis (structural or thermal, static or dynamic, and linear or nonlinear).
2. Element type.
3. Real constants for elements (Cross-sectional area, Moment of Inertia, Shell thickness, and so on).
4. Material Model (Homogeneous, Isotropic, and Anisotropic) and Material properties (Young's Modulus, Poisson's ratio, Spring Constant, Thermal Conductivity, Coefficient of Thermal Expansion, and so on).
5. Geometric model (either created in the FEA software or imported from other CAD packages).
6. FEA model (discretizing the geometric model into small elements).
7. Loading and boundary conditions (defining loads, pressures, moments, temperature, conductivity, convection, constraints (fixed, pinned, or frictionless/symmetrical), and so on.

The input data are preprocessed for the output data and the preprocessor generates the data files automatically with the help of users. These data files are used in the subsequent phase (solution), refer to Figure 1-6.

Solution

The solution phase is completely automatic. The FEA software generates element matrices, computes nodal values and derivatives, and stores the result data in files. These files are further used in the subsequent phase (postprocessor) to review and analyze the results through the graphic display and tabular listings, refer to Figure 1-6.

Postprocessor

The output from the solution phase (result data files) is in the numerical form and consists of nodal values of the field variable and its derivatives. For example, in structural analysis, the output of the postprocessor is nodal displacement and stress in elements. The postprocessor processes the result data and displays them in graphical form to check or analyze the result. The graphical output gives the detailed information about the required result data. The postprocessor phase is automatic and generates graphical output in the specified form, refer to Figure 1-6.

Coordinate Systems

There are three types of commonly used coordinate systems and they are discussed next.

Global Coordinate System

The global coordinate system is used to define the coordinates of points with respect to the single coordinate system in the entire domain under consideration. It is also referred to as Cartesian Coordinate System.

Local Coordinate System

The local coordinate system is used to define elements with respect to the individual coordinate system. Every single element has its own coordinate system and all the corresponding nodes of the element are specified by using the respective local coordinate system.

Natural Coordinate System

The natural coordinate system is used to define the point within the element by a set of dimensionless numbers, whose magnitude varies from -1 to +1. Natural coordinates are defined with respect to the element rather than with reference to the global coordinates. Also, they are dimensionless quantities.

FEA SOFTWARE

There are a variety of commercial FEA software packages available in the market. Every CAE software provides various modules for various analysis requirements. Depending on your requirement, you can select a required module for your analysis. Some firms use one or more CAE software and others develop customized version of commercial software to meet their requirements.

Since 1970s, some well-known commercial FE codes, such as ANSYS, NASTRAN, MARC, ABAQUS, LSDYNA, COMSOL, Radioss, and OptiStruct have been developed to solve the structural problems. Among them, ANSYS software has the most powerful nonlinear solver, and hence it has become the most widely used software in both academia and industry.

Advantages and Limitations of FEA Software

Following are some of the advantages and limitations of FEA software:

Advantages

1. It reduces the amount of prototype testing, thereby saving the cost and time.
2. It gives the graphical representation of the result of analysis.
3. The finite element modeling and analysis are performed in the preprocessor and solution phases, which if done manually would consume a lot of time and in some cases, might be impossible to perform.
4. Variables such as stress and temperature can be measured at any desired point of the model.
5. It helps optimize a design.
6. It is used to simulate the designs that are not suitable for prototype testing.
7. It helps you create more reliable, high quality, and competitive designs.

Limitations

1. It does not provide exact solutions.
2. FEA packages are costly.

3. An inexperienced user can deliver incorrect answers, upon which expensive decisions will be based.
4. Results give solutions but not remedies.
5. Features such as bolts, welded joints, and so on cannot be accommodated to a model. This may lead to approximation and errors in the result.
6. For more accurate results, more hard disk space, RAM, and time are required.

KEY ASSUMPTIONS IN FEA

There are four types of key assumptions that must be considered while performing the finite element analysis. These assumptions are not comprehensive but cover a wide variety of situations applicable to the problem. Moreover, by no means do all the following assumptions apply to all situations. Therefore, you need to consider only those assumptions that are applicable for your analysis problem.

Assumptions Related to Geometry

1. Displacement values will be small so that a linear solution is valid.
2. Stress behavior outside the area of interest is not important. Therefore, geometric simplifications in those areas do not affect the outcome.
3. Only internal fillets in the area of interest will be included in the solution.
4. Local behavior at the corners, joints, and intersection of geometries is of primary interest, therefore, no special modeling of these areas is required.
5. Decorative external features will be assumed insignificant for the stiffness and performance of the part and these external features will be omitted from the model.
6. Variation in the mass due to suppressed features is negligible.

Assumptions Related to Material Properties

1. Material properties will remain in the linear region and the nonlinear behavior of the material property cannot be accepted.
2. Material properties are not affected by the load rate.
3. The component is free from surface imperfections that can produce stress concentration.
4. All simulations will assume room temperature, unless otherwise specified.
5. The effects of relative humidity or water absorption on the material used will be neglected.
6. No compensation will be made to account for the effect of chemicals, corrosives, wears, or other factors that may have an impact on the long term structural integrity.

Assumptions Related to Boundary Conditions

1. Displacements will be small so that the magnitude, orientation, and distribution of the load remains constant throughout the process of deformation.
2. Frictional loss in the system is considered to be negligible.
3. All interfacing components will be assumed rigid.
4. The portion of the structure being studied is assumed as a separate part from the rest of the system, so that any reaction or input from adjacent features is neglected.

Assumptions Related to Fasteners

1. Residual stresses due to fabrication, pre loading on bolts, welding, or other manufacturing or assembly processes will be neglected.

2. All welds between components will be considered as ideal and continuous.
3. The failure of fasteners will not be considered.
4. The load on the threaded portion of the part is supposed to be evenly distributed among the engaged threads.
5. The stiffness of bearings, both in radial and in axial directions, will be considered as infinite or rigid.

APPLICATIONS OF FEA

The Finite Element Analysis (FEA) is a powerful numerical technique used to solve boundary-value problems in a wider range of engineering applications. Some of the major advancements in the field are discussed next.

Automobile Applications

In a vehicle having monocoque construction, the body panels, which are connected to the suspension, are subjected to transient type road loads. Hence, structural stress, strain analysis of these body panels are of interest. An automotive engine cylinder block experiences severe pressure, temperature gradients, and other transient loads. Therefore, it is essential to predict accurately the level of stresses and vibrations to predict the system vibrational characteristics such as natural frequencies and mode shapes. The same FE analysis is also used for crash analysis to determine the crashworthiness of the vehicle body.

Manufacturing Process Applications

A 3-D thermo-mechanical FE simulation can be carried out to study the solidification, thermal field, evaluation of stress, and the cause for failure. This evaluated result information is further used to optimize the processing parameters to reduce high stresses concentration. However, other parameters such as surface finish, surface heat treatment, alloying elements, decarburizing, actual welding pattern, and so on are not considered in the analysis.

Electromagnetics Applications

For the complex configuration of electrodes and dielectric insulating materials, analytical formulations are inaccurate and extremely difficult. Electromagnetics applications such as antennas, phased arrays, electric machines, high-frequency circuits, and crystal photonics are solved by using the FEA tool. FEA can be used for reliability enhancement and optimization of insulation design in high voltage equipment by finding accurately the voltage stresses and corresponding withstands.

Aerospace Applications

In typical aerospace applications, finite element analysis can be effectively used in structural analysis for stresses, strains, deformations, frequencies, mode shapes, impact analysis, aerodynamics, and crash analysis.

TYPES OF ENGINEERING ANALYSES

You can perform different types of analyses using FEA software and these are discussed next.

Structural Analysis

In structural analysis, first the nodal degrees of freedom (displacement) are calculated and then the stress, strains, and reaction forces are calculated from nodal displacements. The classification of structural analysis is shown in Figure 1-7.

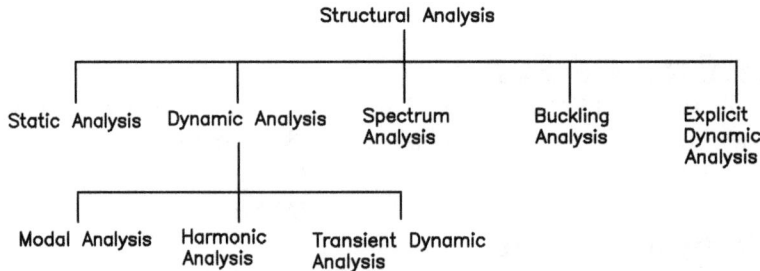

Figure 1-7 Types of structural analysis

Static Analysis

In static analysis, the load or field conditions do not vary with respect to time, and therefore, it is assumed that the load or field conditions are applied gradually, not suddenly. The system under this analysis can be linear or nonlinear. The inertia and damping effects are ignored in structural analysis. In structural analysis, the following matrices are solved:

$$[K] \times [X] = [F]$$

Where,
$$K = \text{Stiffness Matrix}$$
$$X = \text{Displacement Matrix}$$
$$F = \text{Load Matrix}$$

The above equation is called the force balance equation for the linear system. If the elements of matrix [K] are the function of [X], the system is known as the nonlinear system. Nonlinear systems include large deformation, plasticity, creep, and so on. The loads that can be applied in a static analysis include:

1. Externally applied forces and pressures
2. Steady-state inertial forces (such as gravity or rotational velocity)
3. Imposed (non-zero) displacements
4. Temperatures (for thermal strain)
5. Fluences (for nuclear swelling)

The outputs that can be expected from the FEA software are given next.

1. Displacements
2. Strains
3. Stresses
4. Reaction forces

Dynamic Analysis

In dynamic analysis, the load or field conditions vary with the time and are applied suddenly. The system can be linear or nonlinear. The dynamic load includes oscillating loads, impacts, collisions, and random loads. The dynamic analysis is classified into the following three main categories:

Modal Analysis

It is used to calculate the natural frequency and mode shape of a structure.

Harmonic Analysis

It is used to calculate the response of a structure to harmonically time varying loads.

Transient Dynamic Analysis

It is used to calculate the response of a structure to arbitrary time varying loads.

In dynamic analysis, the following matrices are solved:

For the system without any external load:

$$[M] \times \text{Double Derivative of } [X] + [K] \times [X] = 0$$

Where,

M = Mass Matrix
K = Stiffness Matrix
X = Displacement Matrix

For the system with external load:

$$[M] \times \text{Double Derivative of } [X] + [K] \times [X] = [F]$$

Where,

K = Stiffness Matrix
X = Displacement Matrix
F = Load Matrix

The above equations are called the force balance equations for a dynamic system. By solving the above set of equations, you can extract the natural frequencies of a system. The load types applied in a dynamic analysis are the same as that in a static analysis. The outputs that can be expected from a software are natural frequencies, mode shapes, displacements, strains, stresses, and reaction forces. All these outputs can also be obtained with respect to time.

Spectrum Analysis

This is an extension of the modal analysis and is used to calculate stress and strain due to the response of the spectrum (random vibrations). For example, you can use it to analyze how well a structure will perform and survive in an earthquake.

Buckling Analysis

This type of analysis is used to calculate the buckling load and the buckling mode shape. Slender structures (that is thin and long structures) when loaded in the axial direction, buckle under relatively small loads. For such structures, the buckling load becomes a critical design factor.

Explicit Dynamic Analysis

This type of structural analysis is available only in the ANSYS LS-Dyna program and is used to get fast solutions for large deformation dynamics and complex contact problems, for example, explosions, aircraft crash worthiness, and so on.

Thermal Analysis

The thermal analysis is used to determine the temperature distribution and related thermal quantities such as: Thermal distribution, amount of heat loss or gain, thermal gradients, and thermal fluxes.

All primary heat transfer modes such as conduction, convection, and radiation can be simulated. You can perform two types of thermal analysis, steady-state and transient.

Steady State Thermal Analysis

In this analysis, the system is studied under steady thermal loads with respect to time.

Transient Thermal Analysis

In this analysis, the system is studied under varying thermal loads with respect to time.

Fluid Flow Analysis

This analysis is used to determine the flow distribution and temperature of a fluid. The ANSYS/ FLOTRAN program is used to simulate the laminar and turbulent flow, compressible and electronic packaging, automotive design, and so on. The outputs that can be expected from the fluid flow analysis are velocities, pressures, temperatures, and film coefficients.

Electromagnetic Field Analysis

This type of analysis is conducted to determine the magnetic fields in electromagnetic devices. The types of electromagnetic analyses are static analysis, harmonic analysis, and transient analysis.

Coupled Field Analysis

This type of analysis considers the mutual interaction between two or more fields. It is impossible to solve fields separately because they are interdependent. Therefore, you need a program that can solve both the problems by combining them.

For example, if a component is exposed to heat, you may first require to study the thermal characteristics of the component and then the effect of the thermal heating on the structural stability.

Alternatively, if a component is bent in different shapes using one of the metal forming processes and then subjected to heating, the thermal characteristics of the component will depend on the new shape of the component. Therefore, first the shape of the component has to be predicted through structural simulations. This is called as the coupled field analysis.

IMPORTANT TERMS AND DEFINITIONS

Some of the important terms and definitions used in FEA are discussed next.

Strength

When a material is subjected to an external load, the system undergoes a deformation. The material, in turn, offers resistance against this deformation. This resistance is offered by virtue of the strength of the material.

Load

The external force acting on a body is called load.

Stress

The force of resistance offered by a body against the deformation is called stress. The stress is induced in the body while the load is being applied on the body. The stress is calculated as load per unit area.

$$p = F/A$$

Where,

p = Stress in N/mm^2
F = Applied Force in Newton
A = Cross-Sectional Area in mm^2

The material can undergo various types of stresses, which are discussed next.

Tensile Stress

If the resistance offered by a body is against the increase in the length, the body is said to be under tensile stress.

Compressive Stress

If the resistance offered by a body is against the decrease in the length, the body is said to be under compressive stress. Compressive stress is just the reverse of tensile stress.

Shear Stress

Shear stress is the resistance offered when applied external load is tangential to the cross-section of the body. The cross-sectional plane on which shear stress exists is called shear plane.

$$\textbf{Shear Stress} = \textbf{Shear resistance (R) / Shear area (A)}$$

Strain

When a body is subjected to a load (force), its dimension changes. The ratio of change in the dimension of the body to its original dimension is called strain. If the body returns to its original shape on removing the load, the strain is called elastic strain. If the body remains distorted after removing the load, the strain is called plastic strain. The strain can be of two types, normal strain (tensile, compressive, and volumetric) and shear strain.

Strain (e) = Change in dimension / Original dimension

Elastic Limit

The maximum stress that a material can withstand without experiencing any permanent deformation is known as the elastic limit of the material. If the stress is within the elastic limit, the material returns to its original shape and dimension on removal of the stress.

Hooke's Law

It states that the stress is directly proportional to the strain within the elastic limit.

Stress / Strain = Constant (within the elastic limit)

Young's Modulus or Modulus of Elasticity

In case of axial loading, the ratio of intensity of the tensile or compressive stress to the corresponding strain is constant. This ratio is called Young's modulus, and is denoted by E.

E = p/e

where,
p = Stress in N/mm^2
e = Strain (Dimensionless quantity)

Shear Modulus or Modulus of Rigidity

In case of shear loading, the ratio of shear stress to the corresponding shear strain is constant. This ratio is called Shear modulus, and it is denoted by C, N, or G.

Ultimate Strength

The maximum stress that a material withstands without breaking when subjected to an applied load is called its ultimate strength.

Factor of Safety

The ratio of the ultimate strength to the estimated maximum stress in ordinary use (design stress) is known as factor of safety. It is necessary that the design stress is well below the elastic limit, and to achieve this condition, the ultimate stress should be divided by a 'factor of safety'.

Lateral Strain

If a cylindrical rod is subjected to an axial tensile load, the length (l) of the rod will increase (dl) and the diameter (Ø) of the rod will decrease (dØ). In short, the longitudinal stress will not only produce a strain in its own direction, but will cause deformation in a direction perpendicular to the applied load. The strain caused due to this deformation is called lateral strain. The ratio dl/l is called the longitudinal strain or the linear strain, and the ratio dØ$/$Ø is called the lateral strain.

Poisson's Ratio

The ratio of the lateral strain to the longitudinal strain is constant within the elastic limit. This ratio is called the Poisson's ratio and is denoted by *1/m*. For most of the metals, the value of the '*m*' lies between 3 and 4.

Poisson's ratio = Lateral Strain / Longitudinal Strain = 1/m

Bulk Modulus

If a body is subjected to equal stresses along the three mutually perpendicular directions, the ratio of the direct stresses to the corresponding volumetric strain is found to be constant for a given material, when the deformation is within a certain limit. This ratio is called the bulk modulus and is denoted by K.

Stress Concentration

The value of stress changes abruptly in the regions where the cross-section or profile of a structural member changes abruptly. The phenomenon of this abrupt change in stress is known as stress concentration and the region of the structural member that is affected by stress concentration is known as the region of stress concentration. The region of stress concentration needs to be meshed densely to get accurate results.

Bending

When a non-axial force is applied on a structural member, the structural member starts deforming. This phenomenon is known as bending. In case of bending, strains vary linearly from the centerline of a beam to the circumference. In case of pure bending, the value of strain is zero at the centerline. The plane section of the beam is assumed to remain plain even after the bending.

Bending Stress

When a non-axial force is applied on a structural member, some compressive and tensile stresses are developed in the member. These stresses are known as bending stresses.

Creep

At elevated temperature and constant load, many materials continue to deform but at a slow rate. This behavior of materials is called creep. At a constant stress and temperature, the rate of creep is approximately constant for a long period of time. After a certain amount of deformation, the rate of creep increases, thereby causing fracture in the material. The rate of creep depends highly on both the stress and the temperature.

Classification of Materials

Materials are classified into three main categories: elastic, plastic, and rigid. In case of elastic materials, the deformation disappears on the removal of load. In plastic materials, the deformation is permanent. A rigid material does not undergo deformation when subjected to an external load. However, in actual practice, no material is perfectly elastic, plastic, or rigid. The structural members are designed such that they remain in the elastic conditions under the action of working loads. All engineering materials are grouped into three categories that are discussed next.

Isotropic Material
In case of Isotropic materials, material properties do not vary with direction, which means they have the same material properties at each point in all directions. Material properties are defined by Young's modulus and Poisson's ratio.

Orthotropic Material
In case of orthotropic materials, material properties vary with direction and are specified in three orthogonal directions. Such materials have three mutually perpendicular planes of material symmetry. Material properties are defined by separate Young's modulus and Poisson's ratio along each axis.

Anisotropic Material
In case of Anisotropic materials, material properties vary with direction and point location, but there is no plane of material symmetry. This means they do not behave in the same way in all directions.

Aspect Ratio
Aspect ratio is defined as the ratio of the longest side to the smallest side of an element.

Axisymmetry
Model that can be defined by rotating its cross-section by 360-degrees about an axis is known as axisymmetry model.

Degrees of Freedom (DOF)
Degrees of freedom can be defined as the minimum number of variables required to constrain an object physically, chemically, and thermally. The variables can be displacement, rotation, temperature, pressure, enthalpy, chemical composition, and so on.

There are six DOFs for any point in 3-dimensional (3D) space:
3 translational DOFs (one each in the X,Y, and Z directions) and
3 rotational DOFs (one each about the X, Y, and Z axes).

Example 1
Consider the stepped bar as shown in the Figure 1-8. For an element q, length and cross-sectional area are represented as Lq and Aq respectively. The Young's modulus of the material is represented as E. Find the global stiffness matrix.

where, q = 1, 2, 3, ...

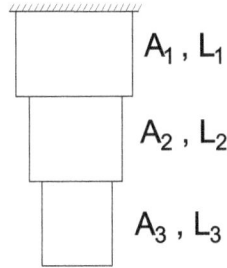

Figure 1-8 *Stepped-Bar*

The problem can be represented as a spring model, as shown in Figure 1-9. In this figure, K1, K2, and K3 represent the elements and 1, 2, 3, and 4 represent the nodal points.

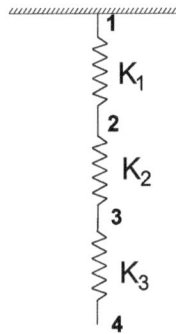

Figure 1- 9 *Spring representation of stepped bar*

So the stiffness matrix for each element can be represented as:

$$K_q = \frac{A_q E}{L_q} \begin{bmatrix} 1 & -1 \\ -1 & 1 \end{bmatrix}$$

Therefore the value of K_1, K_2, and K_3 can be obtained by putting the value of q as 1, 2, and 3.

$$K_1 = \frac{A_1 E}{L_1} \begin{bmatrix} 1 & -1 \\ -1 & 1 \end{bmatrix} \begin{matrix} 1 \\ 2 \end{matrix}$$

$$K_2 = \frac{A_2 E}{L_2} \begin{bmatrix} 1 & -1 \\ -1 & 1 \end{bmatrix} \begin{matrix} 2 \\ 3 \end{matrix}$$

$$K_3 = \frac{A_3E}{L_3}\begin{bmatrix} 1 & -1 \\ -1 & 1 \end{bmatrix}\begin{matrix} 3 \\ 4 \end{matrix}$$

As there are four nodal points, therefore, global stiffness matrix will have 4 rows and 4 columns and this global stiffness matrix can be obtained by equation given next.

$$K_{eq} = K_1 + K_2 + K_3$$

$$K_{eq} = \frac{A_1E}{L_1}\begin{bmatrix} 1 & -1 & 0 & 0 \\ -1 & 1 & 0 & 0 \\ 0 & 0 & 0 & 0 \\ 0 & 0 & 0 & 0 \end{bmatrix} + \frac{A_2E}{L_2}\begin{bmatrix} 0 & 0 & 0 & 0 \\ 0 & 1 & -1 & 0 \\ 0 & -1 & 1 & 0 \\ 0 & 0 & 0 & 0 \end{bmatrix} + \frac{A_3E}{L_3}\begin{bmatrix} 0 & 0 & 0 & 0 \\ 0 & 0 & 0 & 0 \\ 0 & 0 & 1 & -1 \\ 0 & 0 & -1 & 1 \end{bmatrix}$$

So, the final global stiffness matrix for the given problem will be equal to:

$$K_{eq} = E\begin{bmatrix} \frac{A_1}{L_1} & \frac{-A_1}{L_1} & 0 & 0 \\ \frac{A_1}{L_1} & (\frac{A_1}{L_1}+\frac{A_2}{L_2}) & \frac{-A_2}{L_2} & 0 \\ 0 & \frac{-A_2}{L_2} & (\frac{A_2}{L_2}+\frac{A_3}{L_3}) & \frac{-A_3}{L_3} \\ 0 & 0 & \frac{-A_3}{L_3} & \frac{A_3}{L_3} \end{bmatrix}$$

Self-Evaluation Test

Answer the following questions and then compare them to those given at the end of this chapter:

1. A minimum of _____ nodes are required to define a line element.

2. A minimum of _____ nodes are required to define an area element.

3. A minimum of _____ nodes are required to define a volume element.

4. FEA simulates the loading conditions of a model and determines its response under those conditions. (T/F)

5. A linear line element has a maximum of two nodes. (T/F)

6. The nodes define the shape of an element. (T/F)

7. An area element should always be triangular in shape. (T/F)

8. You cannot import an external geometry file into an FEA software. (T/F)

9. 2D elements are also referred as shell elements. (T/F)

Review Questions

Answer the following questions:

1. In dynamic analysis, the boundary conditions are a function of _____.

2. Modal analysis is used to calculate the _____ frequencies of a model.

3. Hooke's law states that stress is directly proportional to _____ within elastic limit.

4. The Finite Element Method gives exact solutions to problems. (T/F)

5. In FEM, the geometry is discretized into small parts, known as elements. (T/F)

6. In space, a rigid body has six degrees of system. (T/F)

Answers to Self-Evaluation Test

1. two, **2.** three, **3.** four, **4.** T, **5.** T, **6.** T, **7.** F, **8.** F, **9.** T

Chapter 2

Introduction to ANSYS Workbench

INTRODUCTION TO ANSYS Workbench

Welcome to the world of Computer Aided Engineering (CAE) with ANSYS Workbench. If you are a new user, you will be joining hands with thousands of users of this Finite Element Analysis software package. If you are familiar with the previous releases of this software, you will be able to upgrade your designing skills with tremendous improvement in this latest release.

ANSYS Workbench, developed by ANSYS Inc., USA, is a Computer Aided Finite Element Modeling and Finite Element Analysis tool. In the Graphical User Interface (GUI) of ANSYS Workbench, the user can generate any dimensional (1D, 2D, and 3D) and FE models, perform analysis, and generate results of analysis. You can perform a variety of tasks ranging from design assessment to finite element analysis to complete product optimization analysis by using ANSYS Workbench. ANSYS also enables you to combine the stand-alone analysis system into a project and to manage the project workflow.

The following is the list of analyses that can be performed by using ANSYS Workbench:

1. Coupled Field Harmonic
2. Coupled Field Modal
3. Coupled Field Static
4. Coupled Field Transient
5. Eigenvalue Buckling
6. Electric
7. Explicit Dynamics
8. Fluid Flow- Blow Molding (Polyflow)
9. Fluid Flow- Extrusion (Polyflow)
10. Fluid Flow (CFX)
11. Fluid Flow (FLUENT)
12. Fluid Flow (Polyflow)
13. Harmonic Acoustics
14. Harmonic Response
15. Hydrodynamic Diffraction
16. Hydrodynamic Response
17. I.C. Engine (Fluent)
18. Magnetostatic
19. Modal
20. Modal Acoustics
21. Random Vibration
22. Response Spectrum
23. Rigid Dynamics
24. Static Acoustics
25. Static Structural
26. Steady-State Thermal
27. Thermal-Electric
28. Topology Optimization
29. Transient Structural
30. Transient Thermal
31. Turbomachinery Fluid Flow

SYSTEM REQUIREMENTS

The following are minimum system requirements to ensure smooth functioning of ANSYS Workbench on your system:

*   Operating System: Windows 64-bit (Windows 10, Windows Server 2016, Windows Server 2019)
*   Platform: Intel i3/i5/ i7/ Xeon
*   Memory: 4 GB of RAM for all applications, 8GB for running CFX and FLUENT
*   DVD drive: For installing the software
*   Graphics adapter: NVIDIA Quadro or AMD Radeon Pro card with at least 1 GB of discrete video memory
*   Browser: Chrome/Microsoft edge 90 or higher, Firefox 89 or higher

STARTING ANSYS Workbench 2023 R2

To start ANSYS Workbench 2023 R2, choose **Start > Ansys 2023 R2 > Workbench 2023 R2** from the taskbar, refer to Figure 2-1. Alternatively, you can start ANSYSWorkbench by double-clicking on its shortcut icon displayed on the desktop of your computer. After the necessary files are loaded and licenses are verified, the **Workbench** window will be displayed on the screen, as shown in Figure 2-2.

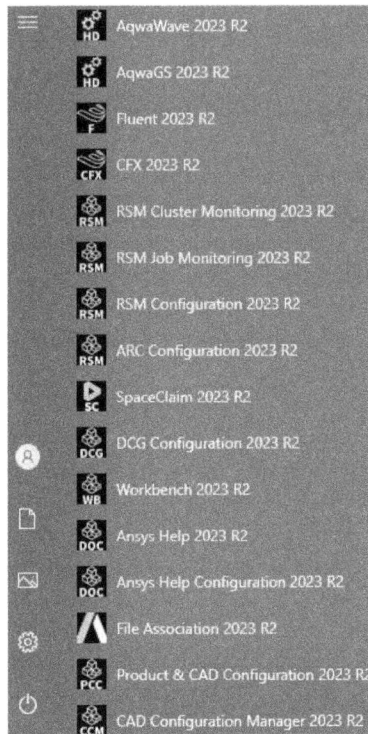

Figure 2-1   *Starting ANSYS Workbench using the taskbar*

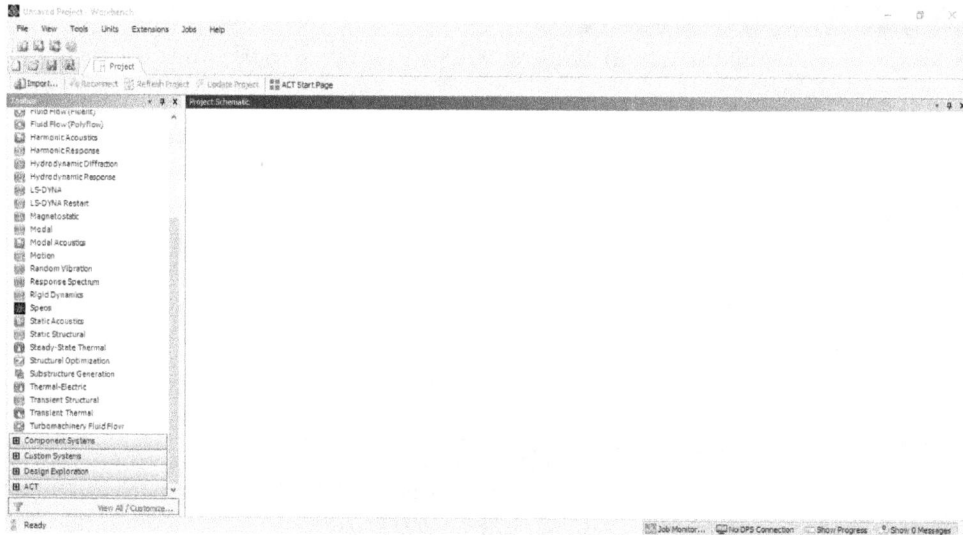

*Figure 2-2 The **Workbench** window*

The **Workbench** window helps streamline an entire project to be carried out in ANSYS Workbench 2023 R2. In this window, one can create, manage, and view the workflow of the entire project created by using Standard analysis systems. The **Workbench** window mainly consists of Menu bar, **Main** toolbar, the **Toolbox** window, **Project Schematic** window, and the **Status bar**, refer to Figure 2-3. Various components of the **Workbench** window are discussed next.

Toolbox Window

The **Toolbox** window is located on the left in the **Workbench** window. The **Toolbox** window lists the standard and customized templates or the individual analysis components that are used to create projects. To create a project, drag a particular analysis or component system from the **Toolbox** window and drop it into the **Project Schematic** window. Alternatively, double-click on a particular analysis or component system in the **Toolbox** window to add it to the **Project Schematic** window and to create the project.

Note

*The double-click action always adds new system to the project, whereas dragging and dropping the system from the **Toolbox** window enables you to specify location of the new system in the **Project Schematic** window. Based on the specified location, you can create data sharing with the existing systems. You will learn more about sharing data between different systems in the **Project Schematic** window in later chapters.*

The **Toolbox** window comprises five toolboxes: **Analysis Systems**, **Component Systems**, **Custom Systems**, **Design Exploration**, and **ACT**. The components of these toolboxes are discussed next.

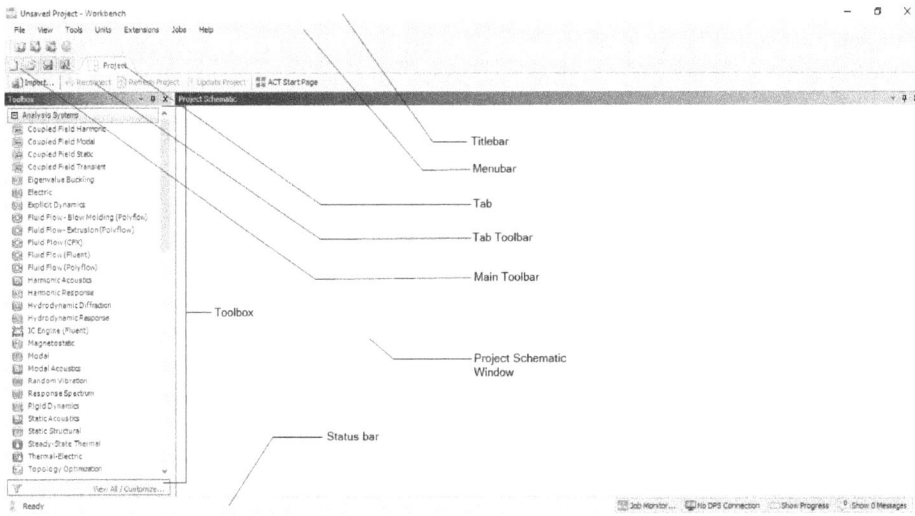

Figure 2-3 *The components of the* **Workbench** *window*

Analysis Systems Toolbox

The **Analysis Systems** toolbox is displayed expanded in the **Toolbox** window by default. It contains predefined templates for different types of analyses that can be carried out in ANSYS Workbench 2023 R2. Each predefined template consists of all the components that are used to perform a particular type of analysis. For example, the **Static Structural** analysis system of the **Analysis Systems** toolbox is used to carry out the Static Structural analysis. When you add this system in the **Project Schematic** window, it contains all the components that are necessary to carry out the Static Structural analysis. Figure 2-4 shows the **Analysis Systems** toolbox with different types of analysis systems available in ANSYS Workbench. These analysis systems are discussed next.

Coupled Field Harmonic

This analysis system, while performing the harmonic analysis, determines the steady-state response of a structure and the surrounding fluid medium to sinusoidally time-varying excited loads. The system supports piezoelectric coupling between electric and structural physics.

Figure 2-4 *The* **Analysis Systems** *toolbox displaying various analysis systems*

Coupled Field Modal

This analysis system, while performing the modal analysis, determines the frequencies and standing wave patterns within a structure and the surrounding fluid medium.

Coupled Field Static

This analysis system is used to determine the response of a structure subjected to steady loads that do not induce significant inertia and damping effects. Using this type of analysis, the displacements, stresses, strains, and forces caused by static loading conditions can be determined. The system supports 2D coupled structural-thermal physics.

Coupled Field Transient

This analysis system is used to determine the responses of a structure under the action of time dependent variables such as displacements, strains, stresses, and forces. The system supports 2D coupled structural-thermal physics.

Electric

This analysis system is used to analyze steady-state electric conduction.

Eigenvalue Buckling

Buckling is failure of structure which is caused when compressive stresses are less than the ultimate compressive stress of the material. Eigenvalue buckling is the theoretical behavior of an ideal elastic structure. However, geometric imperfections and nonlinearity prevent most structures from attaining their theoretical eigenvalue strength. To assess these effects, a nonlinear buckling is needed.

Explicit Dynamics

This analysis system is used to identify the dynamic response of a component under stress wave propagation, or time-dependent loads or impacts. It is also used for modal mechanical phenomena that are highly non-linear.

Fluid Flow - Blow Molding (Polyflow)

This analysis system is used to perform blow-molding simulations that provide only the application-specific capabilities.

Fluid Flow - Extrusion (Polyflow)

This analysis system is used to perform extrusion simulations that provide only the application-specific capabilities.

Fluid Flow (CFX)

This system allows users to carry out flow analysis of compressible and incompressible fluids. It is also used to analyze heat transfer in fluids.

Fluid Flow (FLUENT)

Like Fluid Flow (CFX), Fluid Flow (Fluent) system is also used to carry out fluid flow analysis of compressible and incompressible fluids and their heat transfer analysis.

Fluid Flow (Polyflow)

This analysis system is used to simulate polyflow application-specific capabilities.

Harmonic Acoustics

Harmonic Acoustics is the prediction of structure borne noise and sound propagation that plays a vital role in the design of large variety of products, for example noise caused by vibrating structural components, transmission of sound through thin panels and the acoustic performance of piezoelectric devices. These types of acoustic wave propagation problems can be solved either in a coupled way in which the fluid and structural domains are solved simultaneously, or in an uncoupled way with the help of Harmonic Acoustics analysis system.

Harmonic Response

Harmonic Response is the response of a system under a sustained cyclic load. Harmonic Response analysis system is used to analyze a system working under periodic or sinusoidal loads. This analysis helps in determining whether a particular structure will be able to withstand resonance, fatigue, and other effects of forced vibration.

Hydrodynamic Diffraction

The **Hydrodynamic Diffraction** system is used in the aqwa applications and determines the wave forces and structure motions in any types of wave pattern.

Hydrodynamic Response

The **Hydrodynamic Response** system is used to determine the response of a structure caused by ocean environment forces such as wind, wave, current in regular or irregular waves.

Magnetostatic

This analysis system is used to analyze the magnetic field developed due to the presence of a temporary or permanent magnet.

Modal

Modal analysis is the study of dynamic properties of a model, subjected to vibrations. Modal analysis system in ANSYS Workbench helps in determining the frequencies and mode shapes of a model.

Modal Acoustics

The **Modal Acoustics** system is used to determine frequencies and standing wave patterns within a structure and the surrounding fluid medium.

Random Vibration

This analysis is carried out to determine the reaction of a structure or a component to changing frequencies of vibrations. Many components experience vibrations which are random in nature. This analysis system is used to determine the responses of structures that are exposed to such varying or random vibrations.

Response Spectrum

Response Spectrum analysis system is similar to Random Vibration analysis system and is used after a transient analysis is done.

Rigid Dynamics

Rigid Dynamics analysis system is used to determine the response of a rigid body or a mechanism consisting of rigid bodies. Response of a robot mechanism is an example of rigid body analysis.

Static Acoustics

This system is used as a base analysis (pre-stress environment) for future downstream Modal Acoustics or Harmonic Acoustics analyses where response of a structure is determined for a load that does not induce significant inertia and damping effects.

Static Structural

The Static Structural analysis system is used to determine the response of a structure subjected to static loading conditions. The loads in this case are assumed to produce no or negligible time based loading characteristics. Using this type of analysis, displacement, stresses, and deformations of structures under static loading conditions can be determined.

Steady-State Thermal

Steady-State Thermal analysis system is used to determine the temperature, thermal gradient, heat flow rates and heat fluxes under the influence of thermal loading which remains constant with time and are static in nature.

Thermal-Electric

Thermal-Electric analysis system is used to simulate thermal and electric fields.

Topology Optimization

The **Topology Optimization** system computes an optimal material layout within a selected region of your model with specified design objectives, boundary conditions, and constraints. The topology optimization is a tool that identifies the region that contribute the least to load bearing as per the set of loads and boundary conditions provided by a preceding **Static Structural, Modal** system or combination of both the analyses.

Transient Structural

Transient Structural analysis system is used to determine responses of structures under the action of time dependent variables. Using this analysis, time-varying displacement, stresses and strains can be determined.

Transient Thermal

Transient Thermal analysis system is used to determine the temperature and other thermal variables of a structure that vary over time.

Turbomachinery Fluid Flow

The **Turbomachinery Fluid Flow** system contains cells for carrying out a study with ANSYS CFX with an added Turbo Mesh cell.

Component Systems Toolbox

By default, the **Component Systems** toolbox is displayed in collapsed state in the **Toolbox** window. To expand the **Component Systems** toolbox, click on the plus sign (+) located on the left of the **Component Systems** title bar. The components displayed in the **Component Systems** toolbox are the basic blocks of a project and form only a part of the analysis system, such as **Geometry** (used to create a model for analysis), **Mesh** (used to generate FEA model), **Results** (used to visualize the results of analysis in the desired form), and so on. Figure 2-5 shows the **Components Systems** toolbox with various components displayed in it.

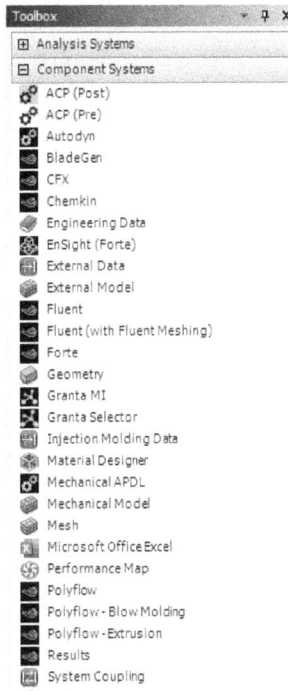

Figure 2-5 *Partial view of the* ***Component Systems*** *toolbox*

Custom Systems Toolbox

By default, the **Custom Systems** toolbox is also displayed in collapsed state in the **Toolbox**. To expand this node, click on the plus sign (+) displayed on the left of the **Custom Systems** title bar, refer to Figure 2-6. The systems in the **Custom Systems** toolbox are used to carry out standard coupled analysis, in which the input and output data of one analysis are used as input for the next analysis. For example, the **Pre-Stress Modal** system is used to carry out Static Structural analysis followed by a Modal analysis. Similarly, the **FSI: Fluid Flow (CFX) -> Static Structural** custom system is used to carry out a Fluid Flow analysis in CFX followed by a Static Structural analysis.

To add a custom system to the **Project Schematic** window, double-click on it in the **Custom Systems** toolbox in the **Toolbox** window. Figure 2-7 shows **FSI: Fluid Flow (CFX) -> Static Structural** custom system added to the **Project Schematic** window. This figure illustrates two different systems sharing the same geometry. This type of sharing is done if a single project requires various analysis types for the same geometry. You will learn about adding systems to the **Project Schematic** window later in this chapter.

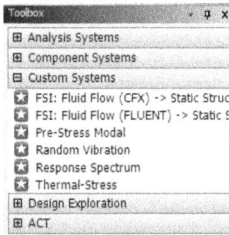

*Figure 2-6 The **Custom Systems** toolbox*

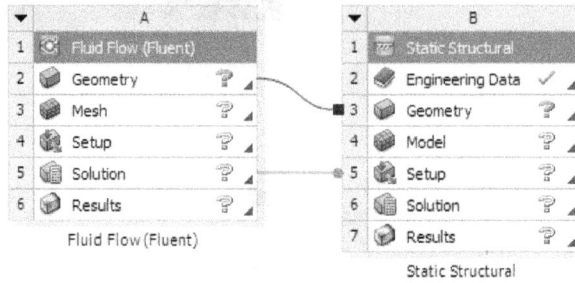

*Figure 2-7 The **FSI: Fluid Flow (CFX) -> Static Structural** custom system added to the **Project Schematic** window*

Design Exploration Toolbox

By default, the **Design Exploration** toolbox is displayed in collapsed state in the **Toolbox** window. Expand this toolbox by following the procedure discussed earlier. The options in the **Design Exploration** toolbox are used to explore a component, so that the design of the component can be further optimized by changing the design variables based on the performance of the product, refer to Figure 2-8.

You can control the display of elements in the **Toolbox**. To do so, choose the **View All / Customize...** button displayed at the bottom of the **Toolbox** window; the **Toolbox Customization** window will be

*Figure 2-8 The **Design Exploration** toolbox*

displayed, as shown in Figure 2-9. In this window, some of the check boxes are selected, indicating that the corresponding element will be displayed in the **Toolbox** window. Clear the check box corresponding to those elements that you do not want to be displayed in the **Toolbox** window.

*Figure 2-9 Partial view of the **Toolbox Customization** window*

Project Schematic Window

The **Project Schematic** window helps manage an entire project. It displays the workflow of entire analysis project. To add an analysis system to the **Project Schematic** window, drag the analysis system from the **Toolbox** window and drop it into the green-colored box displayed in the **Project Schematic** window, as shown in Figure 2-10 and 2-11. Alternatively, double-click on an analysis system in the **Toolbox** window to include it in the **Project Schematic** window. You can also add an analysis system to the **Project Schematic** window by using the shortcut menu displayed on right-clicking in the **Project Schematic** window. The procedure of adding an analysis system by using the shortcut menu is discussed later in this chapter.

In the **Project Schematic** window, when you click on the down arrow available at the top right corner, a flyout is displayed with various options to close, float, restore, minimize, and maximize the **Project Schematic** window, refer to Figure 2-12.

Each time you drag and drop an analysis system or an item into the **Project Schematic** window, a system is formed. Each system, consists of cells which are used to carry out various tasks within a system. You can add more than one systems in the **Project Schematic** window by dragging and dropping them from the **Toolbox** window, as per the requirement. After adding systems to the **Project Schematic** window, you can share the data available in the cells of one system with the corresponding cells of another system. A common example of systems sharing same kind of data among various cells of different systems is shown in Figure 2-13. The different types of systems connection links are discussed later in this chapter.

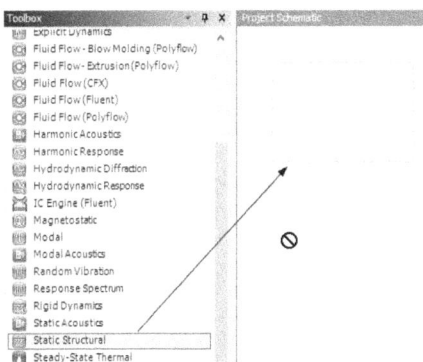

Figure 2-10 Dragging the **Static Structural** analysis system into the **Project Schematic** window

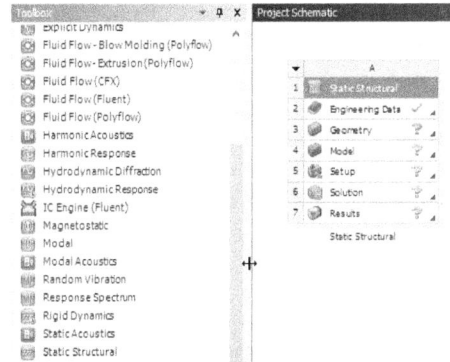

Figure 2-11 The **Static Structural** analysis system open in the **Project Schematic** window

Figure 2-12 Partial view of the **Project Schematic** window with the flyout displayed

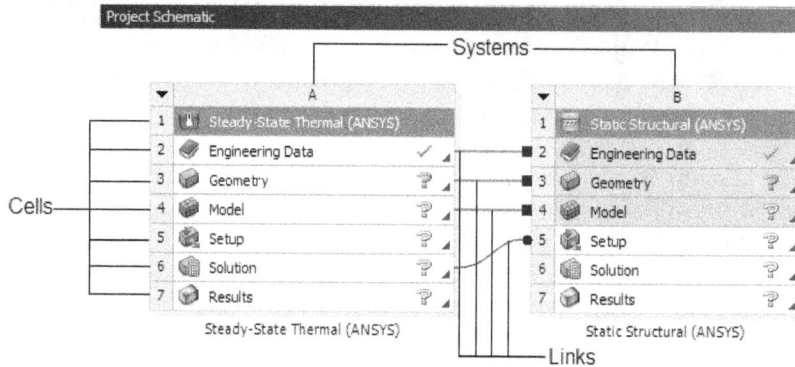

Figure 2-13 *Partial view of the **Project Schematic** window showing sharing of cells among two different analysis systems*

Note
*You will learn about the **Project Schematic** window, systems, and cells in detail later in this chapter.*

Menu Bar

Menu bar is located on the top of the **Workbench** window and contains various options such as **File**, **View**, **Tools**, and so on. These options enable you to control and manage the files of the current project. Figure 2-14 shows the Menu bar in the **Workbench** window. The options available in various menus will be discussed in detail later in this chapter.

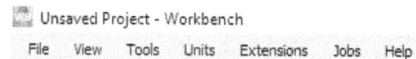

Figure 2-14 *The Menu bar*

Main and Tab Toolbar

The **Main** and **Tab** toolbar is a collection of the frequently used tools in ANSYS Workbench 2023 R2 and is shown in Figure 2-15. The tools available in the **Main** and **Tab** toolbar are also available in the Menu bar.

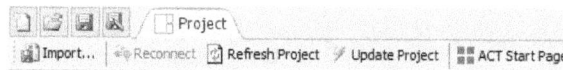

Figure 2-15 *The **Main** and **Tab** toolbar*

The various tools available in the **Main** and **Tab** toolbar are **New**, **Open**, **Save**, **Save As**, **Import**, **Reconnect**, **Refresh Project**, **Update Project**, and so on. You can use these tools to create new projects, open existing ones, save a project, save a project to a different location with a different name, import a project from another source, refresh project status after changes are made to it, update a project to its latest status and so on.

Shortcut Menu

In ANSYS Workbench, you can invoke most of the tools by using a shortcut menu displayed on right-clicking. The shortcut menus displayed are context sensitive, that is, the context in the shortcut menus will change depending upon the place where you right-click to invoke it. You can right-click anywhere in the **Workbench** window to display a shortcut menu. Some of the options in a shortcut menu display an arrow on their right. This arrow indicates that one more menu will be displayed on choosing this option. Figure 2-16 shows the shortcut menu that is displayed by right-clicking on the **Geometry** cell of the **Static Structural** system in the **Project Schematic** window.

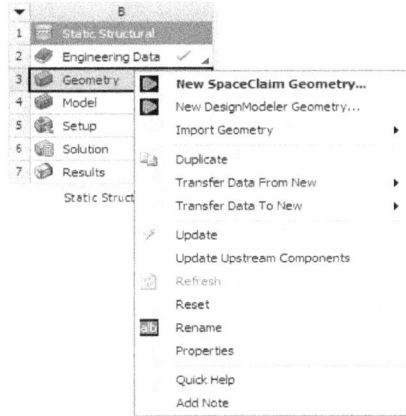

*Figure 2-16 The shortcut menu displayed by right-clicking on the **Geometry** cell in the **Project Schematic** window*

WORKING ON A NEW PROJECT

To start working on a new project, you need to add an appropriate analysis or component system to the **Project Schematic** window.

Adding a System to a Project

After starting a new project, it is necessary to define the tasks to be carried out in ANSYS Workbench 2023 R2. To start a new analysis, you need to add an analysis system to the **Project Schematic** window, as shown in Figure 2-17. There are many ways to add a system to a project. They are discussed next.

Adding a System by Drag and Drop

To add a system to a project by dragging and dropping, pick the required system template from the **Toolbox** window and then drag the cursor to the **Project Schematic** window; the green rectangular area of dash lines will be displayed, representing the location where the picked analysis system can be dropped. Move the cursor over the green rectangular area; the green rectangle will convert into a red rectangle of solid lines, refer to Figure 2-17. Drop the system in the red box; the system will be added to the project and will be displayed in the **Project Schematic** window.

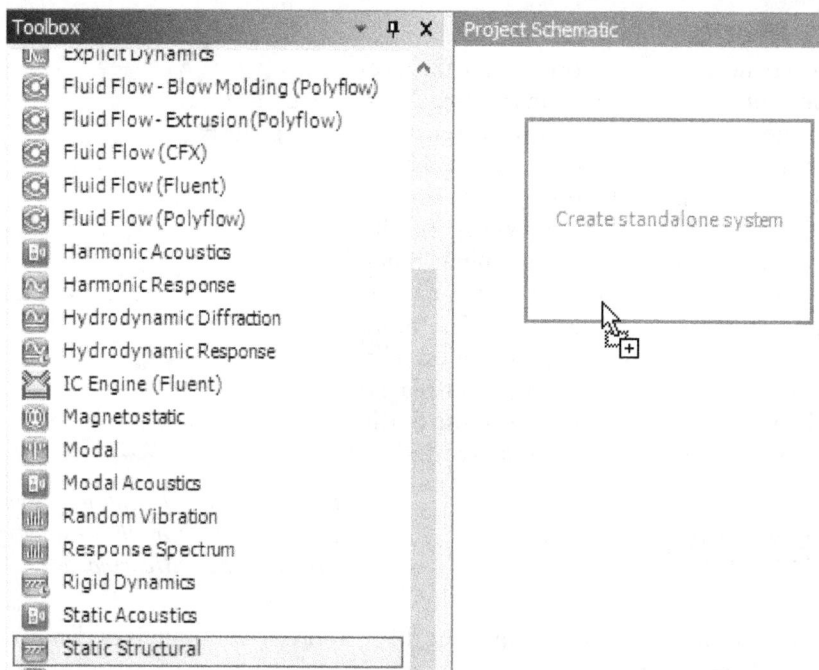

Figure 2-17 Adding an analysis system by using the drag and drop functionality

Note

*After adding the first analysis system into the **Project Schematic** window, when you drag the next analysis system from the **Toolbox** window to add to the **Project Schematic** window, more than one green rectangular areas of dash lines will be displayed, representing the possible locations where you can drop analysis.*

Adding a System by Double-clicking

You can also add an analysis system by double-clicking the left mouse button. To do so, double-click on the system that has to be added to the project; the system will be automatically added to the **Project Schematic** window.

Note

If an analysis system already exists in a project and then you double-click to add a new system, it will be added below the existing one.

Adding a System Using the Shortcut Menu

You can also add an analysis system by using the shortcut menu. To do so, right-click on the **Project Schematic** window; a shortcut menu will be displayed. Using this shortcut menu, you can add analysis, component, and custom systems to the **Project Schematic** window. To add an analysis system, choose the **New Analysis Systems** option from the shortcut menu; a flyout will be displayed. Choose the desired analysis system from the flyout to add it to the project, refer to Figure 2-18.

To add a new component system in the project, choose the **New Component Systems** option from the shortcut menu; a flyout will be displayed. Next, choose the desired component system from the flyout to add to the **Project Schematic** window.

Similarly, to add a new custom system or a new design exploration system into the **Project Schematic** window, choose the **New Custom Systems** or **New Design Exploration** option, respectively from their respective shortcut menu. Next, choose the desired option from the flyout to add to the **Project Schematic** window.

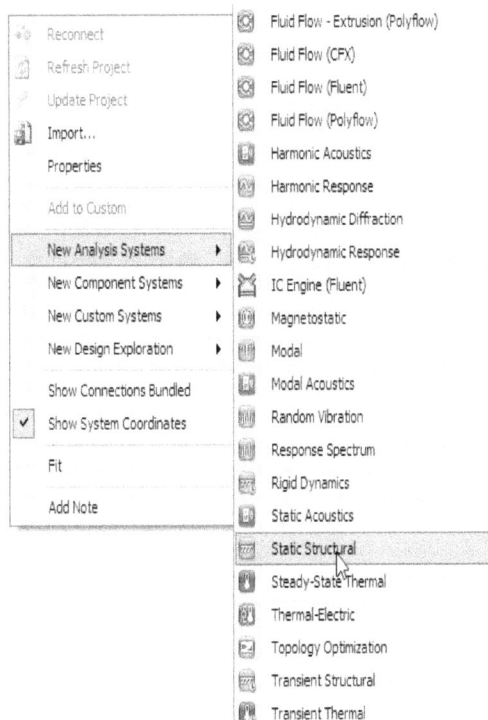

Figure 2-18 *Choosing the* ***Static Structural***
analysis system from the shortcut menu

Tip
After a system is added to the project, it is now important to define the cells that are displayed in respective systems. The most common types of cells that exist in a system are discussed later in this chapter.

RENAMING A SYSTEM

After a system is added to the **Project Schematic** window, its name will be highlighted at the bottom of the system. You can also rename an existing project by double-clicking on the name of the current project. Alternatively, click on the black down-arrow displayed at the upper left corner of the analysis system; a flyout will be displayed. Choose the **Rename** option from this flyout, refer to Figure 2-19; the name of the system will be highlighted. Specify a name to the system.

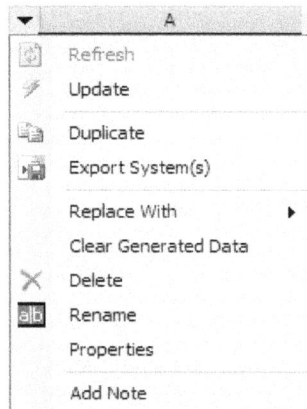

Figure 2-19 *Choosing the **Rename** option from the shortcut menu*

DELETING A SYSTEM FROM A PROJECT

To delete a system from the **Project Schematic** window, right-click on its name displayed in the title bar; a shortcut menu will be displayed, refer to Figure 2-20. Choose **Delete** option from the shortcut menu; the **ANSYS Workbench** message box will be displayed, as shown in Figure 2-21. Choose the **OK** button from this message box; the selected system will be deleted from the project. Alternatively, click on the down-arrow displayed at the upper left corner of the system; a flyout will be displayed. Choose the **Delete** option from this flyout to delete the system from the project.

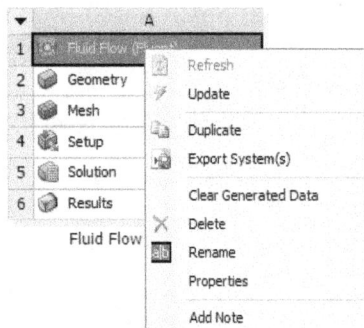

Figure 2-20 *Choosing the **Delete** option from the shortcut menu*

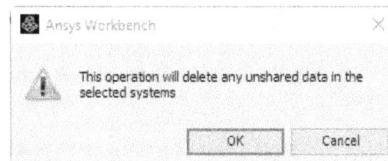

Figure 2-21 *The **ANSYS Workbench** message box*

DUPLICATING A SYSTEM IN A PROJECT

To duplicate a system, select the down arrow available at the top left corner of the selected system; a flyout is displayed, refer to Figure 2-19. Choose the **Duplicate** option from the flyout to duplicate it.

Note

*While duplicating a system, all the cells will be duplicated except the **Result** cell.*

SAVING THE CURRENT PROJECT

Whenever you start a new analysis project, the title bar of the **Workbench** window displays **Unsaved Project - Workbench**. This indicates that the current project is not saved yet. To save the current project, choose the **Save Project** button from the **Main** toolbar. Alternatively, choose the **Save** option from the **File** menu; the **Save As** dialog box will be displayed, as shown in Figure 2-22. You can also invoke the **Save As** dialog box by pressing the CTRL and S keys together. In this dialog box, browse to the location where you want to save the current project and then specify its name in the **File name** edit box. Next, choose the **Save** button; the project will be saved at the specified location. After saving the project, the title bar of the **Workbench** window will display the name that you have specified while saving the project.

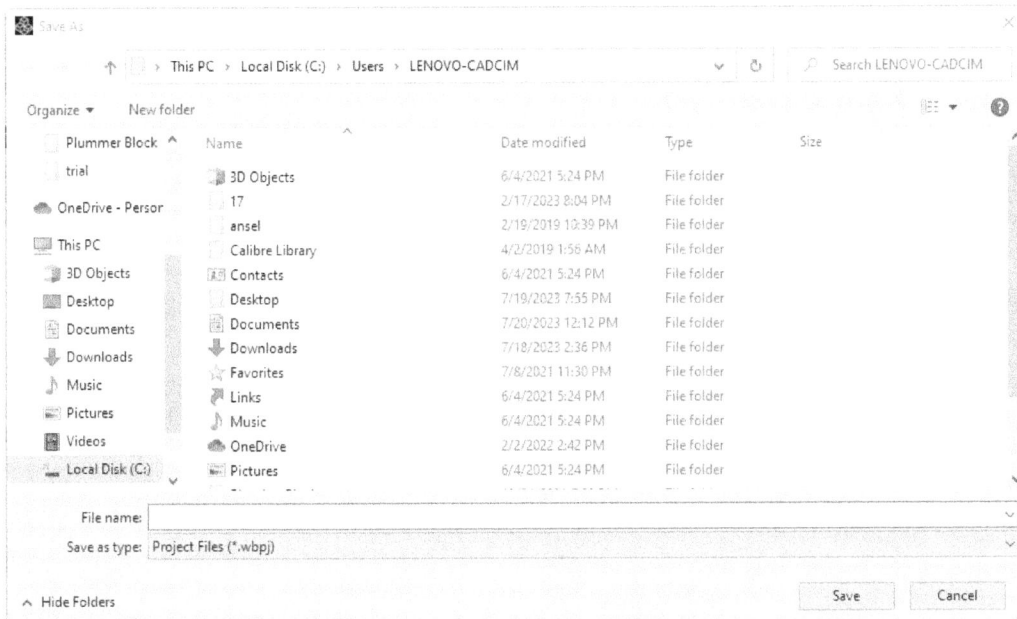

Figure 2-22 The Save As dialog box

Note

*If you have already saved the project, the **Save As** dialog box will not be displayed on choosing the **Save** button.*

If you want to save an opened project with a different name or at a different location, choose the **Save Project As** button from the **Main** toolbar; the **Save As** dialog box will be displayed, refer to Figure 2-22. Specify a new name and then choose the **Save** button from this dialog box; the same project will be saved with the new name and will become the current project.

OPENING A PROJECT

Open... To open an existing project, choose the **Open Project, Archive or Script** button from the **Main** toolbar; the **Open** dialog box will be displayed, as shown in

Figure 2-23. Browse to the location where the project file is saved, select the *.wbpj* file, and then choose the **Open** button from this dialog box. The selected project file will be opened and its name will be displayed on the title bar of the **Workbench** window. Alternatively, choose the **Open** option from the **File** menu.

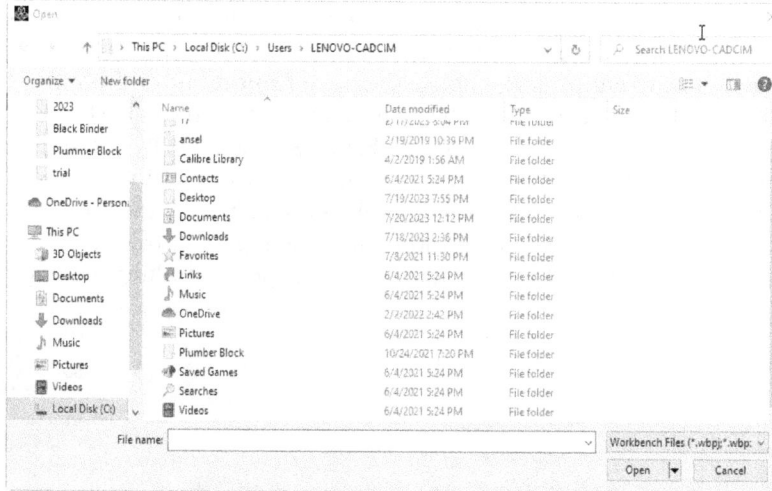

*Figure 2-23 The **Open** dialog box*

ARCHIVING THE PROJECT DATA

If you want to move the project data from one system to another, you can archive all project related data in a single zip file. This zip file contains all files and folders necessary to run the project on another computer, such as project file (*.wbpj*) and project folder (*name_files*). To archive a project, choose **File > Archive** from the Menu bar; the **Save Archive** dialog box will be displayed, refer to Figure 2-24. By default, *name* is displayed as the name of the file in the **File name** edit box, where, *name* is the project name of the current project. If you want to assign a different name to the archive file, specify it in the **File name** edit box.

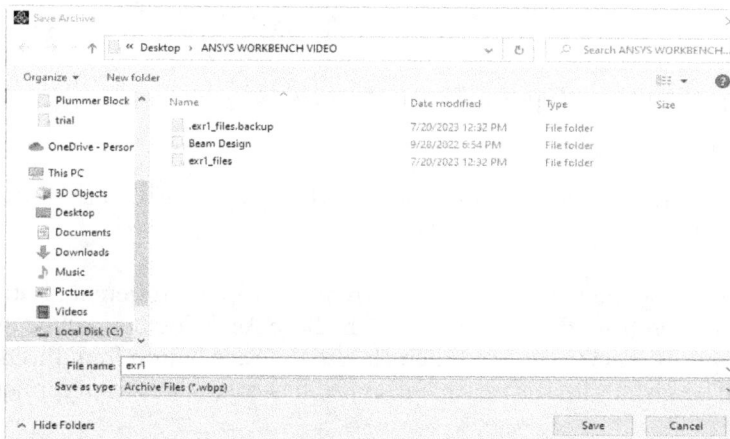

*Figure 2-24 The **Save Archive** dialog box*

Next, specify the location where you want to save the archive file and choose the **Save** button from the **Save Archive** dialog box; the **Archive Options** dialog box will be displayed, as shown in Figure 2-25. If you do not want to include the result and the solution file or some imported file that has not been created in the project, clear the respective check box from the **Archive Options** dialog box. Next, choose the **Archive** button from this dialog box; the zipped archive file will be created at the specified location. Transfer this zip file to any system and restore the project files.

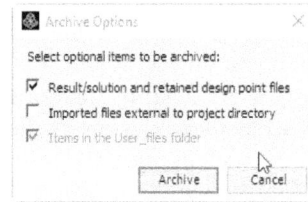

Figure 2-25 *The Archive Options dialog box*

UNITS IN ANSYS Workbench

In ANSYS Workbench, you can use any of the following predefined unit systems:

1. **Metric (kg, m, s, °C, A, N, V)** Unit System

Mass = Kilogram (kg)	Temperature = Degree Celsius (°C)
Length = Meter (m)	Current = Ampere (A)
Time = Second (s)	Force = Newton (N)
Voltage = Volts (V)	

2. **Metric (tonne, mm, s, °C, mA, N, mV)** Unit System

Mass = Tonne	Temperature = Degree Celsius (°C)
Length = Millimeter (mm)	Current = Milliampere (mA)
Time = Second (s)	Force = Newton (N)
Voltage = Millivolt (mV)	

3. **U.S. Customary (lbm, in, s, °F, A, lbf, V)** Unit System

Mass = Pound (lb)	Temperature = degree Fahrenheit (°F)
Length = Inch (in)	Current = Ampere (A)
Time = Second (s)	Force = Pound (lbf)
Voltage = Volts (V)	

4. **SI (kg, m, s, K, A, N, V)** Unit System

Mass = Kilogram (kg)	Temperature = Kelvin (K)
Length = Meter (m)	Current = Ampere (A)
Time = Second (s)	Force = Newton (N)
Voltage = Volts (V)	

5. **U.S. Engineering (lb, in, s, R, A, lbf, V)** Unit System

Mass = Pound (lb)	Temperature = Renkine (R)
Length = Inch (in)	Current = Ampere (A)
Time = Second (s)	Force = Pound (lbf)
Voltage = Volts (V)	

Metric (kg, m, s, °C, A, N, V) unit system is the default unit system that is assigned to a project. However, you can change the unit system for the current project or set the default unit system, to be assigned to any new project.

ANSYS WORKBENCH DATABASE AND FILE FORMATS

When you save a project, a project file is created with a name *name.wbpj*, where *name* can be any user specified name. In addition to the project file, a folder is also created with a name *name_files*. All other files relevant to the project are automatically saved in this folder under various sub folders such as, *dp0*, *user_files*, and so on.

In ANSYS Workbench, various files are created with different file extensions. To view all files associated with the current project, choose **View > Files** from the Menu bar; the **Files** window will be displayed, as shown in Figure 2-26.

	A	B	C	D	E	F
1	Name	Ce...	Size	Type	Date Modified	Location
2	Static Structural_Plate.wbpj		38 KB	Workbench Project File	26-05-2021 17:51:56	D:\TIET\ANSYS\ANSYS_Practice\Exercise
3	act.dat		259 KB	ACT Database	26-05-2021 17:51:50	dp0
4	EngineeringData.xml	A2	26 KB	Engineering Data File	26-05-2021 17:51:53	dp0\SYS\ENGD

*Figure 2-26 The **Files** window*

The **Files** window lists all files of the current project. The information displayed in the **File** window includes the following:

1. Name and extension of the files.
2. The reference of the analysis component cell with which the file is associated.
3. The file size and its location.

To open a folder containing the selected files, right-click on the corresponding cell and then choose the **Open Containing Folder** option from the shortcut menu displayed. To filter in any particular type of file extension, right-click on any cell to display a shortcut menu. Next, choose the **File Type Filter** option from the shortcut menu; the **File Type Filter** dialog box will be displayed, as shown in Figure 2-27. Select the check boxes displayed on the left of the desired file type from the **File Type Filter** dialog box; the list displayed in the **File** window will change dynamically based on the selection made in the **File Type Filter** dialog box. After selecting the desired file types, choose the **Ok** button from the **File Type Filter** dialog box to exit.

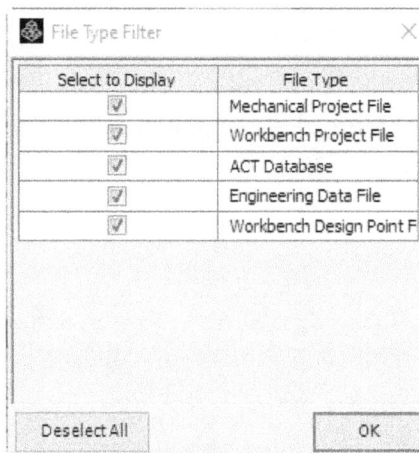

Select to Display	File Type
☑	Mechanical Project File
☑	Workbench Project File
☑	ACT Database
☑	Engineering Data File
☑	Workbench Design Point F

*Figure 2-27 The **File Type Filter** dialog box*

The following are some of the main file extensions used in ANSYS:

*.wbpj   = ANSYS Workbench project database file
*.engd   = Engineering Data
*.agdb   = DesignModeler file
*.fedb   = FE Modeler files
*.cmdb = Meshing file
*.mechdb = Mechanical file

*.rsx   = Mesh Morpher
*.ad   = ANSYS AUTODYN
*.dxdb   = Design Exploration
*.bgd   = BladeGen
*.db   = Mechanical APDL database file
*.cas, **.dat*, **.msh* = FLUENT files
*.cfx, **.def*, **.res*, **.mdef*, and **.mres* = CFX files
*.cmdb = CFX-Mesh files

CHANGING THE UNIT SYSTEMS

You can change the unit system being used in the current project. To do so, choose the **Units** menu from the Menu bar; a menu with various options will be displayed. Select the desired unit system from the menu; a tick mark will be displayed on the left of the selected unit system, refer to Figure 2-28.

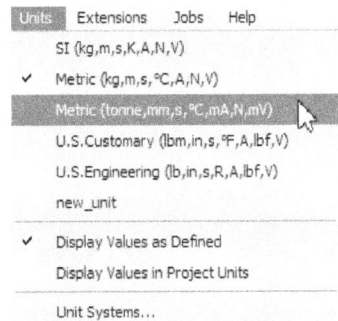

In addition to the unit systems displayed in the Units menu, you can customize to add some more unit systems by using the **Unit Systems** dialog box. To invoke this dialog box, choose **Units > Unit Systems** from the Menu bar; the **Unit Systems** dialog box will be displayed, as shown in Figure 2-29. All unit systems supported by ANSYS Workbench will be displayed under the **Unit System** column in the **Unit Systems** dialog

*Figure 2-28 The **Units** menu*

box. If you select a unit system in this column, the units used for measuring various quantities will be displayed on the right pane in this dialog box.

ANSYS Workbench supports nineteen main unit systems. However, by default only five Main unit systems are displayed under the **Units** menu. This is because the rest of the unit systems are suppressed in the **Unit Systems** dialog box, by default. To unsuppress a unit system, clear the corresponding check box under the **Suppress** column in the **Unit Systems** dialog box. Now onward, the unsuppressed unit system will be displayed in the **Units** menu.

You can change the default unit system that is assigned to a new project. To do so, select the radio button corresponding to the desired unit system, in the **Default** column in the **Unit Systems** dialog box. Now onward, the specified unit system will become the default unit system for new projects. Select the radio button corresponding to the desired unit system under the **Active Project** column to make it the unit system for the current project.

If the Main unit systems given in ANSYS Workbench do not suit your requirement, you can customize a unit system according to your requirement by using the **Unit Systems** dialog box. To do so, select the unit system that closely fits your requirement under the **Unit System** column and then choose the **Duplicate** button from the **Unit Systems** dialog box; a new unit system will be added under the **Unit System** column with the default name *Custom Unit System*. Rename this system. Select the newly defined unit system; the corresponding measuring units will be displayed on the right pane. Click on the down-arrow displayed on the right of the units under the **Unit** column; a drop-down list will be displayed with all feasible units for the quantity to be measured. Select the desired units from this drop-down list. You can also export custom units for the use of other users or import an already saved unit system by using the **Export** or **Import** button, from the **Unit Systems** dialog box. The imported or exported unit system files are saved in *.xml* format. To delete the selected customized unit system, choose the **Delete** button from the **Unit Systems** dialog box.

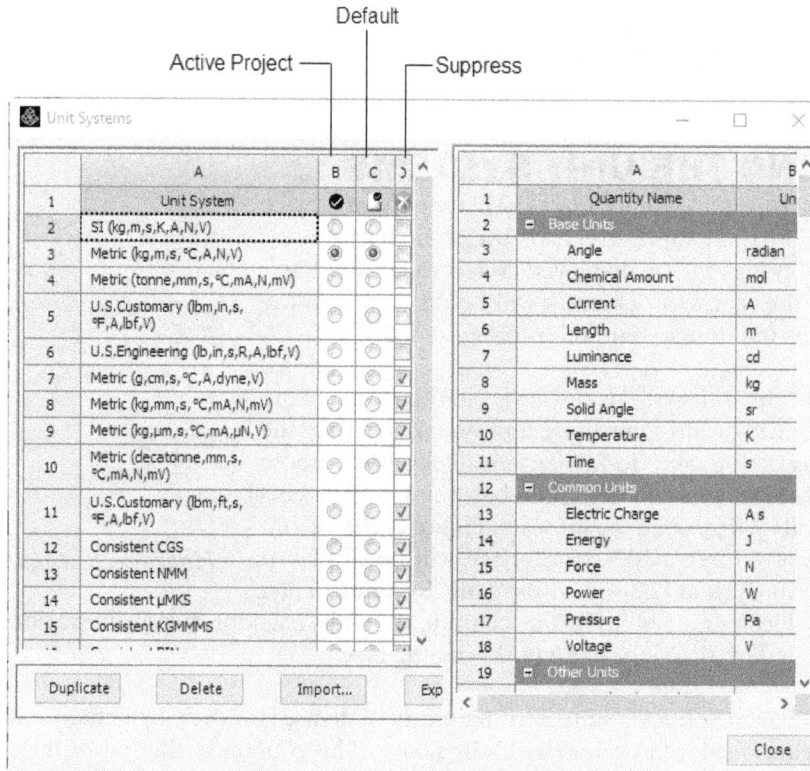

*Figure 2-29 The **Unit Systems** dialog box*

COMPONENTS OF A SYSTEM

An item that is added from the **Toolbox** window to the **Project Schematic** window is known as system and the constituent elements of the system are known as cells. Each cell of a system plays an important role in carrying out a project and are discussed next.

Engineering Data	Geometry	Model/Mesh
Setup	Solution	Results

Engineering Data Cell

The **Engineering Data** cell is used to define the material to be used in the analysis. To define the material, double-click on the **Engineering Data** cell; the workspace corresponding to this the **Engineering Data** cell will be displayed, as shown in Figure 2-30. Alternatively, you can also invoke the Engineering Data workspace to specify the material by right-clicking on this cell and then choosing the **Edit** option from the shortcut menu displayed.

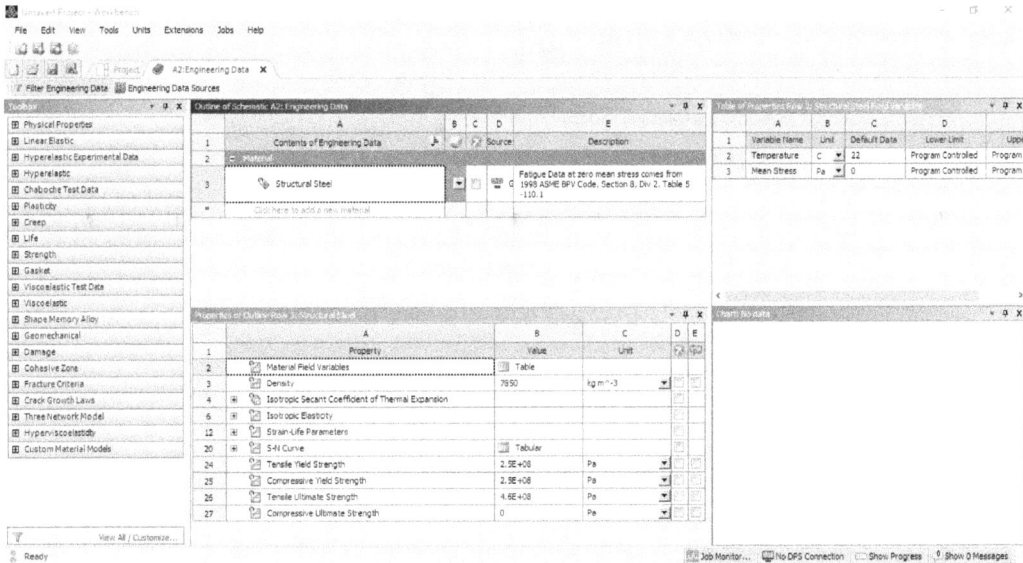

Figure 2-30 The Engineering Data workspace

Note
When you right-click on **Engineering Data** *cell, the* **Edit** *option will be highlighted in the shortcut menu displayed. This indicates that this is the most suitable action that can be taken. You can select an option from the shortcut menu as per the requirement. However, when you double-click on the* **Engineering Data** *cell, the most suitable action will be initiated directly.*

Geometry Cell

The **Geometry** cell is used to create, edit, or import the geometry that is used for analysis. To create a geometry for analysis, right-click on this cell and then choose either the **New SpaceClaim Geometry** or **New DesignModeler Geometry** option from the shortcut menu displayed, refer to Figure 2-31. Choose the **New SpaceClaim Geometry** option in the shortcut menu; the **SpaceClaim** window will be opened. Alternatively, if you double-click on the **Geometry** cell; the **SpaceClaim** window will be displayed which is used to create geometry. You can choose the **New DesignModeler Geometry** option which uses Parasolid geometry modeling kernel to create geometry.

Some other important options in the shortcut menu that are specific to the **Geometry** cell are discussed next. These options will only be available if you have not defined a geometry. The **Import Geometry** option is used to import any existing geometry to the current analysis system. You can also import the geometry created in other CAD packages. The **Edit Geometry** option will only be displayed if a geometry is already associated with the current analysis system. When you choose this option, the geometry opens in the **SpaceClaim** window for modification. The **Replace Geometry** option is used to replace the existing geometry that is associated with the analysis system with another geometry.

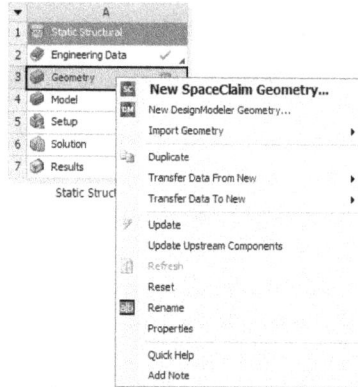

*Figure 2-31 The shortcut menu displayed on
right-clicking on the **Geometry** cell*

Note
*To exit the **SpaceClaim** window, choose **File** > **Close SpaceClaim** from the Menu bar.*

Model Cell

The **Model** cell will be displayed for mechanical analysis systems and is used to discretize geometry into small elements, apply boundary and load conditions, solve the analysis, and so on. On double-clicking on this cell, the **Mechanical** window will be displayed. In other words, this cell is associated with the **Mechanical** window.

Mesh Cell

The **Mesh** cell will be displayed for fluid flow analysis systems and is used to mesh the geometry. On double-clicking on this cell, the **Meshing** window will be displayed. In other words, this cell is associated with the **Meshing** window.

Setup Cell

The **Setup** cell is used to define the boundary conditions of an analysis system, such as loads and constraints. This cell is also associated with the **Mechanical** workspace.

Solution Cell

The **Solution** cell is used to solve the analysis problem based on the conditions defined in the cells above the **Solution** cell. This cell is also associated with the **Mechanical** workspace.

Results Cell

The **Results** cell is used to display the results of the analysis in the user specified formats. This cell is also associated with the **Mechanical** workspace.

STATES OF A CELL IN AN ANALYSIS SYSTEM

All cells display some visual symbols on their right to indicate their current status. This helps the user understand the next step in the analysis process as well as understanding the overall status of an analysis, refer to Figure 2-32. The explanation of these symbols is given in Table 2-1.

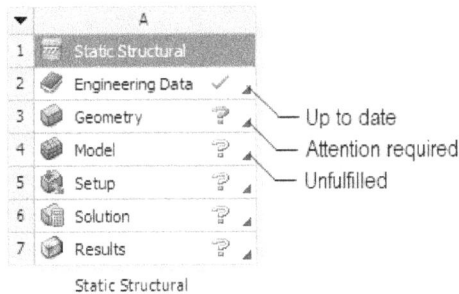

Figure 2-32 *Symbols for cell status*

Table 2-1 *Symbols and their meaning*

Symbol	Meaning	Function
?	Attention required symbol	indicates an immediate requirement of action for a cell and a user cannot proceed further without fixing the cell
?	Unfulfilled	indicates that the previous cells do not have sufficient or appropriate data
✓	Up to date	indicates that the cell is up-to-date and the data in the cell is ready to be shared with other cells
⟳	Refresh required	indicates that the data of the previous cells has been changed since last update and you need to refresh the cell with this symbol.
✓	Input changes pending	indicates that the data of the current cell will change on updation, if changes are done in the previous cells
⚡	Update required	indicates that the input data of the cell has been changed and the output data needs to be updated
⟳ₓ	Refresh failed	indicates that the last refresh process for the cell has failed
⚡ₓ	Update failed, update needed	indicates that the last update process for the cell has failed and you need to update it again
⚡ₓ	Update failed, attention needed	indicates that the last update process has failed and the cell needs immediate attention.

Figure 2-33 shows an analysis system in which the geometry of the analysis model has been changed after performing the complete analysis. Therefore, the refresh required symbol is displayed against the **Model** cell.

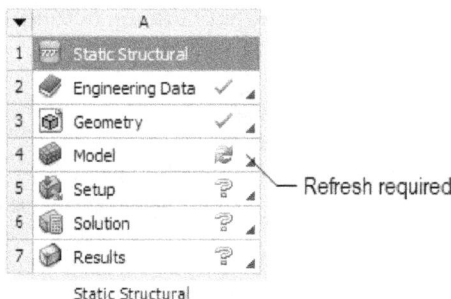

Figure 2-33   The Static Structural analysis system with the refresh required symbol annotated

REFRESHING AND UPDATING A PROJECT

Refresh is a process in which ANSYS Workbench reads the entire modified data. To refresh the entire project, choose the **Refresh Project** button from the **Tab** toolbar. Alternatively, choose **Tools > Refresh Project** from the Menu bar.

Note
When you refresh the entire project, the output data will not be recalculated based on the modifications made in the project.

Update is a process in which ANSYS Workbench refreshes the input data and recalculates the output data. To update the entire project, choose the **Update Project** button from the **Tab** toolbar. Alternatively, choose **Tools > Update Project** from the Menu bar.

You can also refresh or update a particular cell by choosing the respective option from the shortcut menu that is displayed on right-clicking on that particular cell. These options will only be available in the shortcut menu if that particular cell needs refreshing or updating.

If you refresh a particular cell, you can get information about its effect on the cells below the specified cell, and if needed, you can make the required changes before updating the cell. Refreshing a cell before updating is useful in complex analysis systems, where updating can take more time compared to refreshing.

ADDING SECOND SYSTEM TO A PROJECT

The second or the subsequent system can be added in the **Project Schematic** window in two ways, either as a stand-alone system or as a connected system that shares data with the existing analysis system. In a project, multiple stand-alone systems are preferred in case of performing analysis for various components of an assembly. In this way, analysis of all components of an assembly can be kept and managed in a single analysis project. The connected analysis systems are preferred more for conducting coupled analysis, where data from the first analysis system can be used or shared with the next analysis system.

As discussed earlier, to add a stand-alone system, double-click on the analysis system template in the **Toolbox** window; the new analysis system will get added in the **Project Schematic** window below the existing analysis system, as shown in Figure 2-34. Alternatively, use the shortcut menu or the drag and drop method to add a stand-alone system to the **Project Schematic** window.

Adding Connectors

There are two ways in which you can add connectors between cells of different systems in the **Project Schematic** window. In the first method, drag the cell from one system and drop it in the corresponding cell of another system; connectors will be added between the systems , which indicates that the data is shared. In the second method, drag the system from the **Toolbox** window to the **Project Schematic** window and move the cursor over the cell of the existing system with which you want to share the data; a red rectangle will be displayed on the right of the existing system with text written inside the rectangle. The text will give you information about the cells with which the data will be shared by the new system, as shown in Figure 2-35. Drop the selected analysis system over the desired cell of the existing analysis system; the new system will be added to the right of the existing analysis system and will share data from the specified cells. Figure 2-36 shows a new analysis system, the **Eigenvalue Buckling** analysis system, dropped over the **Solution** cell of the **Static Structural** analysis system. Note that the new system will automatically share the necessary data from the cells that are above the cell over which you dropped the new analysis system.

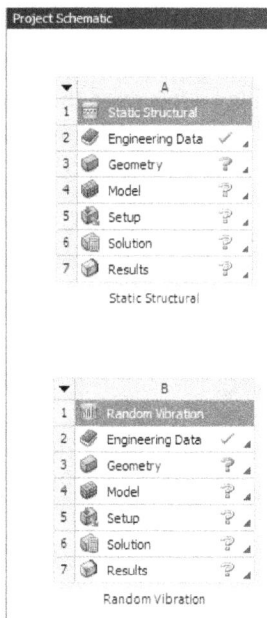

Figure 2-34 *Two stand-alone systems added in the **Project Schematic** window*

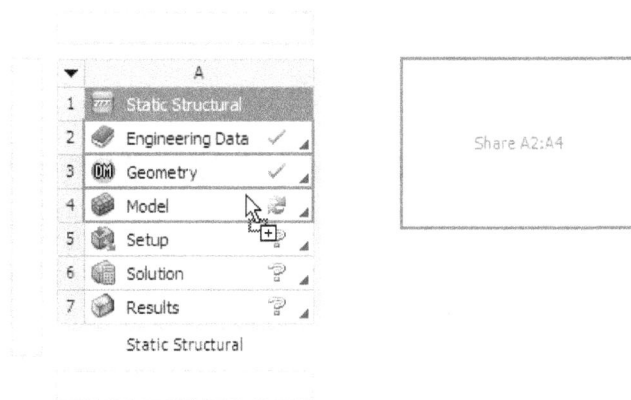

Figure 2-35 *Text reference of the cells to be shared*

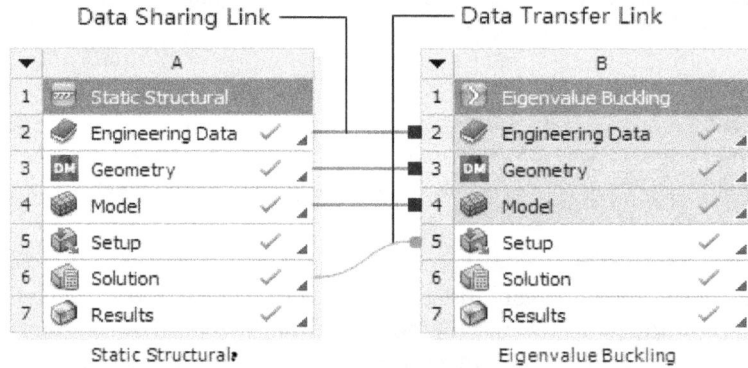

Figure 2-36 *The **Eigenvalue Buckling** analysis system added as a connected system to the **Static Structural** analysis system*

Connector/Link Types

Links connecting systems represent data sharing or data transfer between the cells of systems in the **Project Schematic** window. ANSYS Workbench supports two different types of connectors, as shown in Figure 2-36. In the first type, line with a square on its right (target) side is used to share data when the inputs and outputs of the two connected cells are identical. Shared data links can only be created between two cells of the same type. In the second type, a curve with a circle on its right (target) side is used to transfer data when the output of one cell is used as the input to the connected cell. Transfer data links are usually created between two cells of different types.

Deleting Connectors/Links

In most cases, you can delete links by right-clicking the link and choosing **Delete** from the shortcut menu. The data associated with the cell in the upstream system is copied to the downstream system so the cells can be edited independently.

In some cases (for example, links between two mechanical geometry cells), you cannot delete shared data links. In such cases, the linked cells in the downstream system have a gray background, refer to Figure 2-36.

SPECIFYING A GEOMETRY FOR ANALYSIS

In ANSYS Workbench, geometry can be specified in two different ways. In the first method, you can import a geometry created by using any other solid modeling software. In the second method, you can create a new geometry in the **SpaceClaim** or **DesignModeler** application of ANSYS Workbench. The procedure to create a new geometry is described next.

Creating a Geometry

When a new system is added in the **Project Schematic** window, the next step is to create the geometry. To create a new geometry, right-click on the **Geometry** cell of the system; a shortcut menu will be displayed, refer to Figure 2-37. Choose **New SpaceClaim Geometry** or

New DesignModeler Geometry from the shortcut menu; the status bar will display the message **Starting SpaceClaim** or **Starting DesignModeler**. If you choose **DesignModeler**, then after sometime, the **DesignModeler** window will be displayed, refer to Figure 2-38.

Next, in the **DesignModeler** window, create a model according to your design requirements by using the tools available in this window. These tools will be discussed in detail in later chapters. After creating the model, close this window. On doing so, the model will be automatically saved in the ANSYS Workbench database. Also, a green color check mark will be displayed on the right of the **Geometry** cell of the analysis system in the **Project Schematic** window, indicating that the geometry requirement is satisfied for the current system.

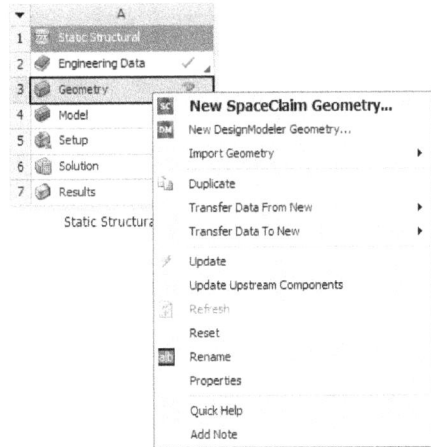

Figure 2-37   *The shortcut menu displayed on right-clicking on the **Geometry** cell*

After the geometry is specified for the analysis, you need to fullfil the requirement of other cells in the system. The cell types in the system are context specific and are solely dependent on the type of system selected.

In a system tree, you need to update the cells from top to bottom. For example, in a Static Structural analysis system, you need to satisfy the requirement of the **Geometry** cell before you move to the **Model** cell.

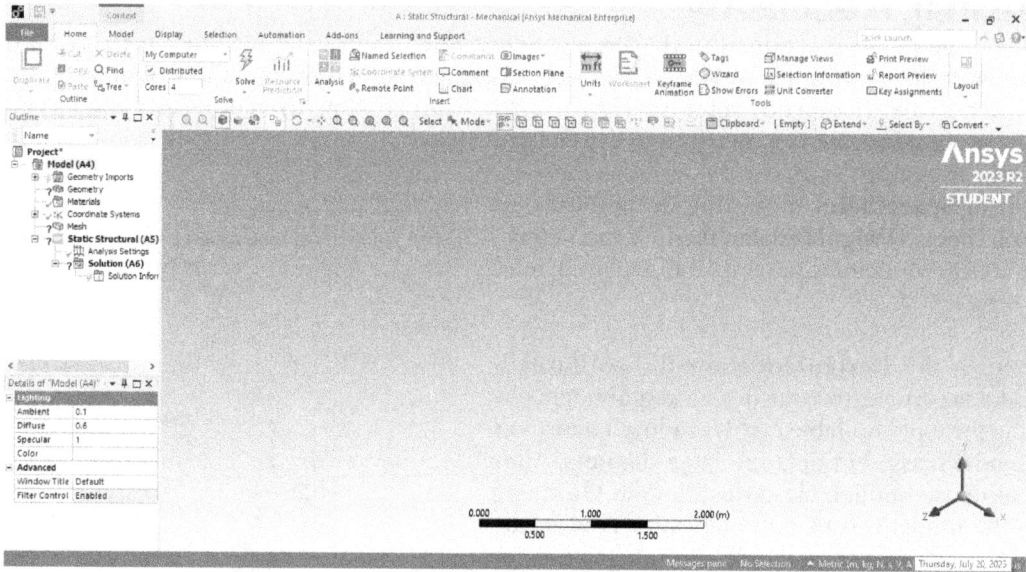

Figure 2-38   *A typical **DesignModeler** window*

Note
The detailed procedure of working through analysis component cells in an analysis system is given in later chapters.

USING HELP IN ANSYS WORKBENCH

In ANSYS Workbench, there are different ways in which you can access help. These methods are discussed next.

ANSYS Workbench Help

You can get online help and documentation while working on ANSYS Workbench. To access the help, choose ANSYS Workbench Help from the Help menu of the Menu bar; you will be redirected to ANSYS help web page of the Ansys website(ansys.com), as shown in Figure 2-39. This Help page is divided into two parts: Navigation pane (left pane) and Document pane (right pane). The Navigation pane consists of the Contents and the Document pane contains the information related to the topic selected in the Navigation pane. When a particular topic is selected in the left pane, corresponding information or help is displayed in the Document pane. You can open multiple help web pages on your browser by clicking at the desired topics. Apart from the ANSYS Help website, ANSYS Workbench offers two more ways to access help and they are discussed next.

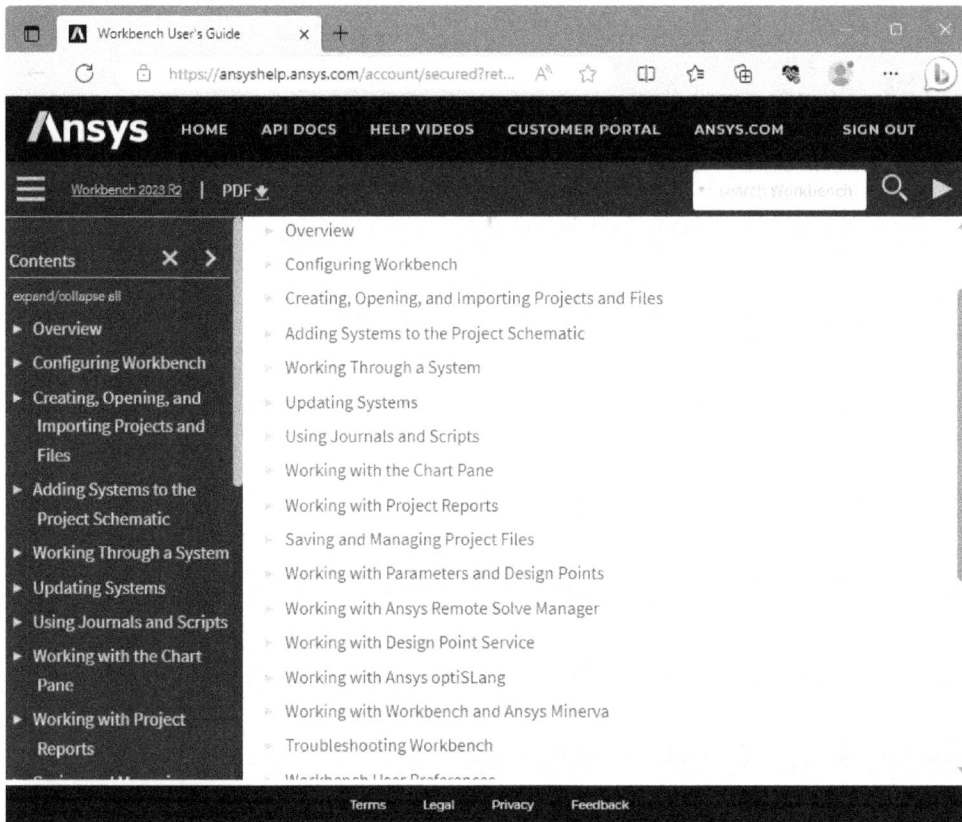

Figure 2-39 The ANSYS Help website

Quick Help

Quick Help is available for all the cells of an analysis or component system in the **Project Schematic** window. Click on the blue inclined arrow at the bottom right corner of any cell; a flyout will be displayed with a short description on the current status of the cell, refer to Figure 2-40. This flyout contains some relevant links that when clicked, will directly open the related document in the **ANSYS Help** window.

Context Sensitive Help

Context sensitive help lists the help topics of the **ANSYS Help** window, related to the currently active option. To open the **ANSYS Help** window, choose **Help > Show Context Help** from the Menu bar; the **Sidebar Help** window will be displayed on the right of the **Project Schematic** window, as shown in Figure 2-41. Alternatively, press the F1 key to open the **Sidebar Help** window.

2 🗐 Engineering Data ✓ ▲	
3 🗐 Geometry ? ▲	
4 🗐 Model	
5 🗐 Setup	Either no data has been
6 🗐 Solution	specified, or the model
7 🗐 Results	contains errors in

Either no data has been specified, or the model contains errors in DesignModeler. Right-click the Geometry cell and choose **Import Geometry** to load an existing geometry, choose **New Geometry** to open DesignModeler and create a new geometry, or choose **Edit** to open the current geometry in DesignModeler to correct the error.

Geometry Properties

Systems and Cells

ANSYS Workbench

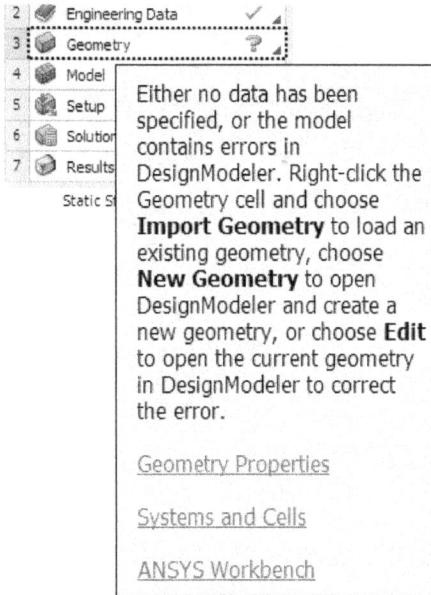

Figure 2-40 The Quick Help

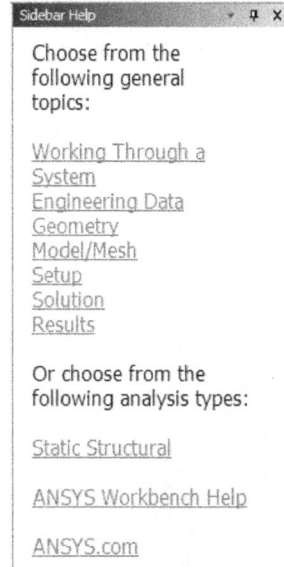

Sidebar Help ⚞ 📌 ✕

Choose from the following general topics:

Working Through a System
Engineering Data
Geometry
Model/Mesh
Setup
Solution
Results

Or choose from the following analysis types:

Static Structural

ANSYS Workbench Help

ANSYS.com

*Figure 2-41 The **Sidebar Help** window*

EXITING ANSYS WORKBENCH

To exit the **Workbench** window, choose **File > Exit** from the Menu bar. If the current project is not saved or if you have made any changes after the project was last saved, then on choosing the **Exit** option, the **ANSYS Workbench** message box will be displayed, as shown in **Figure 2-42**. Choose the **Yes** button from this message box to save the changes made in the current project and exit the **Workbench** window. Choose the **No** button to discard changes and close the **ANSYS Workbench** window. Choose the **Cancel** button from this message box to cancel the exit process.

⚠ Ansys Workbench ✕

The current project has been modified. Do you want to save it?

Yes   No   Cancel

*Figure 2-42 The **ANSYS Workbench** message box*

TUTORIAL

Tutorial 1

In this tutorial, you will start ANSYS Workbench and add a new Static Structural analysis system to the **Project Schematic** window. After adding the analysis system, save the project with the name *c02_ansWB_tut01* at the location *C:\ANSYS_WB\c02\Tut01*.

(Expected time: 15 min)

The following steps are required to complete this tutorial:

a. Create the project folder.
b. Start ANSYS Workbench 2023 R2.
c. Add the Static Structural analysis system to the **Project Schematic** window.
d. Save the project and exit the ANSYS Workbench session.

Creating the Project Folder

Before you start ANSYS Workbench, you need to create a project folder in which you will save all your projects.

1. Create a folder with the name **ANSYS_WB** at the location *C:*.

You will save all your projects in the folder *ANSYS_WB*.

Starting ANSYS Workbench 2023 R2

First, you need to start ANSYS Workbench and then add the analysis system to the project.

1. Choose **Start > Ansys 2023 R2 > Workbench 2023 R2** from the Start menu; the **Workbench** window is displayed.

2. Choose the **Save** button from the **Main** toolbar; the **Save As** dialog box is displayed.

3. Browse to the location *C:\ANSYS_WB* and then create a folder with the name **c02**.

4. Now, create another folder with the name **Tut01** under the *c02* folder and then open the newly created folder by double-clicking on it.

5. Enter **c02_ansWB_tut01** in the **File name** edit box and then choose the **Save** button from the **Save As** dialog box; the project is saved with the specified name.

Adding Static Structural Analysis System to the Project

Next, you need to add the **Static Structural** analysis system to the project.

1. Double-click on the **Static Structural** option displayed in the **Analysis Systems** toolbox in the **Toolbox** window; the **Static Structural** system is added to the project and is displayed in the **Project Schematic** window.

Note

*The **Static Structural** analysis system can also be added by dragging it from the **Toolbox** window and then dropping it on the **Project Schematic** window. You can also add it by using the options in the shortcut menu that is displayed on right-clicking in the **Project Schematic** window.*

2. Once the project is added to the **Project Schematic** window, its name is highlighted at the bottom of the analysis system in blue. If it is not highlighted, double-click on the name and enter **c02_ansWB_tut01**, as shown in Figure 2-43. The analysis system is renamed.

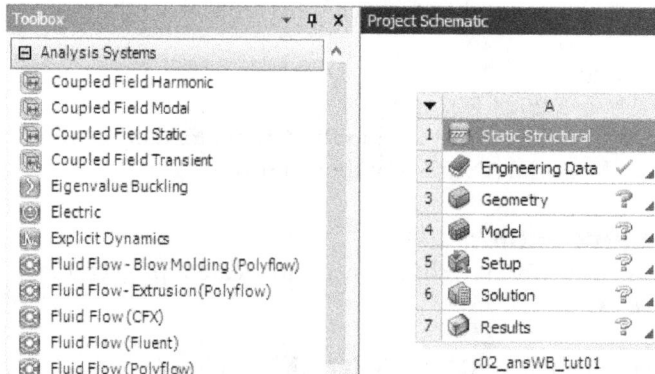

Figure 2-43   *The **Static Structural** analysis system added to the project*

Saving the Project and Exiting ANSYS Workbench

Now, you need to save the project and exit ANSYS Workbench.

1. Choose the **Save** button to save the project with the name *c02_ansWB_tut01*.

2. Choose **Exit** from the **File** menu; the current ANSYS Workbench session is closed

Self-Evaluation Test

Answer the following questions and then compare them to those given at the end of this chapter:

1. The archived project files are saved in _____ format.

2. To view all files related to the current project, choose _____ from the Menu bar.

3. Press the _____ key to open the **Sidebar Help** window.

4. In the **Workbench** window, you can change the units by using the options available in the _____ menu.

5. You can add an analysis system to the **Project Schematic** window by double clicking on the desired component in the **Analysis Systems** toolbox. (T/F)

6. The geometry for an analysis can not be imported from an external software package. (T/F)

7. The **Results** cell is used to display the contents of an analysis system. (T/F)

8. If you choose an option that has 3 dots (...) on its right then a dialog box corresponding to that option is displayed. (T/F)

Review Questions

Answer the following questions:

1. Which one of the following unit systems is assigned to a project by default?

    (a) **Metric (kg,m,s,°C,A,N,V)**
    (b) **SI (kg,m,s,K,A,N,V)**
    (c) **U.S. Engineering (lbm,in,s,R,A,lbf,V)**
    (d) **U.S. Customary (lbm,in,s,°F,A,lbf,V)**

2. An ANSYS project file is saved with _____ extension.

3. The _____ environment is used to create the geometry to be used in an analysis.

4. The customized analysis systems are added under the _____ toolbox in the **Toolbox** window.

5. To create a connected system for coupled analysis, double-click on an analysis system in the **Toolbox**. (T/F)

6. The *.wbpj* file is the default extension for the project files in ANSYS Workbench 2023 R2. (T/F)

7. A project consists of systems and a system consists of cells. (T/F)

8. You can specify a unit system from the Menu bar. (T/F)

9. The **Engineering Data** cell is used to define the geometry to be used in an analysis project. (T/F)

EXERCISE

Exercise 1

Start ANSYS Workbench and add a new thermal analysis system to the project, as shown in Figure 2-44. Save the project with the name *c02_exr01* at the location *C:\ANSYS_WB\c02\Exr*.

(Expected time: 10 min)

Hint: Use the **Thermal-Stress** analysis system displayed under the **Custom Systems** toolbox in the **Toolbox**.

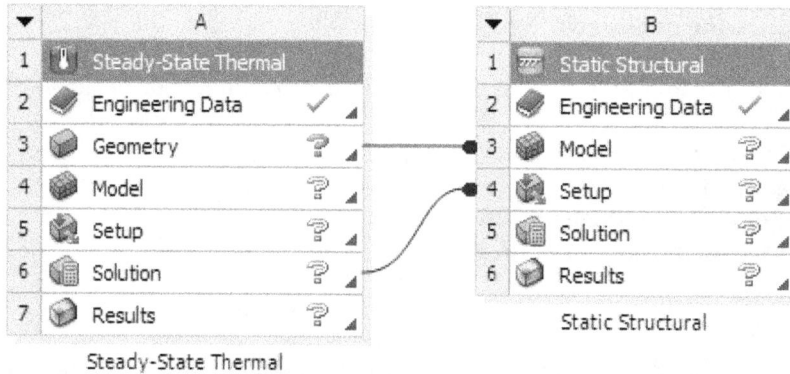

*Figure 2-44 The **Thermal-Stress** analysis system added to the **Project Schematic** window*

Chapter 3

Part Modeling - I

Learning Objectives

After completing this chapter, you will be able to:
- *Understand the DesignModeler Workspace*
- *Draw sketches*
- *Convert sketches into 3D models*
- *Understand views of the model in 3D space*
- *Apply constraints and relations*
- *Create new sketching planes*

INTRODUCTION TO PART MODELING

For conducting an FEA analysis, a part model is mandatory. In ANSYS Workbench, the next step after defining the material properties is to define the geometry to which the material properties will be applied. Like most of the other CAD modeling packages, the part models created in ANSYS Workbench are parametric and feature based. The parametric nature of an application is defined as its ability to use the standard properties or parameters (dimensions) in determining the shape and size of the geometry. You can also modify the shape and size of the model using these parameters. Feature is defined as the smallest building block of the model that can be modified individually.

Most of the 3D models created using ANSYS Workbench are a combination of sketched and placed features. The placed features are created without drawing a sketch, but the sketched features require a sketch to be drawn first. Generally, the base feature of any 3D model is a sketched feature and is created using a sketch. Therefore, while creating any design, you first need to draw the sketch for the base feature. Once you have drawn the sketch, you can convert it into the base feature and then add the other sketched and placed features to it to complete the 3D model.

In general terms, a sketch is defined as the basic contour for the feature. For example, consider the spanner shown in Figure 3-1. It is created using a single sketch drawn on the XY plane, as shown in Figure 3-2.

Figure 3-1 *Base feature of the spanner*

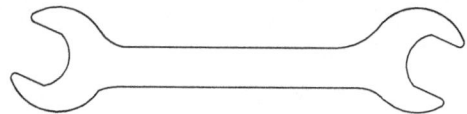

Figure 3-2 *Sketch for the base feature of the spanner*

INTRODUCTION TO DesignModeler WINDOW

In ANSYS Workbench, the part models and their sketches are defined in the **DesignModeler** window. You can define part models either by importing the CAD model created in some other CAD applications such as Pro/E, Solidworks, and so on, or by creating the model in the **DesignModeler** window of ANSYS Workbench 2023 R2.

In any system, the **DesignModeler** window is associated with the **Geometry** cell. The **Geometry** cell can be added to any analysis project as a standalone component system or as a part of any mechanical analysis system. Figure 3-3 displays system **A** as a standalone component system. In system **B**, the standalone system **A** is used as a part of an analysis system. In analysis system **C**, the model will be defined in the **DesignModeler** window that is associated with the **Geometry** cell of the same system.

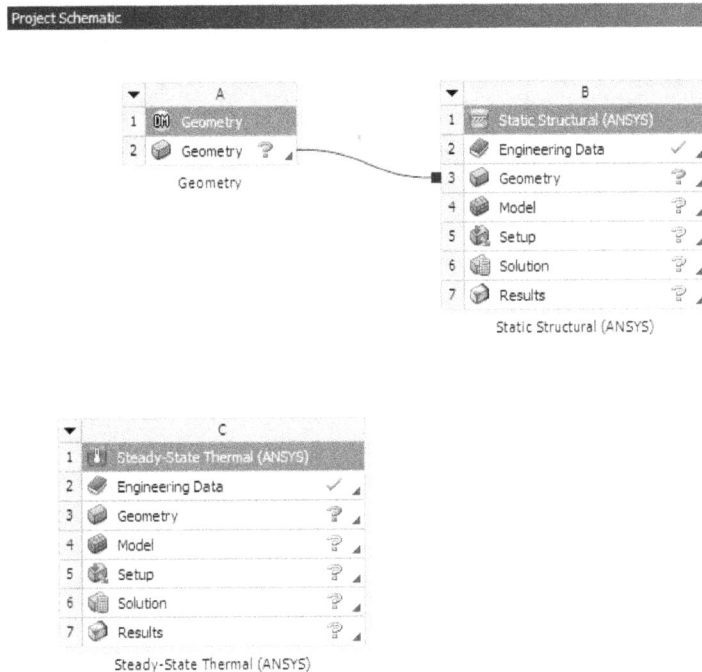

*Figure 3-3 The **Project Schematic** window displaying various standalone systems in it*

To open the **DesignModeler** window, right-click on the **Geometry** cell of any analysis system and choose the **New DesignModeler Geometry** option from the shortcut menu displayed; the **DesignModeler** window will be displayed.

When the **DesignModeler** window is opened, you can choose the desired unit from the **Units** menu. However, you can also use the unit system specified for the project in the **Workbench** window by selecting the **Display values in Project Units** from the **Units** menu. The **DesignModeler** window is shown in Figure 3-4.

The **DesignModeler** window can be used in two basic modes that are discussed next.

Sketching Mode

The **Sketching** mode is used to draw 2D sketches. Later on, these sketches can be converted into 3D models using the **Modeling** mode. To invoke the **Sketching** mode, choose the **Sketching** tab from the **DesignModeler** window, refer to Figure 3-4. The **Sketching** mode displays the **Sketching Toolboxes** window which contains five toolboxes. These toolboxes are used to create, modify, and dimension the sketches, refer to Figure 3-5.

Modeling Mode

The **Modeling** mode is used to generate the part model using the sketches drawn in the **Sketching** mode. By default, the **Modeling** mode is active when the **DesignModeler** window is invoked. If not, choose the **Modeling** tab from the **DesignModeler** window, refer to Figure 3-4. The **Modeling** mode in the **DesignModeler** window contains three default planes. Apart from three default planes, the list of all operations that are used to create any model in the **Modeling** mode will be listed in the Tree Outline as they are created, refer to Figure 3-6.

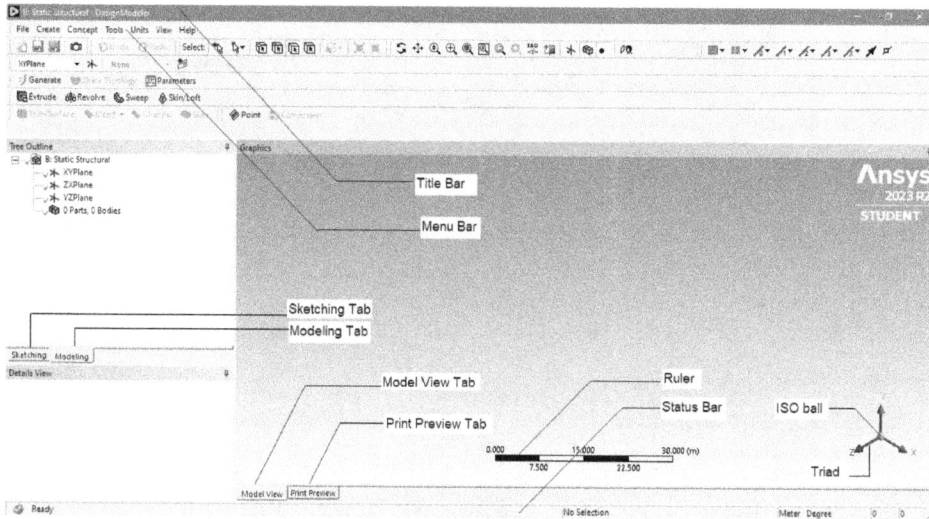

*Figure 3-4 The **DesignModeler** window*

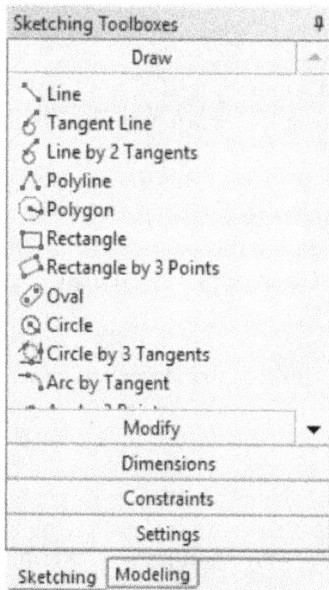

*Figure 3-5 The **Sketching Toolboxes** window*

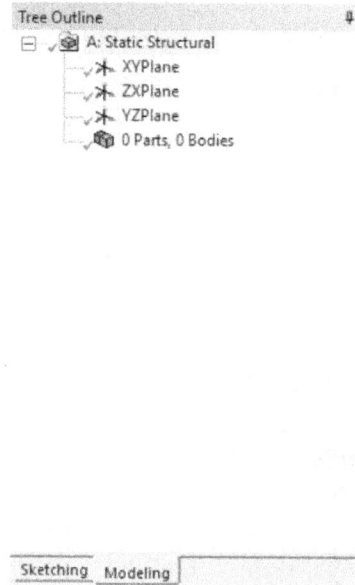

*Figure 3-6 The **Tree Outline***

SCREEN COMPONENTS OF THE DesignModeler WINDOW

The various screen components of the default **DesignModeler** window and important terms related to this window are discussed next.

Tree Outline

In ANSYS Workbench three planes **XYPlane**, **YZPlane**, and **ZXPlane** corresponding to the XY, YZ, and ZX planes of cartesian coordinate system are created by default and are displayed in the Tree Outline. These planes are used as sketching planes for drawing the sketches to be used for generating the part model.

A sketch is a collection of 2D drawing entities which is used for generating the features of a part model. The sketch can be created on any one plane. However, a single plane can have multiple sketches associated with it. The plane for creating the sketch can be specified by selecting it from the Tree Outline. When you click on a plane in the Tree Outline, it is displayed in the **Graphics** window. Figure 3-7 shows the plane that will be displayed on selecting **XYPlane** from the Tree Outline.

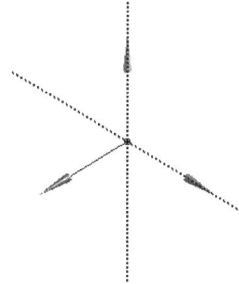

Figure 3-7 The XY plane displayed when **XYPlane** *is selected in the Tree Outline*

Note

1. The last sketching plane you worked on will act as the active plane for any future operation, unless and until you change the plane by selecting it from the Tree Outline.

2. Apart from these three default planes, the user can also create new planes at the specified location and orientation in the **Graphics** *window. The method of creating new planes will be discussed in detail in the later chapters.*

3. To start a new sketch, choose the **New Sketch** *tool from the* **Active Plane/Sketch** *toolbar. On doing so, a new sketch instance will be added under the desired plane in the Tree Outline. Alternatively, you can select the desired plane in the Tree Outline and then switch to the* **Sketching** *mode and draw a sketch. On doing so, a sketch instance will automatically be added under the selected plane.*

Details View Window

The **Details View** window located near the **Graphics** window is contextual in nature and changes its content according to the selection made in the Tree Outline. Figure 3-8 shows the **Details View** window which is displayed when the **Extrude1** is selected in the Tree Outline. You can also modify the selected entity by editing its parameters in the **Details View** window.

Details View	
− **Details of Extrude1**	
Extrude	Extrude1
Geometry	Sketch1
Operation	Add Material
Direction Vector	None (Normal)
Direction	Normal
Extent Type	Fixed
FD1, Depth (> 0)	5 m
As Thin/Surface?	No
Merge Topology?	Yes
− **Geometry Selection: 1**	
Sketch	Sketch1

Figure 3-8 The **Details View** *window*

Model View/Print Preview

The **Model View** and the **Print Preview** tabs are located at the lower left corner of the **Graphics** window, refer to Figure 3-4. By default, the **Model View** tab is chosen in the **DesignModeler** window. Subsequently, the sketches and the part model are displayed in this interface. To preview the current view of the model, choose the **Print Preview** tab. After previewing the model, choose **File > Print** from the Menu bar to print it. Note that the option to print the model will be available only in the **Print Preview** mode.

Ruler

The Ruler is displayed at the bottom of the **Graphics** window, refer to Figure 3-9. The Ruler helps the user to visualize and compare the actual size of the model with the size of the model displayed. The number displayed at the end of each block in the Ruler represents the cumulative length of the blocks on the left of the number. The quantity shown in brackets at the extreme right of the ruler represents current unit of the length. To toggle the display of the Ruler in the **Graphics** window, choose **View > Ruler** from the Menu bar.

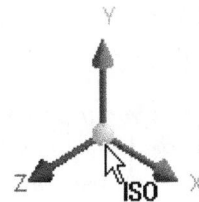

Triad

The Triad is displayed at the lower right corner of the **Graphics** window, refer to Figure 3-10. Triad helps the user to visualize the X, Y, and Z directions in the **Graphics** window. To toggle the display of the Triad in the **Graphics** window, choose **View > Triad** from the Menu bar. Move the cursor to any axis, its name will be displayed attached to the cursor. Moving the cursor in the negative direction of the three orthogonal axis system displays the temporary view of the axis and its name with a negative symbol (-). If you click on any axis system, the view will get oriented normal to the selected axis. Click on the ISO ball displayed at the center of the triad to orient the model in the isometric view.

Figure 3-9 The Ruler   *Figure 3-10* The Triad

Status Bar

The Status bar is located at the bottom of the screen, refer to Figure 3-4. The instructions for the currently active tool and its status are displayed on the left of the Status bar. The middle portion of the Status bar displays the information about the currently selected object. The right portion of the Status bar displays the current unit system and the coordinate value of the cursor location.

TUTORIALS

Tutorial 1

In this tutorial, you will create the I-section solid model shown in Figure 3-11. The dimensions of the model are shown in Figure 3-12. Save the project with the name *c03_ansWB_Tut01* at the location *C:\ANSYS_WB\c03\Tut01*.   **(Expected time: 30 min)**

Figure 3-11 *Model for Tutorial 1*

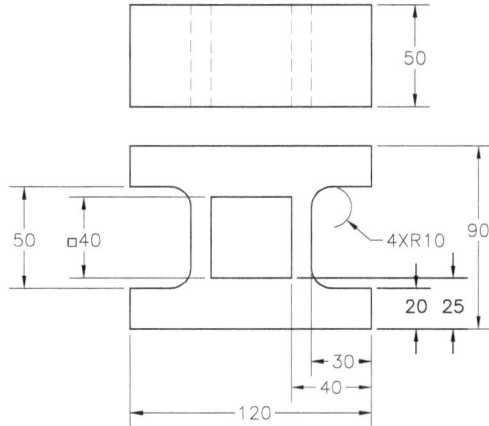

Figure 3-12 *Dimensions of the I-section*

The following steps are required to complete this tutorial:

a. Start ANSYS Workbench 2023 **R2** and add the **Geometry** component system.
b. Create the sketch for the outer loop of the I-section.
c. Apply constraints to the sketch.
d. Generate dimensions and edit them to achieve the required size of sketch.
e. Create the base feature.
f. Create the second feature.
g. Create the blend feature.
h. Rotate the view of the model dynamically.
i. Save the project and exit the ANSYS Workbench session.

Starting ANSYS Workbench and Adding the Component System

First you need to start ANSYS Workbench and then add an analysis system to the **Project Schematic** window.

1. Start ANSYS Workbench 2023 **R2.** The Workbench window is displayed.

2. Double-click on **Geometry** displayed under the **Component Systems** toolbox in the **Toolbox** window; the **Geometry** component system is added and displayed in the **Project Schematic** window, refer to Figure 3-13.

After the **Workbench** window is displayed and an analysis system is added to the **Project Schematic** window, the first step in any analysis is to define the geometry. There are two ways to define a geometry: by creating a new geometry and by importing an already created geometry from any solid modeling software. The **DesignModeler** window is used to create and edit geometries which are used in ANSYS Workbench. In ANSYS Workbench 2023 R2, a standalone system known as **Geometry** component system, is available to create geometries. Figure 3-13 shows a **Geometry** component system added to the **Project Schematic** window. Now, you need to save the project.

3. In the **Workbench** window, choose the **Save Project** button from the **Main** toolbar; the **Save As** dialog box is displayed.

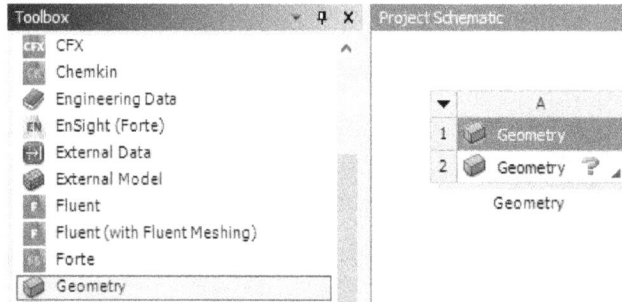

Figure 3-13 *The **Geometry** component system added to the **Project Schematic** window*

4. Browse to the location *C:\ANSYS_WB*.

5. Create a folder with the name **c03** in the *ANSYS_WB* folder and then double-click on the newly created *c03* folder to open it. Next, create a sub folder with the name **Tut01** under the *c03* folder and then choose the **Open** button from the **Save As** dialog box.

6. Enter **c03_ansWB_Tut01** in the **File name** edit box and choose the **Save** button from the **Save As** dialog box; the project is saved with the name specified.

Creating the Sketch

You now need to invoke the **DesignModeler** window to create the sketch.

1. Right-click on the **Geometry** cell of this component system to display a shortcut menu. Next, choose **New DesignModeler Geometry** from the shortcut menu; the **Starting DesignModeler** message is displayed in the status bar. After sometime, the **DesignModeler** window is displayed.

2. Choose the **Millimeter** option from the **Units** menu of the Menu bar.

   Next, you need to select the plane in which you want to create the sketch for the tutorial.

3. Select **XYPlane** in the Tree Outline that is displayed in the left pane of the **DesignModeler** window.

   The Tree Outline in the **DesignModeler** window lists all the operations that are carried out on the model. The Tree Outline for **Geometry** component system is shown in Figure 3-14. Note that the operations in the Tree Outline are listed in the sequence in which they were created.

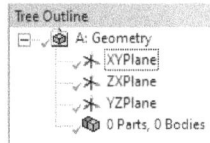

Figure 3-14 *The Tree Outline displayed in the* **DesignModeler** *window*

4. Choose the **Sketching** tab available at the bottom of the Tree Outline to display all the tools available for creating sketches.

   Since the XY plane has already been selected for drawing the sketch, as shown in Figure 3-15, you need to orient the plane normal to the viewing direction.

5. Right-click anywhere in the **Graphics** window and choose the **Look At** tool from the shortcut menu; the view is oriented as required, refer to Figure 3-16.

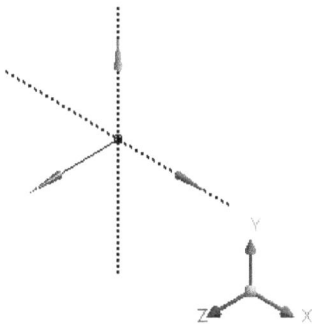

Figure 3-15 *The default Isometric view of the XY plane*

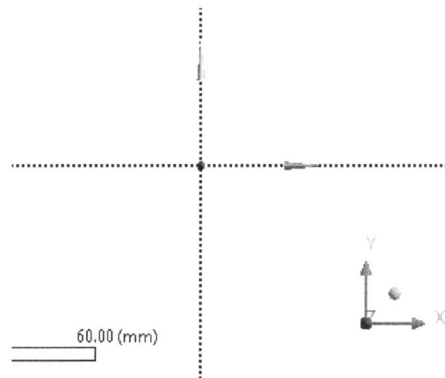

Figure 3-16 *XY plane after choosing* **Look At** *tool from the* **Graphics** *window*

The **Look At** tool is used to orient the plane on which the sketch is drawn, normal to the viewing direction. You can invoke this tool from the **Graphics** toolbar, refer to Figure 3-17. Alternatively, invoke this tool from the shortcut menu which is displayed on right-clicking anywhere in the **Graphics** window.

Figure 3-17 *The Graphics toolbar*

Next, you need to create the sketch, refer to Figure 3-18. For ease of creating the sketch, the entities of the sketch have been numbered.

6. From the **Draw** toolbox, choose the **Line** tool; the Status bar displays the message **Line -- Click, or Press and Hold, for start of line**.

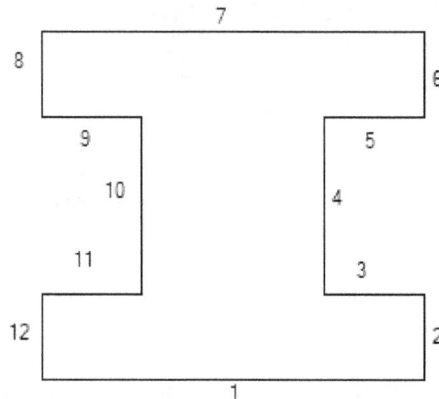

Figure 3-18 Sketch for Tutorial 1

A line is defined as the shortest distance between two points. A line is one of the basic sketching entities available in the **DesignModeler** window. To draw a line, choose the **Line** tool from the **Draw** toolbox in the **Sketching Toolboxes** window. On doing so, you will be prompted to specify the start point of the line. Click anywhere in the Graphics window to specify the start point; you will be prompted to specify the end point of the line. Next, click to specify the end point of the line; a line will be created.

After specifying the start point of the line, if you realize that the specified start point is wrong, right-click in the **Graphics** window and choose the **Back** option from the shortcut menu displayed. As a result, the specified start point will get nullified without exiting the **Line** tool.

7. Move the cursor close to the origin in the **Graphics** window and click to specify the start point of the line when the symbol of Coincident Point constraint (**P**) is displayed, refer to Figure 3-19.

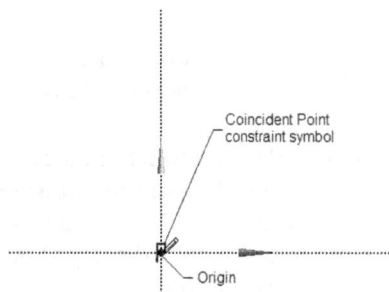

Figure 3-19 Specifying the start point of the line

After specifying the start point of the line, if you move the cursor horizontally, an H symbol will be displayed on the line. This symbol indicates that if you specify the end point of the line at the current location, a horizontal line will be drawn. Similarly, while drawing a line close to an existing line, if C symbol is displayed, it indicates that the end point specified at the current location of the cursor will be coincident with the existing line.

Table 3-1 *List of auto constraints*

Auto Constraint Name	Symbol	Description
Horizontal	H	Makes the entity horizontal
Vertical	V	Makes the entity vertical
Parallel	//	Makes the entity parallel to another entity
Perpendicular	⊥	Makes the entity perpendicular to another entity
Tangent	T	Makes the entity tangent to another entity
Equal Radius	R	Makes the entity equal to another entity
Coincident	C	Makes the end point of an entity coincident with another entity
Coincident Point	P	Makes the end point of the current drawing entity coincident with a point.

8. Move the cursor toward right to some distance and click to specify the end point of line when the symbol of Horizontal constraint (**H**) is displayed, as shown in Figure 3-20. Line 1 is created and its blue color indicates that the line is fully constrained, refer to Figure 3-21.

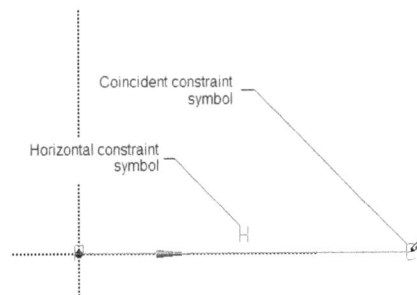

Figure 3-20 *Specifying the end point of the line*

9. Next, move the cursor close to the end point of line created in the last step; the symbol of Coincident Point constraint (**P**) is displayed. Click to specify the start point of the second line, refer to Figure 3-21.

10. Move the cursor vertically upward; the symbol of Vertical constraint (**V**) is displayed. Click to specify the end point of the line; line 2 is drawn, as shown in Figure 3-22.

12. Now, move the cursor close to the end point of line 2 created in the last step; the symbol of Coincident Point constraint (**P**) is displayed. Click at that point to specify the first point of line 3, as shown in Figure 3-23.

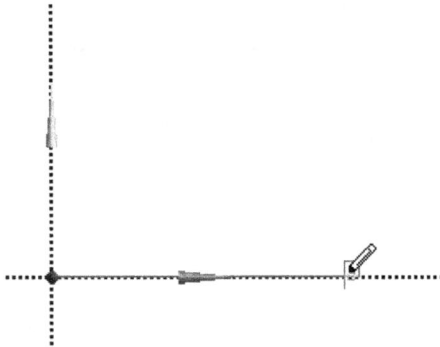

Figure 3-21 *Specifying the startpoint of line 2*   ***Figure 3-22*** *Specifying the endpoint of line 2*

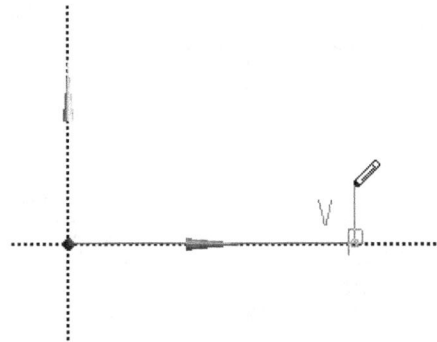

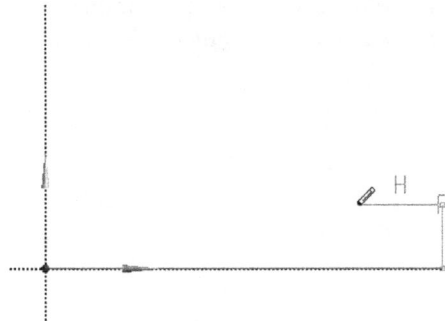

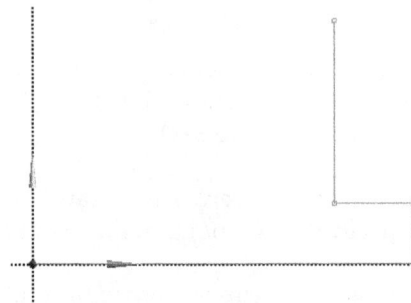

13. Move the cursor toward left to some distance; the symbol of Horizontal constraint (**H**) is displayed. Next, Click to specify the end point of line 3, as shown in Figure 3-24.

14. Move the cursor close to the end point of line3; the symbol of Coincident Point constraint (**P**) is displayed. Click at that point to specify the start point of line 4, as shown in Figure 3-25.

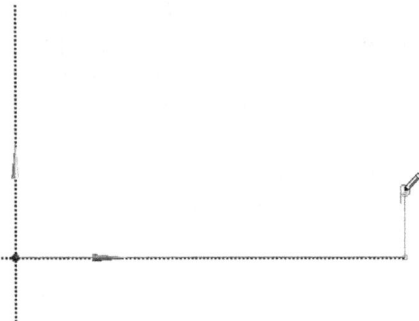

Figure 3-23 *Specifying the startpoint of the line 3*   ***Figure 3-24*** *Specifying the endpoint of line 3*

15. Move the cursor upward to some distance; the symbol of Vertical constraint (**V**) is displayed along the cursor. Click to specify the end point of line 4, refer to Figure 3-26.

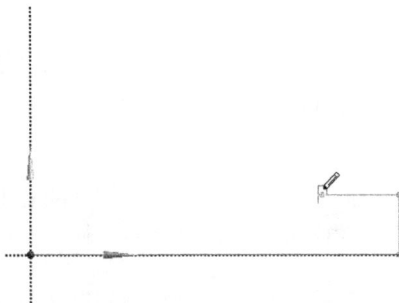

Figure 3-25 *Specifying the startpoint of the line 4*   ***Figure 3-26*** *Specifying the endpoint of line 4*

16. Similarly, draw all the lines specified in the sketch. The sketch drawn is just a representation of the final sketch to be drawn based on dimensions. The final sketch before applying the constraints and dimensions is shown in Figure 3-27.

Note

1. The sketch shown in Figure 3-27 is just for reference and is not created based on dimensions.

*2. While drawing the sketch, some of the constraints like Horizontal, Vertical and Coincident are automatically applied. You can also apply constraints manually after drawing the sketch, by choosing the desired constraint tool from the **Constraints** toolbox.*

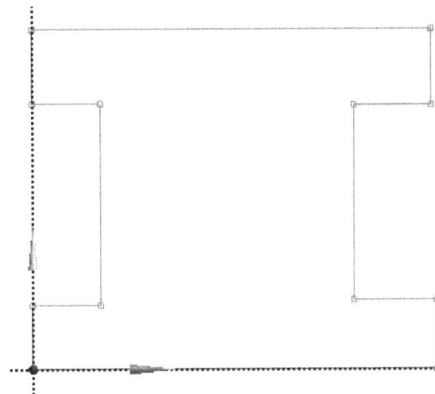

Figure 3-27 *The sketch drawn using automatic constraints*

Applying Constraints to the Sketch

After a sketch for the model is drawn, refer to Figure 3-27, you need to constrain the entities of the sketch to restrict some degrees of freedom.

1. Expand the **Constraints** toolbox in the **Sketching Toolboxes** window.

2. Choose the **Horizontal** tool from the **Constraints** toolbox; the **Horizontal -- Select line or ellipse for horizontal constraint** message is displayed in the Status bar indicating that the Horizontal constraints can be applied to entities of the sketch.

    Horizontal constraint is used to make a line or the major axis of an ellipse horizontal.

3. Select lines 1, 3, 5, 7, 9, and 11 one by one to make them horizontal, if not already horizontal, refer to Figure 3-20.

4. Next, choose the **Vertical** tool from the **Constraints** toolbox; the **Vertical -- Select line or ellipse for vertical constraint** message is displayed in the Status bar indicating that you can now apply vertical constraints to entities.

    Vertical constraint is used to make a line or the major axis of an ellipse vertical.

5. Select the lines 2, 4, 6, 8, 10, and 12 one by one to apply vertical constraints to them; all these lines become vertical now, refer to Figure 3-28.

    The next step is to apply Equal Length constraints to lines which are equal in length.

6. Choose the **Equal Length** tool from the **Constraints** toolbox of the **Sketching Toolboxes** window. Use the down arrow available next to the **Settings** toolbox tab to scroll down and view the **Equal Length** tool, if it is not visible by default; you are prompted to select the first line to apply Equal Length constraint.

7. Move the cursor close to line 1; a rectangular box is attached to the cursor, which means that you can now select the line to apply Equal Length constraint. Select line 1; it turns yellow and you are prompted to select the second line.

8. Select 7; the Equal Length constraint is applied and both the lines become equal in length.

9. Similarly, apply Equal Length constraint between the lines 2 and 12, 6 and 8, and 2 and 6 to make these lines equal. In addition, make the lines 3, 5, 9, and 11 equal in length.

10. Next, apply Equal Length constraint to lines 4 and 10.

Applying Dimensions to the Sketch

Now you need to apply dimensions to the sketch. To apply dimensions to the sketch, you need to invoke the desired tool from the **Dimensions** toolbox tab.

1. Choose the **Dimensions** toolbox from the **Sketching Toolboxes** window to expand it.

2. Choose the **General** tool from the expanded **Dimensions** toolbox; the **Details View** window is displayed. Also, you are prompted to select first point or 2D edge for applying Horizontal dimension.

   The **General** tool is used to create dimensions depending upon the entities selected. After the **General** tool is invoked, select an entity from the **Graphics** window to dimension it.

3. Move the cursor close to line 1; a rectangle is attached to the cursor. Select the line 1; the cursor changes to pencil shape. Click to place dimension. As the line is horizontal, H1 is placed above the line 1.

   Note

   *In **DesignModeler**, the dimensions are placed as symbols. For example, when you invoke the **Horizontal** tool from the **Dimensions** toolbox and then dimension an entity, the dimension that would be placed is **H1**. The values of all the placed dimensions in the sketch can be changed by changing the corresponding dimensional values under the **Dimensions** node in the **Details View** window.*

4. Similarly, select the line 8 and place dimension V2 to the left of the line.

5. Select line 9, and place dimension H3 below the line.

6. Select line 10 and place dimension V4 below V2, as shown Figure 3-28.

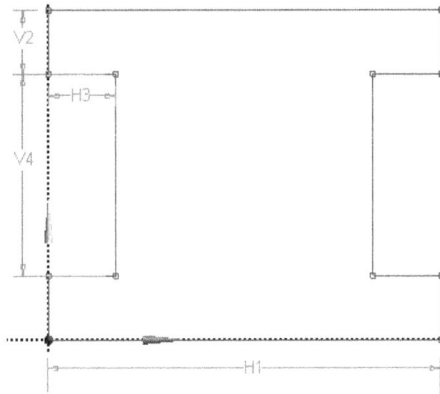

Figure 3-28 *Sketch after applying dimensions*

Note
*1. When line 11 is selected for dimensioning, the **A:Geometry - DesignModeler** warning message box is displayed, as shown in Figure 3-29. This message states that placing this dimension will over-constrain the sketch and you can edit and place this dimension as a reference. On choosing the **OK** button from the warning message box, the dimension of line 9 will turn red, indicating that the sketch is over-constrained, as shown in Figure 3-30.*

2. If you want to place the over-constrained dimension, place it on the sketch, as shown in Figure 3-31.

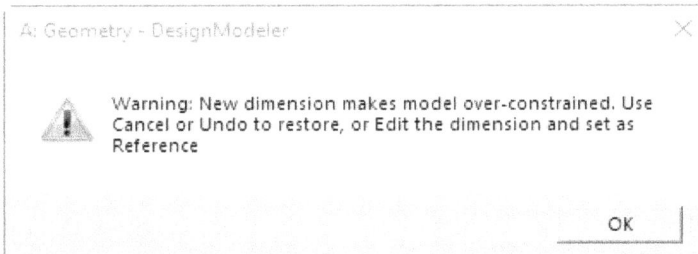

Figure 3-29 *The **A: Geometry-DesignModeler** warning message box*

To undo changes, click anywhere in the **Graphics** window and then press Ctrl+Z keys; over-constrained dimensions will vanish.

Tip
*By default, the name of the dimension is displayed on the dimension line. However, if you want to display the value of the dimension or both the name and value of the dimension on the dimension line, choose the **Display** tool from the **Dimensions** toolbox. On doing so, the **Name** and **Value** check boxes will be displayed on the right of the **Display** tool. Select the respective check box, the corresponding parameter will be displayed on the dimension lines in the **Graphics** window. Note that you cannot clear both the check boxes at the same time.*

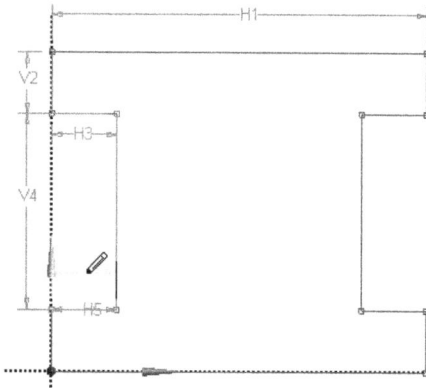

Figure 3-30 *Over-constrained sketch*

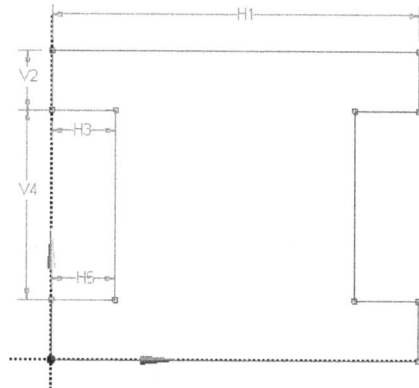

Figure 3-31 *Placing dimension on sketch*

Assigning Dimensional Values to the Sketch

After dimensions are assigned to the sketch, you need to change their values to get the exact dimensions.

1. In the Tree Outline, choose the **Modeling** tab to view the Tree Outline. On doing so, the **Details View** window is also activated with various nodes active in it.

2. Expand the Tree Outline, if it is not already expanded and then expand the **XYPlane** node to view **Sketch1**. Select **Sketch1**; the **Details View** window with corresponding parameters is displayed.

3. Expand the **Dimensions** node in the **Details View** window, if it is not already expanded.

    There are four dimensions present under this node : **H1**, **H3**, **V2**, and **V4**. These dimensions refer to the dimensions of lines 1, 9, 8, and 10, respectively.

4. To assign a different dimension, click on the edit box to the right of **H1**; the corresponding dimension is highlighted in yellow in the sketch. Enter **120** as the new dimension value and then press Enter; the lengths of lines 1 and 7 are changed.

Note

*1. As you specify the dimensional values to a dimension in the **Details View** window, the length of the entity is changed and the length of the entities which are constrained along with this entity are also modified.*

*2. While changing the dimensions, the complete sketch may not fit in the screen of your computer. To fit the sketch in the screen, choose the **Look At** tool from the **Graphics** toolbar.*

5. Next, in the **Details View** window, click on the edit box next to **H3**; the corresponding dimension is highlighted in the sketch. Enter **30** as the new dimension value and then press Enter; the length of lines 3, 5, 9, and 11 is changed.

6. Click on the edit box on the right of **V2**; the corresponding dimension is highlighted in yellow in the sketch. Enter **20** in the edit box and then press Enter; the length of lines 2, 6, 8, and 12 is changed in the sketch.

7. Click on the edit box on the right of **V4**; the corresponding dimension is highlighted in yellow in the sketch. Enter **50** in the edit box and then press Enter; lines 4 and 10 are modified.

You will notice that as soon as **H1**, **V2**, **H3**, and **V4** are modified in the **Details View** window, the whole sketch is modified. Therefore, changing the dimension of one entity will ensure that the dimensions of all other entities which are equal in length are also modified.

Extruding the Sketch

After creating the sketch of the outer profile of the I-section, you need to convert it into the base feature by using the **Extrude** tool. The **Extrude** tool is used to add or remove material from the specified sketch along a straight line in the specified direction.

1. Click on the ISO ball placed at the bottom right in the **Graphics** window; the view is changed to Isometric, as shown in Figure 3-32.

2. Choose the **Extrude** tool from the **Features** toolbar; the preview of the extruded feature with default values is displayed in the **Graphics** window, as shown in Figure 3-33. Also, **Extrude1** is added under the three default planes in the Tree Outline, as shown in Figure 3-34.

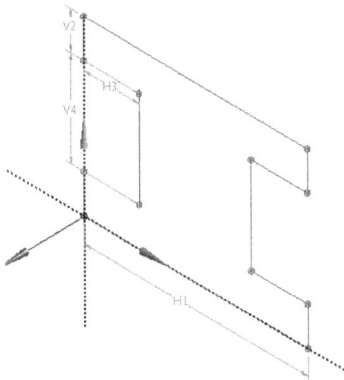

Figure 3-32 *The Isometric view of the sketch*

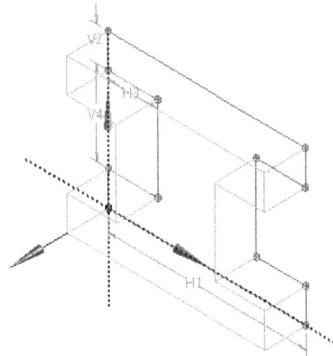

Figure 3-33 *The preview of feature after choosing the **Extrude** tool*

Note
1. The sketch plane on which you worked last acts as an active plane for any future operation, until you change the plane.

*2. To remove material from a feature, you need to have the base feature. Therefore, you cannot use the **Extrude** tool for material removal before you create the base feature.*

The default parameters used for generating the preview of the extruded feature are displayed in the **Details View** window. To get the required shape of the base feature, you need to edit parameters specified in each node of the **Details View** window. The **Both-Symmetric** option is used to add material on both the sides of the sketch with same values of the depth of material addition.

3. Based on the design requirements, the material should be added normal and symmetrically on both sides of the sketch. To add material to the extrusion, click on **Direction** drop-down list; a down arrow is displayed on the right of **Direction** in the **Details View** window. Next, click on the down arrow and select the **Both-Symmetric** option from the list displayed, as shown in Figure 3-35. The preview of the extruded feature is changed in the **Graphics** window and now it displays the same amount of material added on both sides of the sketch.

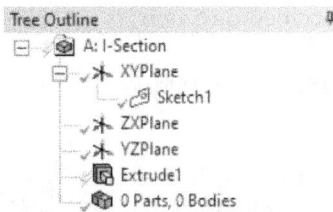

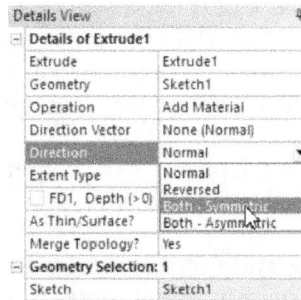

Figure 3-34 *The Tree Outline with*   **Figure 3-35** *Selecting the **Both-symmetric** option from*
Extrude 1 *added to it*   *the **Direction** drop-down list in the **Detail View** window*

The **Reversed** option in the **Direction** drop-down list is used to reverse the direction of material addition. The **Both-Asymmetric** option is used to add material on both the sides of sketch with different values of the depth of material addition.

Next, you need to specify the depth of material addition.

4. Enter **25** in the **FD1, Depth (>0)** edit box of the **Details View** window, refer to Figure 3-36.

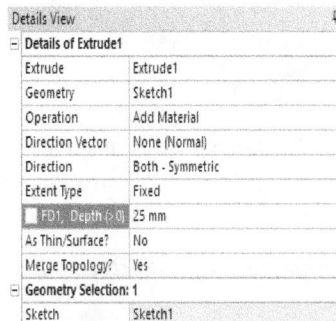

Figure 3-36 *Specifying the depth of material addition in the*
Details View *window*

The total depth of material addition is 50 mm, but the material will be added symmetrically by the same depth on both the sides of the sketch. Therefore, 25 mm has been specified as the depth value.

After all the parameters in the **Details View** window are specified, the preview of the extrude feature is displayed in the **Graphics** window. To create the solid model, you need to generate the extrude feature. Also, notice the thunderbolt symbol displayed before **Extrude 1** in the Tree Outline, which indicates that the feature needs to be generated.

Next, you need to complete the extrusion process with the specified values.

5. Choose the **Generate** tool from the **Features** toolbar; the base feature is created with the specified settings, refer to Figure 3-37. Also, the thunderbolt symbol is changed to green tick mark indicating that the feature is updated.

The **Generate** tool is available in the **Features** toolbar and is used to update the model after any changes are made in it. This tool can also be invoked from the shortcut menu which is displayed on right-clicking anywhere in the **Graphics** window.

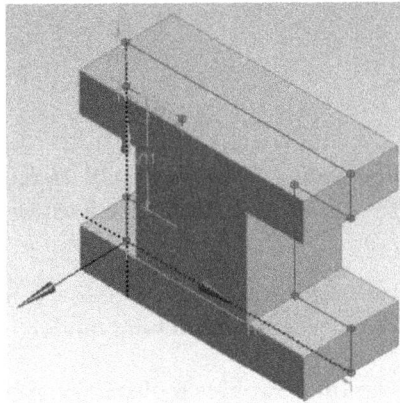

Figure 3-37 Resulting base feature

Tip
Instead of choosing the Generate tool from the Features toolbar, you can press the F5 key to generate a feature.

Creating the Second Feature

Next you need to create a rectangular cutout in the I-section, refer to Figure 3-11. To create a rectangular cutout, you will draw a sketch on the XY plane. Next, you will remove material by extruding he sketch.

1. Select **XYPlane** from the Tree Outline; the XY plane becomes the active plane.

2. Choose the **New Sketch** tool from the **ActivePlane/Sketch** toolbar; the new entry with the name **Sketch2** is added under the **XYPlane** node in the Tree Outline, refer to Figure 3-38.

The **New Sketch** tool is available in the **Active Plane/Sketch** toolbar located just above the **Graphics** window. This tool is used to create new sketches for the models.

Notice that the Sketch 1 and the dimensions are still displayed in the **Graphics** window. Since this sketch is not required, you can hide its display.

3. Right-click on **Sketch1** in the Tree Outline and then choose the **Hide Sketch** option from the shortcut menu displayed, refer to Figure 3-39.

Note

*If needed, you can display the sketch and its dimensions again. To do so, right-click on **Sketch1** in the Tree Outline and then choose the **Show Sketch** option from the shortcut menu displayed.*

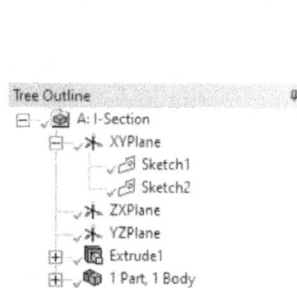

Figure 3-38 The Tree Outline with Sketch2 added to it

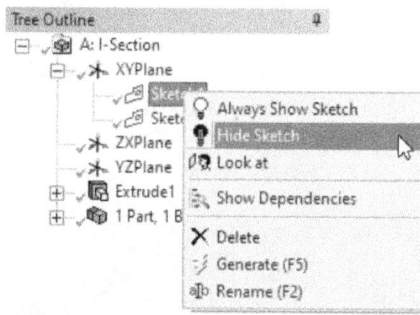

*Figure 3-39 Hiding **Sketch1** by choose the **Hide Sketch** option from the shortcut menu*

4. Select **Sketch 2** from the Tree Outline and then choose the **Sketching** tab displayed at the bottom of the Tree Outline to invoke the **Sketching** mode.

5. Choose the **Look At** tool from the **Graphics** toolbar; the sketching plane is oriented normal to the viewing direction. Orienting the sketching plane enables you to easily draw the sketch.

6. Choose the **Rectangle** tool from the **Draw** toolbox; the cursor changes into the Draw cursor and you are prompted to specify the first corner point of the rectangle.

In the **DesignModeler** window, there are two methods of drawing rectangle: by specifying two diagonally opposite points of the rectangle and by specifying the three corners of the rectangle. The rectangle created by using the first method can be either horizontal or vertical. The rectangle created by the second method is placed at an orientation specified by the user. You can use any of the two methods for drawing the rectangle using the **Draw** toolbox.

7. Specify the first corner point, refer to Figure 3-40; the preview of the rectangle is attached to the cursor and you are prompted to specify the diagonally opposite corner point of the rectangle.

8. Click to specify the second point of the rectangle, refer to Figure 3-40; the rectangle is created, refer to Figure 3-41.

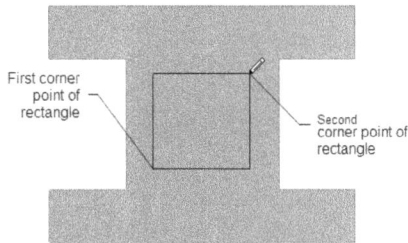

Figure 3-40 Specifying the first and second corner points of the rectangle

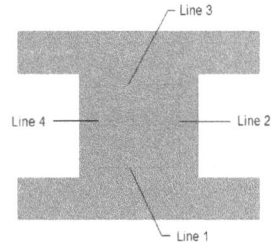

Figure 3-41 The rectangle created

9. Press the Esc key to exit the **Rectangle** tool.

   After creating the rectangle, you need to generate its dimensions so that you can specify the size of the rectangle and place it at the desired location.

10. Expand the **Dimensions** toolbox in the **Sketching Toolboxes** window.

11. Choose the **General** tool from the **Dimensions** toolbox, if not chosen by default.

12. Select Line 1; the preview of dimensional constraint attached to the cursor is displayed.

13. Click below the Line 1 to place the dimension; the dimension is generated and its value is displayed on the dimension line.

    Next, you need to generate a dimension between Line 4 and the Y axis.

14. Move the cursor near the lower end point of Line 4 and click when the cursor changes into the Point selection cursor.

15. Click on the Y axis; the preview of the dimension is attached to the cursor.

16. Move the cursor downward and click to place the dimension, refer to Figure 3-42.

17. Similarly, generate other two dimensions, refer to Figure 3-42.

    After the dimensions are placed on the sketch, you need to specify their exact values in the **Details View** window.

18. Edit the value of the second dimension **L5** to 40 in the **Details View** window; the length of the line is changed to 40 mm and is displayed in the **Graphics** window, refer to Figure 3-42.

19. Click on the edit box next to the second dimension **L5** and specify the dimensional value as 40; the dimension of the line is changed to 40, refer to Figure 3-42.

20. Similarly change the dimensional values of **V7** and **V8** dimensions to 25 and 40, respectively. Figure 3-42 shows the final sketch of the cutout feature.

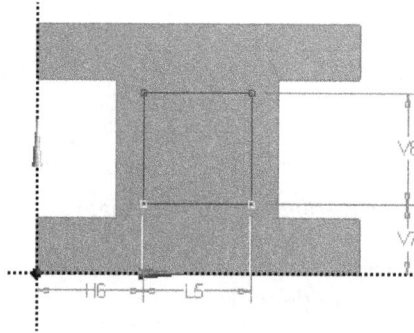

Figure 3-42 *Final sketch of the cutout feature*

Note

*The name of the dimension displayed on the **Details View** window may be different in your system.*

21. Choose the **Modeling** tab displayed at the bottom of the **Sketching Toolboxes** window to switch to the **Modeling** mode; the Tree Outline is displayed.

    After you exit the **Sketching** mode, the sketching plane still remains normal to the viewing direction. To proceed further with the feature creation operation, it is advised that you change the view of the sketching plane to isometric view.

22. Right-click in the **Graphics** window, and then choose the **Isometric View** option from the shortcut menu displayed; the sketch is displayed in the Isometric view.

Tip
*You can also click on the ISO ball (cyan color) displayed on the Triad to change the view of the sketching plane to the Isometric view. The Triad is displayed at the lower right corner of the **Graphics** window, refer to Figure 3-43.*

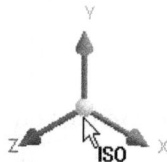

Figure 3-43 *The Triad with the ISO ball*

After the sketch is created, you need to cut the material from the sketch to create the cutout feature. You will use the **Extrude** tool to cut the material.

23. Select **Sketch2** from the Tree Outline and then choose the **Extrude** tool [🔲 Extrude] from the **Features** toolbar; the preview of the extruded feature with default values is displayed in the **Graphics** window. Also, a node for the extruded feature with the name **Extrude2** is added below the **Extrude1** node in the Tree Outline.

The default parameters used for generating the preview of the extruded feature are displayed in the **Details View** window. You need to edit values in the **Details View** window to get the desired shape of the cutout.

In this tutorial, material should be removed in the normal direction and symmetrically from both sides of the sketch.

24. Select the **Cut Material** option from the **Operation** drop-down list in the **Details View** window, refer to Figure 3-44; the preview of the material to be removed from the existing base feature is displayed in the **Graphics** window.

25. Select the **Both-Symmetric** option from the **Direction** drop-down list in the **Details View** window, refer to Figure 3-45; the same amount of material is removed from both sides of the sketch.

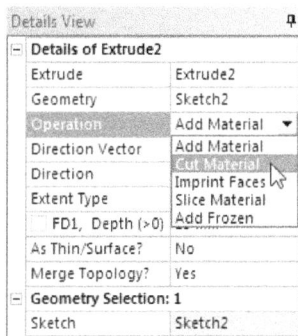

Figure 3-44 *Selecting the* **Cut Material** *option from the* **Operation** *drop-down list*

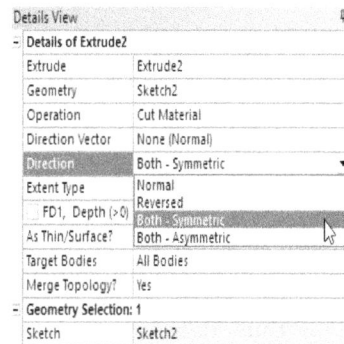

Figure 3-45 *Selecting the* **Both-Symmetric** *option from the* **Direction** *drop-down list*

Next, you need to select the method to specify the depth of material removal.

26. Select the **Through All** option from the **Extent Type** drop-down in the **Details View** window, refer to Figure 3-46.

The **Through All** option is used to remove material throughout the model in the specified direction. Instead of selecting this option, you can also select the **Fixed** option and then specify the exact depth of material removal, as you did for the base feature.

Notice the yellow thunderbolt symbol displayed at the upper-right corner of the **Extrude2** node in the Tree Outline, which indicates that the feature needs to be generated.

27. Next, to complete the process of material removal, choose the **Generate** tool [⚡ Generate] from the **Features** toolbar; the cutout feature is created, as shown in Figure 3-47.

Also, the yellow thunderbolt symbol is changed to green check mark, indicating that the feature is updated.

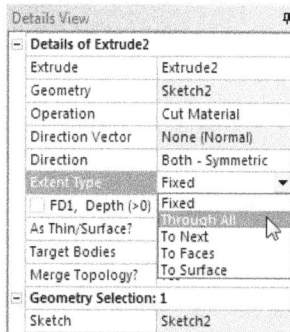

Figure 3-46 *Selecting the **Through All** option from the **Extent Type** drop-down list*

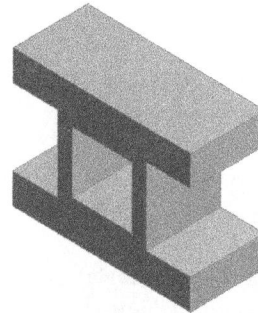

Figure 3-47 *The model after creating the cutout*

> **Note**
> *In Figure 3-47, the display of plane has been turned off to see the cutout feature clearly.*

In the Tree Outline, certain symbols are displayed on the upper right corner of each feature node. These symbols and their meanings are given in Table 3-2.

Table 3-2 *Various symbols and their meaning*

Symbol	Meaning
✓ Green tick mark	The feature creation succeeded and can be used for further processes.
⚡ Yellow thunderbolt mark	The feature has been modified and it needs to be updated.
✓ Yellow tick mark	The feature has been generated, but some warnings are associated with it.
❶ Red exclamation mark	The feature has failed to generate and you may need to redefine the feature.
✕ Blue cross mark	The feature is suppressed and has no influence on the final model.

Creating the Blend Feature

Next you need to create the blend feature (fillet) with the radius of 10 mm at the sharp corners of I-section, refer to Figure 3-11. You will create the fillet on the four edges of the model by using the **Fixed Radius** option from the **Blend** drop-down in the **Features** toolbar.

1. Choose the **Fixed Radius** tool from the **Blend** drop-down in the **Features** toolbar, refer to Figure 3-48; **FBlend1** is attached to the Tree Outline. Also, you are prompted to select the edges to be blended.

The **Blend** tool is used to remove sharp edges in a 3D model. The **Fixed Radius** option is used to create blends with a radius that is constant throughout the edge on which the blend is applied.

2. Press the CTRL key and select the four edges of the model, refer to Figure 3-49. You can use the tools available in the **Graphics** toolbar to rotate the model to view its respective edges.

Note

*To select the edges of a model without rotating it for generating the blend feature, change the display mode to wireframe. To change the display of the model to wireframe mode, choose the **Wireframe** option from the **View** menu. To change the display of the model back to the shaded mode, choose the **Shaded Exterior and Edges** or **Shaded Exterior** option from the **View** menu.*

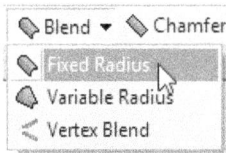

*Figure 3-48 Choosing the **Fixed Radius** option from the **Blend** drop-down*

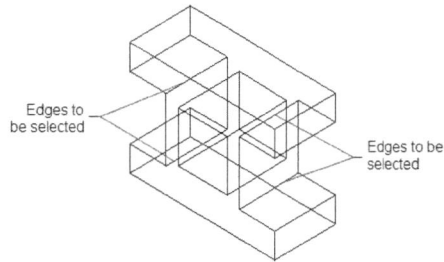

Figure 3-49 Edges to be selected for creating the blend feature

3. Choose the **Apply** button from the **Geometry** selection box in the **Details View** window to accept the selection of the edges to be blended, as shown in Figure 3-50.

4. Enter **10** in the **FD1, Radius (>0)** edit box to specify the radius of the blend feature.

5. Next, choose the **Generate** tool from the **Features** toolbar to finish creating the blend with the specified radius. The final model after creating the blend is shown in Figure 3-51.

Note

If you select a face for creating blend, all the edges of the selected face will be blended.

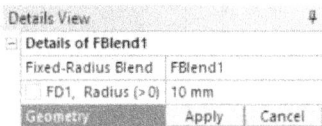

*Figure 3-50 Choosing the **Apply** button from the **Details View** window*

Figure 3-51 The model after generating the blend feature

Rotating the View of the Model Dynamically

You can rotate the view of the model dynamically in 3D space so that it can be viewed from all directions. This allows you to view all features clearly.

1. Choose the **Rotate** tool from the **Graphics** toolbar; the cursor changes into the Rotate cursor. Alternatively, right-click in the **Graphics** window and then choose the **Cursor Mode > Rotate** option from the shortcut menu displayed.

    The **Rotate** tool is used to rotate the view of the model freely in the **Graphics** window. You can invoke this tool from the **Graphics** toolbar of the **DesignModeler** window. Like other tools of the **Graphics** toolbar, the **Rotate** tool is also a transparent tool, implying that the **Rotate** tool can be invoked even when you are using some other tools.

2. Next, press and hold the left mouse button and drag the cursor to rotate the model in 3D space.

3. Choose the **Rotate** tool from the **Graphics** toolbar to exit it.

> **Tip**
> *You can also rotate the view of the model without invoking the **Rotate** tool. To do so, press and hold the middle mouse button in the **Graphics** window and drag the cursor.*

> **Note**
> *After rotating the model dynamically, you can restore the Isometric view again by clicking on the ISO ball (cyan color) in the Triad.*

4. Exit the **DesignModeler** window; the **Workbench** window is displayed.

> **Note**
> *You can close the **DesignModeler** window even without saving the project. However, in this case, the model will be saved automatically but the project will remain unsaved. You need to save the project by choosing the **Save Project** button from the **Standard** toolbar of the **Workbench** window. If the project is saved then its name is displayed on the title bar of the **Workbench** window. Otherwise it will display **Unsaved Project** on the title bar after closing the unsaved project. Also, the model data that was saved automatically will be lost forever.*

Apart from freely rotating the model, you can also view it using the standard orthographic projections. To view the model in the orthographic direction, move the cursor over the Triad on any axis, the name of the axis will be displayed attached to the cursor. Click on any axis; the view will be oriented normal to the selected axis. Move the cursor in the negative direction of the three orthogonal axes, the system will display the temporary view of the axis and its name with a negative symbol (-). You can also use these negative axes to orient the view of the model. Table 3-3 lists the various orthographic views and axes that need to be selected for achieving the corresponding view and the shortcut key to get it.

Table 3-3 *Various orthographic views and axes that are selected*

Orthographic Views	Triad	Shortcut Key (Num pad)
Right View	+X	3
Left View	-X	7
Top View	+Y	8
Bottom View	-Y	2
Front View	+Z	1
Back View	-Z	9
Default Isometric	ISO ball (cyan color)	5

Saving the Project and Exiting ANSYS Workbench

After visualizing the model and restoring the default Isometric view, you need to save the project and exit ANSYS Workbench.

1. Choose the **Save Project** button from the **Standard** toolbar; the project is saved.

2. Close the **Workbench** window to close the ANSYS Workbench session.

Note

1. Instead of creating the I-section with cutout as three separate features, you can create it as single feature. To do so, create the sketch as shown in Figure 3-52 and extrude it symmetrically by 25 mm on both sides. If the sketch consists of some closed loops inside the outer loop, they will automatically be subtracted from the outer loop while extruding.

2. In this tutorial, the model has been created as three separate features to explain the various tools and options available in the software. Also, creating the model consisting of various small features makes the editing work easier in case any changes are to be made in the design at later stages.

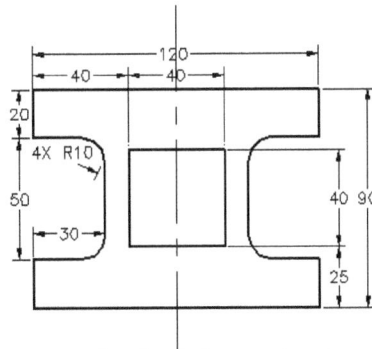

Figure 3-52 The sketch for creating the complete model as single feature

Tutorial 2

In this tutorial, you will create the solid model of the Spring Plate shown in Figure 3-53. The dimensions of the model are shown in Figure 3-54. Thickness and width of the Spring Plate are 2 mm and 20 mm respectively. Save the project with the name *c03_ansWB_Tut02* at the location *C:\ANSYS_WB\c03\Tut02*.   **(Expected time: 30 min)**

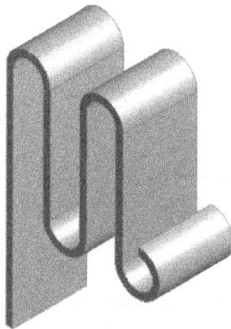

Figure 3-53 Model for Tutorial 2   *Figure 3-54 Sketch and dimensions for Tutorial 2*

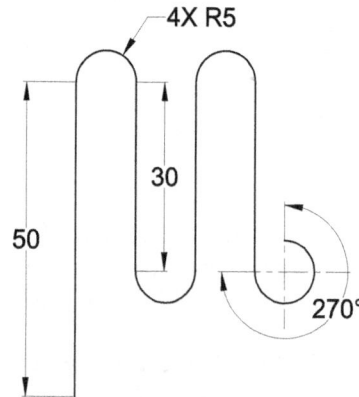

For the ease of creating the model, the sketch has been divided into small segments, as shown in Figure 3-55.

The following steps are required to complete this tutorial:

a.   Start ANSYS Workbench and add the **Geometry** component.
b.   Create the sketch.
c.   Apply constraints to the sketch.
d.   Apply dimension to the sketch.
e.   Extrude the sketch.
f.   Save the model.

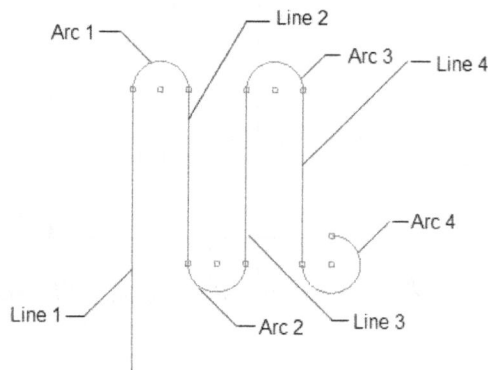

Figure 3-55 Various entities of the sketch

Starting ANSYS Workbench and Adding Geometry Component System

In this section, you need to start ANSYS Workbench and then add a component system to the project.

1. Start ANSYS Workbench 2023 R2. The Workbench window is displayed.

    After invoking the **Workbench** window, you have to add appropriate analysis system or the component system to the **Project Schematic** window. In this tutorial you will create a solid model using the **Geometry** component system.

2. Right-click in the **Project Schematic** window and choose **New Component Systems > Geometry** from the shortcut menu displayed; the **Geometry** component system is added to the **Project Schematic** window, as shown in Figure 3-56.

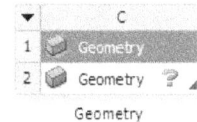

*Figure 3-56 The **Geometry** component system added to the project*

By adding the **Geometry** component system to the **Project Schematic** window, you can have a stand-alone system for the model to be analyzed. When any change is made on the geometry using the **Geometry** component system, the changes are displayed in all analysis systems with which the geometry is shared.

Note
*A component system can also be added by dragging it from the **Toolbox** and then dropping it in the **Project Schematic** window or by double-clicking on the **Geometry** option displayed under the **Component Systems** toolbox in the **Toolbox** window. But in this tutorial, you will use the method mentioned in Step 2.*

3. Double-click on the name field of the **Geometry** component system and enter **Spring Plate** to rename it.

Note
*Once the **Geometry** component system is added to the **Project Schematic** window, you can rename it when the name field gets highlighted at the bottom of the component system in blue.*

4. Choose the **Save Project** button from the **Standard** toolbar; the **Save As** dialog box is displayed.

5. Browse to the location *C:\ANSYS_WB\c03* and then create a sub folder with the name **Tut02** in the *c03* folder.

6. Enter **c03_ansWB_Tut02** in the **File name** edit box in the **Save As** dialog box and then choose the **Save** button in it; the project is saved with the specified name.

Creating the Sketch

After the component system is added to the project, you need to create the sketch for the Spring Plate model. To do so, invoke the **DesignModeler** window and perform the following steps to complete the sketch.

1. Right-click on the **Geometry** cell of the **Spring Plate** component system; a shortcut menu is displayed.

2. Choose the **New DesignModeler Geometry** option from the shortcut menu; the **DesignModeler** window is displayed.

3. Choose the **Millimeter** option from the **Units** menu of the **Menu** bar.

4. In the Tree Outline, expand the **A: Spring Plate** node, if not already expanded; the components of the Tree Outline are displayed.

5. Select **XYPlane** from the Tree Outline; the XY plane becomes the active plane.

6. Choose the **New Sketch** tool from the **Active Plane/Sketch** toolbar; the new entry with the name **Sketch1** is added under the **XYPlane** node in the Tree Outline, refer to Figure 3-57.

7. Right-click on **Sketch1** node under the **XYPlane** node to display a shortcut menu.

8. Choose the **Look At** tool from the shortcut menu displayed; the sketching plane is oriented normal to the viewing direction.

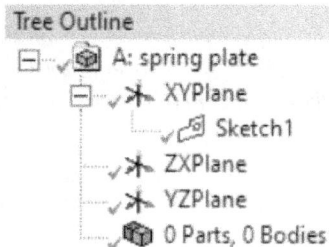

Figure 3-57   The Tree Outline with Sketch1 added to it

9. Choose the **Sketching** tab from the bottom of the Tree Outline to switch to the **Sketching** mode.

10. Next, click on the **Draw** toolbox to expand it, if it is not already expanded.

11. In the **Draw** toolbox, choose the **Line** tool; you are prompted to specify the start point of the line. Also, the cursor is replaced with Draw cursor.

12. Move the cursor close to the origin; the symbol of Coincident constraint (P) is displayed, as shown in Figure 3-58.

13. Click on the origin to specify the start point of the line; you are prompted to specify the end point of the line.

14. Move the cursor upward along the axis such that the symbol of Vertical constraint (**V**) is displayed along the path of the cursor. Move the cursor to some distance along the Y axis and then click to specify the end point of line, as shown in Figure 3-59. The line is created and is displayed in blue color indicating that it is fully constrained.

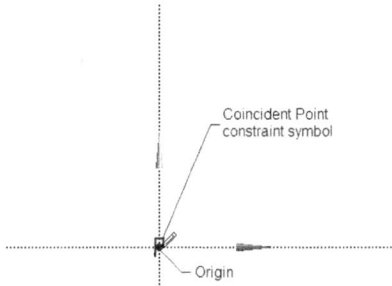

Figure 3-58 Specifying the start point of the line

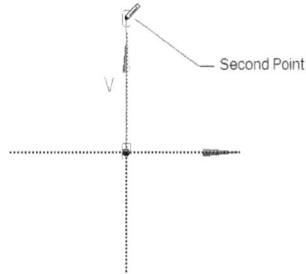

Figure 3-59 Specifying the end point of the line

After Line 1 is created, you now need to create the arc, Arc 1 (refer to Figure 3-55 for naming conventions used in this tutorial).

15. Now, invoke the **Arc by Tangent** tool from the **Draw** toolbox.

The **Arc by Tangent** tool is used to create an arc tangent to a line. Choose the **Arc by Tangent** tool from the **Draw** toolbox and specify a point on any existing sketched entity, the arc to be created will maintain tangency with the specified point. Also, the symbol of Tangent constraint (**T**) is displayed when you move the cursor close to the line. Next, specify the second point to define the end point of the arc.

16. Move the cursor close to end point of Line 1; the symbol of Point Coincident (P) is displayed. Next, click to specify the start point for arc 1, as shown in Figure 3-60.

17. Move the cursor toward right to some distance and click to specify the end point of arc 1; the arc is created, as shown in Figure 3-61.

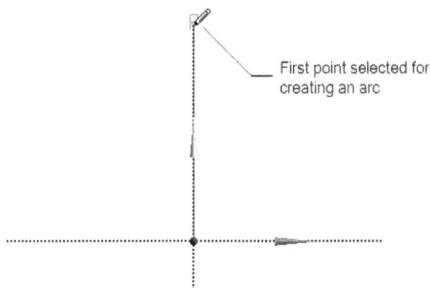

Figure 3-60 Specifying the start point of the arc

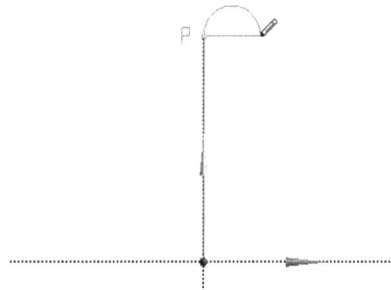

Figure 3-61 Specifying the endpoint of the arc

Note

*In case the arc is created in a direction opposite to the desired direction, right-click to display a shortcut menu and then choose **Reverse** from it.*

18. Now you need to create a line tangent to the arc 1. Invoke the **Tangent Line** tool from the **Draw** toolbox and move the cursor close to the end point of arc 1; the symbols of Tangent constraint (**T**) and the Point Coincident constraint (**P**) are displayed. Click to specify the start point of the line 2, as shown in Figure 3-62.

19. Specify the end point for line 2, as shown in Figure 3-63.

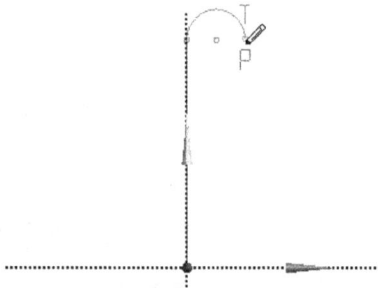

Figure 3-62 Specifying the startpoint of the line

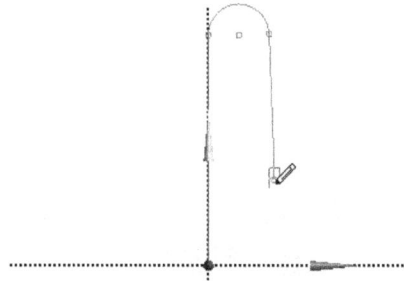

Figure 3-63 Specifying the endpoint of the line

You can create a tangent line by using the **Tangent Line** and **Line by 2 Tangents** tools. To create a tangent line using the **Tangent Line** tool, choose this tool from the **Draw** toolbox; the cursor will change to the Draw cursor. Next, select a curved sketch; the preview of the line will be displayed attached to the Draw cursor. Next, specify the end point of the line to define the length of the line, refer to Figure 3-62. You can also create lines by using the **Line by 2 Tangents** tool. To do so, invoke this tool from the **Draw** toolbox and select two existing curve sketched entities; the tangent line will be created. In case more than one tangent locations are available on the selected entities, the tangent line will be generated using the tangent location nearest to the point of selection.

Note

*The lines created using the **Tangent Line** tool may not always be vertical. They can be made vertical by applying Vertical constraints.*

Now, you need to create an arc.

20. Invoke the **Arc by Tangent** tool from the **Draw** toolbox of the **Sketching** tab; you are prompted to specify the start point of the arc 2.

21. Move the cursor close to the end point of line 2; the symbol of Coincident constraint (**P**) sign is displayed. Click to specify the start point of arc 2, as shown in Figure 3-64.

22. Move the cursor to the right till the symbol of Equal Radius constraint (R) is displayed. Then click to specify the end point of arc 2; arc 2 is created, as shown in Figure 3-65.

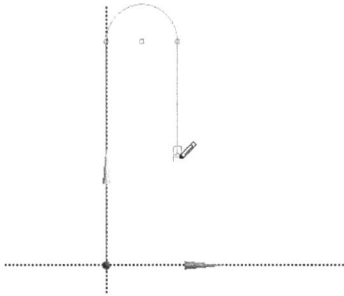

Figure 3-64 *Specifying the start point of the arc*

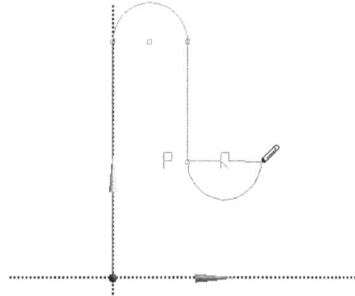

Figure 3-65 *Specifying the end point of the arc*

Now you need to draw the line 5.

23. Invoke the **Tangent Line** tool from the **Draw** toolbox; you are prompted to specify the start point of line. Move the cursor close to the end point of arc 2 and click to specify the start point of line 3 when the symbols of Tangent constraint (T) and Point Coincident constraint (P) are displayed, as shown in Figure 3-66.

24. Move the cursor upward to some distance. While moving the cursor, the symbol of Vertical constraint (**V**) is displayed along with the preview of the line. Click to specify the end point of line 3, as shown in Figure 3-67.

25. Next, invoke the **Arc by Tangent** tool; you are prompted to select a 2D edge or end point of a line.

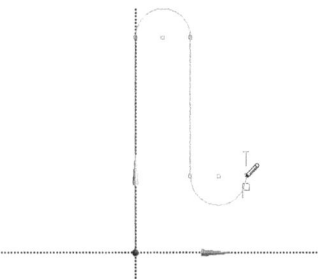

Figure 3-66 *Specifying the start point of the line 3*

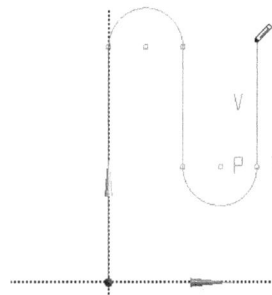

Figure 3-67 *Specifying the endpoint of the line 3*

26. Select the end point of line 3, as shown in Figure 3-68; you are prompted to specify the end point of arc 3.

27. Move the cursor toward right till the symbol of Equal Radius constraint (R) is displayed. Click to specify the end point of arc 3, as shown in Figure 3-69.

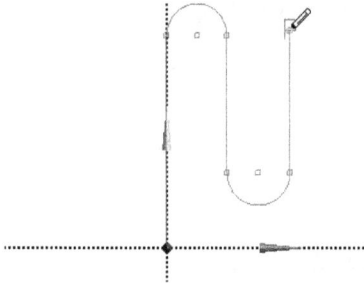

Figure 3-68 Specifying the endpoint of the line 3

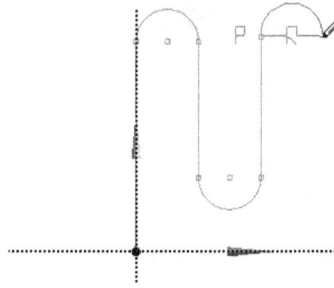

Figure 3-69 Specifying the endpoint of the arc 3

Next, you need to draw the line 4 and arc 3 to finish the sketch.

28. Invoke the **Tangent Line** tool from the **Draw** toolbox; you are prompted to specify the start point of the line.

29. Move the cursor close to end point of arc 3; the symbols of Tangent constraint (T) and the Point Coincident constraint (P) are displayed attached to the cursor.

30. Select the end point of arc 3 to specify it as the start point of line 4, as shown in Figure 3-70. Also, you are prompted to specify the end point of line 4.

31. Move the cursor downward; the preview of the line will be displayed. Also, the symbol of Vertical constraint (V) gets attached to the preview. After moving the cursor to some distance, click to specify the end point of line; line 4 is drawn, as shown in Figure 3-71.

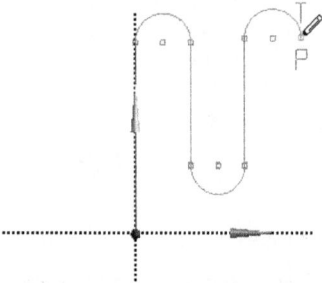

Figure 3-70 Specifying the start point of the line

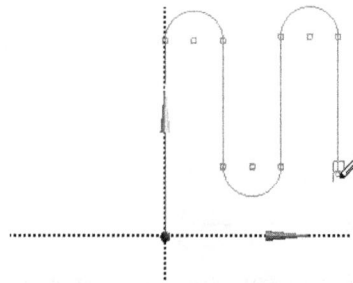

Figure 3-71 Specifying the endpoint of the line

Now, you need to draw the arc 4 to complete the sketch. Note that the arc to be drawn should be made with an angle of 270 degrees.

32. Invoke the **Arc by Tangent** tool from the **Draw** toolbox; you are prompted to select an edge or end point of line to specify the start point of arc. Select the end point of line 4 as the start point of arc 4, as shown in Figure 3-72.

33. Next, move the cursor toward right and then create an arc similar to the one shown in Figure 3-73.

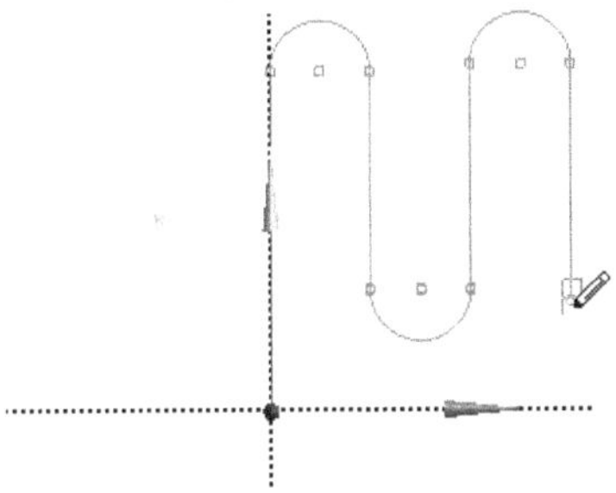

Figure 3-72 Specifying the start point of arc 4 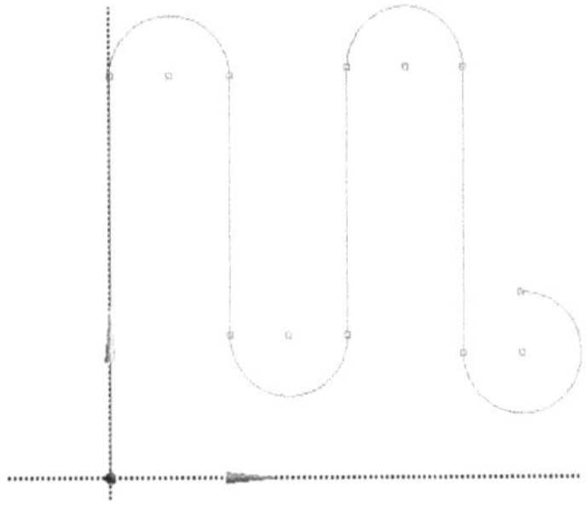   *Figure 3-73* Final sketch after the arc 4 is drawn

Applying Constraints to the Sketch

After the sketch is drawn, you need to apply constraints to the entities.

1. Expand the **Constraints** toolbox from the **Sketching Toolboxes** window.

2. Choose the **Equal Length** tool from the **Constraints** toolbox; you are prompted to select the first line on which Equal Length constraint is to be applied.

   The **Equal Length** tool is used to force two linear entities to maintain the same length.

3. Select line 2; you are prompted to select second line for equal constraint.

4. Select line 3; line 2 and line 3 become equal in length.

5. As the **Equal Length** tool is still active, select line 3 and then line 4; both the lines become equal in length.

   Now, you need to apply the Equal Radius constraint between all the arcs.

6. Choose the **Equal Radius** tool from the **Constraints** toolbox; you are prompted to select the first arc or circle to apply the constraint.

   The **Equal Radius** tool is used to force two circular entities to become equi-radius.

7. Select arc 1; you are prompted to select the second arc to apply the Equal Radius constraint.

8. Select arc 2; arcs 1 and 2 become equal in radius.

9. Similarly, apply the Equal Radius constraint between arcs 2 and 3 and then between arcs 3 and 4; all the arcs become equal in radius.

Assigning Dimensions to the Sketch

After the sketch is drawn and the constraints are applied, you need to assign dimensions to the entities.

1. Choose the **General** tool from the **Dimensions** toolbox; you are prompted to select a 2D edge or line that you want to dimension.

2. Select line 1; the shape of the cursor changes to Draw cursor. Also, the preview of the dimension is attached to the cursor.

3. Click on the left of line 1 to place the dimension, as shown in Figure 3-74.

4. Place all other dimensions to their respective places, as shown in Figure 3-75.

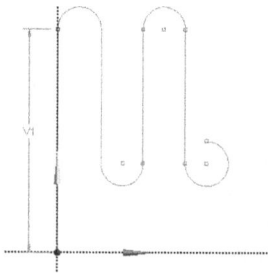

Figure 3-74 Placing the dimension of line 1

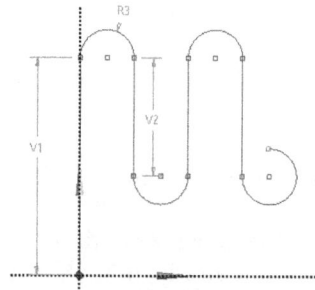

Figure 3-75 Sketch after all the dimensions are placed

You need to place only three dimensions in the sketch, remaining dimensions will be applied automatically as the Equal Radius and Equal Length constraints have been applied to them earlier in this tutorial. After the dimensions are placed, you need to assign values to each of them.

5. Choose the **Modeling** tab located at the bottom of the **Sketching Toolboxes** window; the **Modeling** mode becomes active.

6. Expand the **XYPlane** node in the Tree Outline, if not already expanded.

7. Select **Sketch1** under the expanded **XYPlane** node; the corresponding **Details View** window is displayed.

8. In the **Details View** window, expand the **Dimensions** node if not already expanded.

   The **Dimensions** node in the **Details View** window displays the dimensions that are placed on the sketch in the **Graphics** window.

9. Click on the **R3** edit box in the **Details View** window and enter **5**; the radius of all arcs is changed to 5 mm instantaneously.

Note
*When you click on any edit box under the **Dimensions** node in the **Details View** window, the corresponding dimension in the **Graphics** window is highlighted in yellow color.*

10. Next, click on the **V1** edit box and then enter **50**; the length of Line1 is changed.

11. Similarly, modify dimension **V2** to 30.

    Figure 3-76 shows the **Dimensions** node of the **Details View** window.

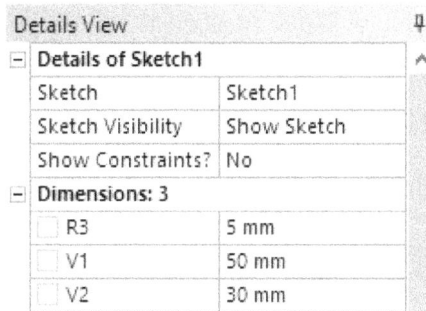

Details View		⊣
⊟ **Details of Sketch1**		⌃
Sketch	Sketch1	
Sketch Visibility	Show Sketch	
Show Constraints?	No	
⊟ **Dimensions: 3**		
☐ R3	5 mm	
☐ V1	50 mm	
☐ V2	30 mm	

*Figure 3-76 The **Dimensions** node in the **Details View** window*

Creating the Extrude Feature

After the sketch is fully constrained and dimensions are applied, you now need to add material to the sketch. This is done by using the **Extrude** tool.

1. Choose the **Extrude** tool from the **Features** toolbar; **Extrude1** is attached to the Tree Outline. Also, the preview of extrusion is displayed in the **Graphics** window.

2. Click on the ISO ball available on the bottom right corner of the **Graphics** window; the view is changed to isometric. Figure 3-77 shows the ISO ball with the Triad and Figure 3-78 shows the Isometric view of the model.

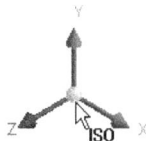

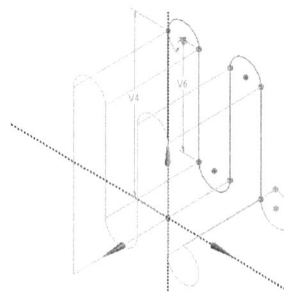

Figure 3-77 The Triad with the ISO ball   *Figure 3-78 The Isometric view of the model*

Now, you need to set the parameters for extrusion in the **Details View** window.

3. Select **Geometry** in the **Details View** window to display the **Apply** and **Cancel** buttons, if not already displayed.

4. Choose the **Apply** button in the **Geometry** selection box to specify the sketch as the sketch to be extruded.

5. In the **Operation** drop-down list in the **Details View** window, select **Add Material** if it is not already selected.

6. In the **Direction** drop-down list, select **Both - Symmetric**; the material is added to both sides of the plane.

7. In the **FDI, Depth (>0)** edit box, enter **10** as the depth of extrusion.

8. In the **As Thin/Surface?** drop-down list, select **Yes**; the **FD2, Inward Thickness (>=0)** and **FD3, Outward Thickness (>=0)** edit boxes are activated.

    The **Thin/Surface** tool is used to create surface out of sketches or create shell features in models.

9. Click on the **FD2, Inward Thickness (>=0)** edit box and enter **1** as the value of thickness in the inward direction.

10. Click on the **FD3, Outward Thickness (>=0)** edit box and enter **1** as the value of thickness in the outward direction.

11. After specifying all the parameters in the **Details View** window, choose the **Generate** tool from the **Features** toolbar; the geometry is extruded, as shown in Figure 3-79.

    The final model for Tutorial 2 is shown in Figure 3-80. The axes and the sketch have been turned off for a better visibility.

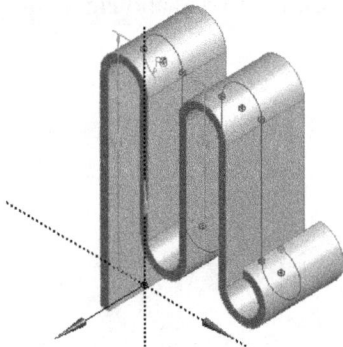

Figure 3-79 *The generated feature*   *Figure 3-80* *The final model for Tutorial 2*

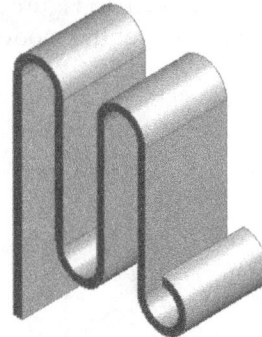

12. Close the **DesignModeler** window; the **Workbench** window is displayed.

Saving the Model

After the model is created, you now need to save your work.

1. Choose the **Save Project** button from the **Standard** toolbar to save the model.

2. Exit the **Workbench** window.

Tutorial 3

In this tutorial, you will create the solid model of the clamp shown in Figure 3-81. The sketch of the model and its dimensions are shown in Figure 3-82. Save the project with the name *c03_ansWB_Tut03* at the location *C:\ANSYS_WB\c03\Tut03*. **(Expected time: 30 min)**

Figure 3-81 Model for Tutorial 3

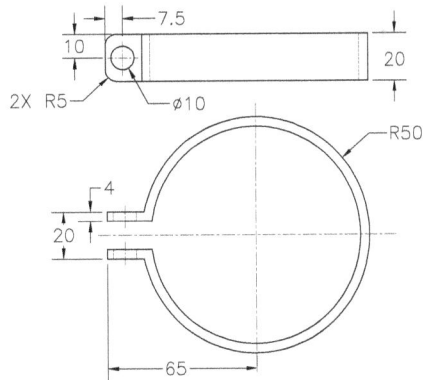

Figure 3-82 Dimensions of the clamp

The following steps are required to complete this tutorial:

a. Start ANSYS Workbench 2023 **R2**.
b. Add the **Geometry** component system to the project.
c. Start **DesignModeler** window and specify unit system.
d. Draw the sketch for the base feature on the XYPlane.
e. Create the base feature.
f. Create the circular cutout.
g. Create the blend feature.
h. Save the project and exit the ANSYS Workbench session.

Starting ANSYS Workbench and Adding Geometry Component System

First, you need to start ANSYS Workbench 2023 R2 and then add a component system to the project.

1. Start ANSYS Workbench 2023 **R2.** The Workbench window will be displayed.

2. After invoking the **Workbench** window, you have to add appropriate analysis system or a component system to the **Project Schematic** window. In this tutorial, you will create a solid model using the **Geometry** component system.

3. Right-click in the **Project Schematic** window and choose **New Component Systems > Geometry** from the shortcut menu displayed, as shown in Figure 3-83; the **Geometry** component system is added to the project and is displayed in the **Project Schematic** window.

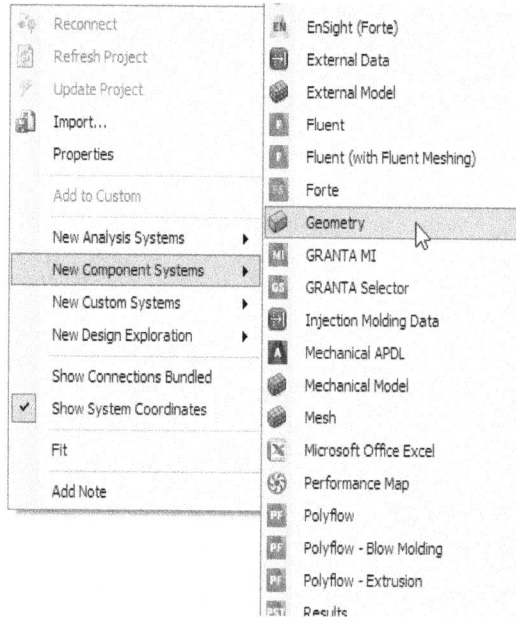

Figure 3-83 *Choosing the **Geometry** option from the shortcut menu displayed on choosing the **New Component Systems** option*

4. Once the **Geometry** component system is added to the **Project Schematic** window, the name field at the bottom of the component system is highlighted in blue. If it is not highlighted, double-click on the name field and enter **Clamp**, refer to Figure 3-84. The component system is renamed as **Clamp**.

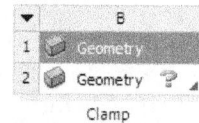

Figure 3-84 *The **Geometry** component system added to the project*

The blue question mark on the right of the **Geometry** cell indicates that an immediate action is required for this cell and the user cannot proceed further without fixing this cell.

5. In the **Workbench** window, choose the **Save Project** button from the **Standard** toolbar; the **Save As** dialog box is displayed.

6. Browse to the location *C:\ANSYS_WB\c03* and then create a sub folder with the name **Tut03** in the *c03* folder and then choose the **Open** button from the **Save As** dialog box.

7. Enter **c03_ansWB_Tut03** in the **File name** edit box in the **Save As** dialog box and then choose the **Save** button in it; the project is saved with the specified name.

Starting DesignModeler Window and Specifying Unit System

To define the geometry, you need to start the **DesignModeler** window associated with this cell.

1. Right-click on the **Geometry** cell in the **Clamp** component system and
    then select **New DesignModeler Geometry**; the **DesignModeler** window
    is displayed.

2. Choose the **Millimeter** option from the **Units** menu of the **Menu** bar.

Drawing the Sketch for the Base Feature

You need to create the sketch for the base feature on the XY plane which is the default plane.
Therefore, you need not specify the plane.

1. Choose the **Sketching** tab displayed in the lower left corner of the Tree Outline to invoke
    the **Sketching** mode.

    Note

    *1. To select a plane other than the default plane (XY), select the **New Plane** from
    the **Active Plane/Sketch** toolbar.*

    *2. To insert a sketch instance or create a new sketch on a plane other than the default plane, you
    can right-click on the plane node in the Tree Outline to display a shortcut menu. Next, choose
    Insert Sketch Instance from it; a sketch instance will be displayed under the desired node.*

    Now, you need to orient the sketching plane normal to the viewing direction, so that you
    can easily draw the sketch on the specified plane.

2. Choose the **Look At** tool from the **Graphics** toolbar to orient the model normal to the
    viewing direction.

    Note

    *You can also orient a plane normal to the viewing direction by choosing the **Look At** tool from the
    shortcut menu displayed on right-clicking on the sketch instance.*

3. Choose the **Circle** tool from the **Draw** toolbox; you are prompted to specify
    the center of the circle.

    In **DesignModeler**, you can draw circles by using two different methods. In the first method,
    specify the center point of the circle and then define its radius. In the second method, you
    need to specify the three existing drawing entities in the **Graphics** window with which the
    new circle to be created must maintain the tangency relation. This type of circle is known
    as tri-tangent circle. You can choose any of the two methods for drawing circles.

4. Move the cursor close to the origin in the **Graphics** window and click; the symbol of
    Coincident Point constraint (**P**) is displayed. After specifying the center point of the circle,
    the preview of the circle is displayed attached to the Draw cursor. Also, you are prompted
    to specify the radius of the circle.

5. Move the cursor away from the center and click; a circle is created, as shown in Figure 3-85.

6. Press the Esc key to exit the **Circle** tool.

After creating the first entity of any sketch, it is better to generate its dimensions first. This gives you a fair idea about the graphics space required to complete the sketch. Also, it helps you to decide the comparative size of other sketched entities to complete the outer profile. Now, you will generate the radius dimension of the circle that you created in the previous step and change its value to 50mm.

7. Expand the **Dimensions** toolbox in the **Sketching Toolboxes** window.

8. Choose the **Radius** tool from the **Dimensions** toolbox; you are prompted to select the entity to place the dimension.

    The **Radius** tool is used to generate dimensions for circles, arcs, or ellipses. When you select an arc for dimensioning, the radius dimension is generated, and when you select an ellipse, the major and minor dimensions are generated.

9. Move the cursor over the circle and select it; the preview of dimension is attached with the cursor.

10. Move the cursor away from the circle and click to place the dimension. The dimension is generated and its name is displayed on the dimension line.

    Other details of the dimension are displayed in the **Details View** window.

11. In the **Details View** window, click in the edit box displayed on the right of the dimension name (**R1**) under the **Dimensions: 1** node.

Note

The name of the dimension displayed on the dimension line can be different in different systems. To avoid such confusion and to facilitate the proper explanation, refer to the corresponding screen captures of the dimensions.

12. Enter **50** in this edit box and press the Enter key; the radius of circle changes to 50 mm and is displayed in the **Graphics** window, refer to Figure 3-86.

    Now, you need to complete the remaining part of sketch for the base feature.

13. Click on the **Modify** toolbox in the **Sketching Toolboxes** window; the **Modify** toolbox expands.

    The **Modify** toolbox contains various tools such as **Fillet**, **Chamfer**, **Trim**, **Extend**, **Split**, and so on. These tools are used to edit sketched entities in the **Sketching** mode.

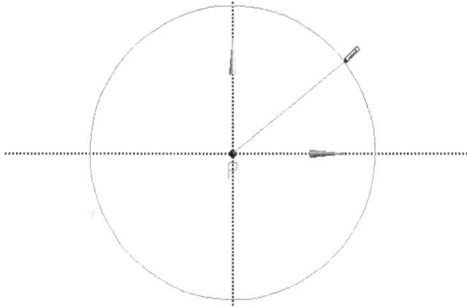

Figure 3-85 *Specifying the radius of the circle*

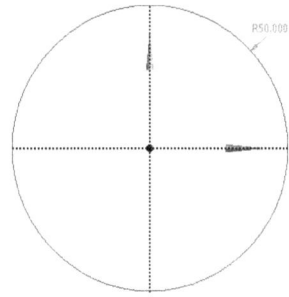

Figure 3-86 *The circle after editing the dimension value*

14. Scroll down the **Modify** toolbox to display other tools, refer Figure 3-87. Next, choose the **Offset** tool from the **Modify** toolbox; you are prompted to select the line or arc to offset.

    The **Offset** tool is used to draw multiple parallel lines, parallel polylines, concentric circles, concentric curves, concentric arcs, and so on. When you choose the **Offset** tool from the **Modify** toolbox, you will be prompted to select the entities to be offset. The entities selected for offsetting must be connected end to end and should form open or closed profile.

15. Select the circle and right-click in the **Graphics** window; a shortcut menu is displayed.

16. Choose the **End selection / Place offset** option from the shortcut menu; the preview of the entity to be offset is displayed attached to the cursor.

    Note
    *If you have made a wrong selection by mistake, then choose the **Clear Selection** option from the shortcut menu and select again the correct entity to be offset.*

17. Move the cursor inside the circle and click to place an offset circle, refer to Figure 3-88.

18. Right-click in the **Graphics** window and choose the **End** option from the shortcut menu displayed; the **Offset** tool is deactivated.

    After editing the sketch, if you want to exit the current selection and select other entity to offset, choose the **Clear Selection** option from the shortcut menu.

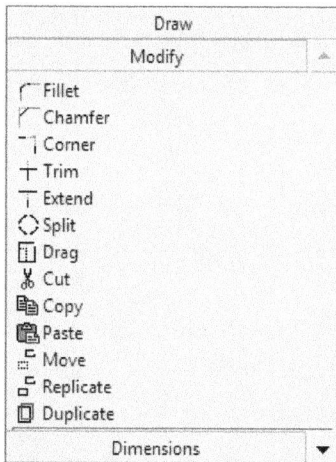

Figure 3-87 *Tools in the **Modify** toolbox*

Figure 3-88 *Specifying the offset distance*

After the circular entities are created, you need to create the linear entities.

19. Expand the **Draw** toolbox and invoke the **Polyline** tool; you are prompted to specify the start point of the line.

    You need to define the start point and end point of the line, each time you want to create a line using the **Line** tool. But if you want to create a continuous connected line where the start point of the next line is automatically defined as the end point of the previous line, choose the **Polyline** tool from the **Draw** toolbox. Specify the start and end points of the first line; the first line will be created and the preview of another line whose start point is the end point of the first line will be attached to the Draw cursor. Specify the third point; the second line will be created and the preview of the third line whose start point will be the end point of the second line will be displayed attached to the Draw cursor. Keep on specifying the points to create continuous lines. To stop creating the polyline and exit the **Polyline** tool, right-click in the **Graphics** window and choose the **Open End** option from the shortcut menu.

20. Move the cursor near the circumference of the outer circle and click when the symbol of Coincident constraint (C) is displayed, refer to Figure 3-89.

21. Move the cursor toward left and click to draw a horizontal polyline.

22. Draw the vertical second entity of the polyline and then draw the horizontal third entity of the polyline. Make sure that the end point of the third entity is coincident with the inner circle, refer to Figure 3-90.

23. After drawing the three entities of the polyline, right-click in the **Graphics** window and choose the **Open End** option from the shortcut menu displayed.

24. Press the Esc key to exit the **Polyline** tool.

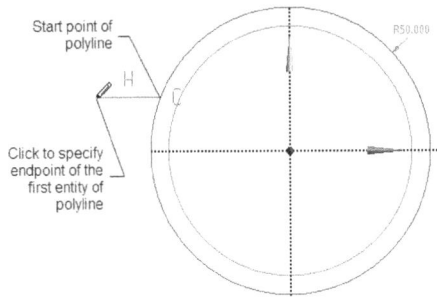

Figure 3-89 *Creating the first entity of the polyline*

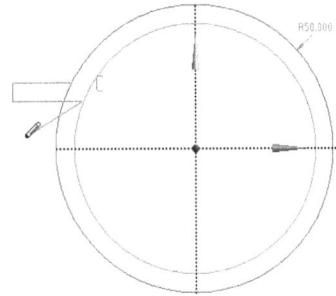

Figure 3-90 *Sketch after creating the third entity of the polyline*

Next, you need to create a similar sketch on the other side of the X axis such that this sketch becomes the mirror copy of the sketch already created.

25. Expand the **Modify** toolbox and choose the **Replicate** tool; you are prompted to select points or edges to replicate.

The **Replicate** tool is used to copy entities from an existing sketch and paste them wherever required.

26. Select the three entities of the polyline created in the previous steps and right-click in the **Graphics** window; a shortcut menu is displayed, as shown in Figure 3-91.

27. Choose the **End / Use Plane Origin as Handle** option from the shortcut menu, refer to Figure 3-91; the preview of the entities to be replicated along with the paste handle is displayed, refer to Figure 3-92.

The paste handle is used to set a reference point while replicating entities. This reference point is used while placing the entities to be replicated. To replicate the entities, select the entities and then right-click to display a shortcut menu. This shortcut menu contains options such as **Clear Selection**, **End / Set Paste Handle**, **End / Use Plane Origin as Handle**, and **End / Use Default Paste Handle**.

The **Clear Selection** option is used to deselect the entities that were selected to replicate earlier. The **End / Set Paste Handle** option is used to specify the paste handle by specifying a point in the **Graphics** window. The **End / Use Plane Origin as Handle** option is used to specify the origin of the sketching plane as the paste handle. The **End / Use Default Paste Handle** option is used to specify a system specified point of the selected entity as the paste handle.

Since the entities to be replicated are the mirror copies of the selected entities, you have to flip them about the X axis.

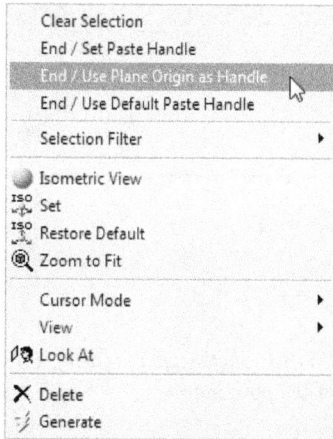

Figure 3-91 *The shortcut menu displayed while the **Replicate** tool is active*

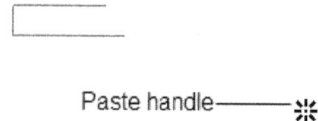

Figure 3-92 *Preview of the entities to be replicated and the paste handle*

28. Next, select the origin; you are prompted to specify the location to paste the entities, as shown in Figure 3-93.

29. Right-click in the **Graphics** window and choose the **Flip Vertical** option from the shortcut menu displayed, refer to Figure 3-94; the preview of the flipped entity is displayed, as shown in Figure 3-95 displayed.

    After specifying the location of paste handle, instead of replicating the entities directly, you can rotate them by the desired angle, scale them by desired scale factor, and flip them along the horizontal and vertical directions. Place the entities at the desired locations, by using the options from the shortcut menu.

    To rotate the selected entities before replicating them, enter the required angle of rotation in the **r** edit box, displayed on the right of the **Replicate** tool. Right-click in the **Graphics** window to display the shortcut menu, refer to Figure 3-94. Next, choose the **Rotate by r Degrees** (Counterclockwise rotation) or **Rotate by -r Degrees** (Clockwise rotation)option from the shortcut menu.

    To scale the selected entities before replicating them, enter the required value for scale factor in the **f** edit box, displayed on the right of the **Replicate** tool. Next, choose the **Scale by factor f** or **Scale by factor 1/f** option from the shortcut menu according to the requirement, refer to Figure 3-94.

30. Move the paste handle to the origin and click when the symbol of Coincident Point constraint (P) is displayed, refer to Figure 3-95; the mirror copies of the selected entities are replicated at their required locations, refer to Figure 3-96.

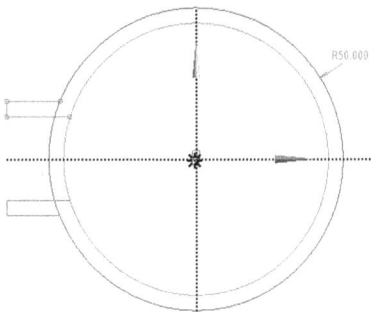

Figure 3-93 *Specifying the location of paste handle*

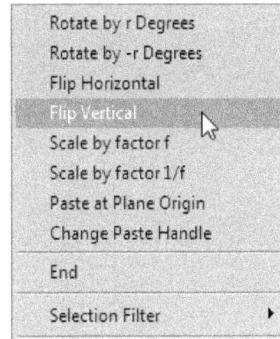

Rotate by r Degrees
Rotate by -r Degrees
Flip Horizontal
Flip Vertical
Scale by factor f
Scale by factor 1/f
Paste at Plane Origin
Change Paste Handle
End
Selection Filter ▶

Figure 3-94 *Specifying the option to flip the entities vertically*

Figure 3-95 *Preview of the entities to be replicated after flipping them vertically*

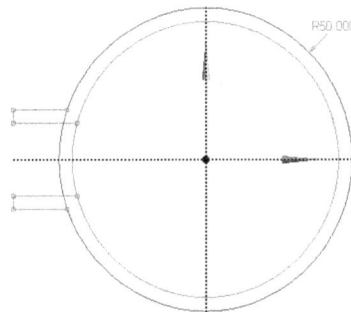

Figure 3-96 *The sketch after replicating the selected entities*

31. Press the Esc key to exit the **Replicate** tool.

    Next, you need to trim the unwanted portion of the sketch using the **Trim** tool from the **Modify** toolbox.

32. Choose the **Trim** tool from the **Modify** toolbox; you are prompted to select edges to trim.

    The **Trim** tool is used to trim the objects that extend beyond a required point of intersection. While creating a sketch, there are a number of places where you need to remove the unwanted and extending edges. Choose the **Trim** tool from the **Modify** toolbox; the Draw cursor will be displayed and you will be prompted to select the edges to be trimmed. Select the sketched entity to be trimmed; the selected sketched entity is trimmed to its nearest point of intersection with any other sketched entity or axis.

33. Select the **Ignore axis** check box displayed on the right of the **Trim** tool and click on the segments one by one, marked in Figure 3-97; the selected segments is trimmed and you get the sketch shown in Figure 3-98.

The **Ignore Axis** check box is selected to ignore the intersection of the segment of circles with the X axis while trimming.

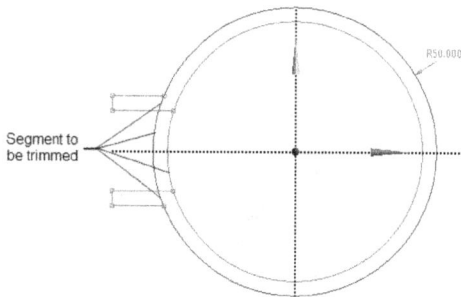

Figure 3-97 *Segments of the circle to be trimmed*

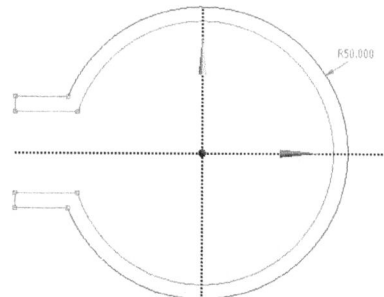

Figure 3-98 *Sketch after trimming the entities*

Applying Geometric Constraints and Dimensions to the Sketch

The entities of the sketch should be fully specified in terms of size, shape, orientation, and location. This is achieved by setting geometric constraints and dimensions.

Geometric constraints are the logical operations that are performed to add relationship (such as tangent or perpendicular) between the sketched entities, planes, axes, edges, or vertices. The constraints applied to the sketched entities are used to capture the design intent. By using constraints in a sketch, you can reduce the number of dimensions that are required in that sketch. The geometric constraints are applied using the tools available in the **Constraints** toolbox.

1. Expand the **Constraints** toolbox and choose the **Equal Length** tool from it; you are prompted to select lines to apply the constraints.

2. Select any one of the two vertical lines from the sketch; you are prompted to select the lines to apply the Equal Length constraint.

3. Select the second vertical line from the sketch; the Equal Length constraint is applied to the two vertical lines and you are prompted to select the first line for applying the Equal Length constraint.

4. Select the top most horizontal line of the sketch and then select the bottom most horizontal line of the sketch to make them equal in length.

5. Select the **Symmetry** tool from the **Constraints** toolbox; you are prompted to specify the axis of symmetry. Select the X axis; you are prompted to select a point or edge to apply the Symmetry constraint.

The **Symmetry** tool is used to make entities symmetric about a centerline. After this tool is invoked, select a centerline and then select the entities which are to be made symmetric.

6. Select the horizontal line from the sketch that is just above the X axis; you are prompted to select the second point or edge to apply the Symmetry constraint.

7. Select the horizontal line from the sketch that is just below the X axis; the two horizontal lines become symmetric about the X axis.

8. Choose the **Concentric** tool from the **Constraints** toolbox and select the two circular arcs from the sketch; the selected arcs become concentric.

The **Concentric** tool is used to force two circular entities share the same center.

Next, you need to generate the dimensions and edit their values to get the sketch of desired size.

9. Expand the **Dimensions** toolbox from the **Sketching Toolboxes** windows and then choose the **General** tool.

10. Generate all dimensions shown in Figure 3-99 and edit the value of dimensions in the **Details View** window, refer to Figure 3-100.

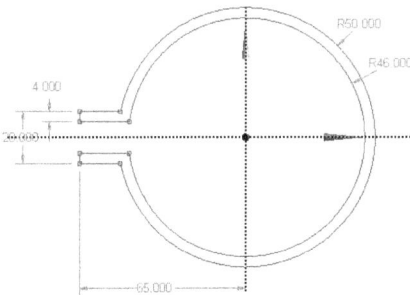

Figure 3-99 Dimensions to be generated for the sketch of base feature

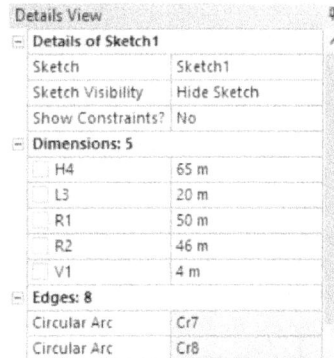

Figure 3-100 Value of dimensions in the Details View window

Note
*The names of the dimensions displayed in the **Details View** window can be different in your system.*

After applying the required geometric constraints and generating the dimensions, the color of the sketch will change to blue indicating that the sketch is fully constrained and is ready to be used for feature creation operations.

After completing the sketch, you need to exit the **Sketching** mode.

11. Choose the **Modeling** tab located at the bottom of the **Sketching Toolboxes** window; the **Sketching** mode is activated and the Tree Outline is displayed. Also, **Sketch 1** is added under the **XYPlane** node.

After exiting the **Sketching** mode, the sketching plane is still normal to the viewing direction. To proceed further with the feature creation operation, it is advised to change the view of the sketching plane to Isometric view.

12. Right-click in the **Graphics** window, and then choose the **Isometric View** option from the shortcut menu displayed; the sketch is displayed in Isometric view.

Creating the Base Feature

Next, you need to create the base feature using the **Extrude** tool from the **Features** toolbar.

1. Choose the **Extrude** tool from the **Features** toolbar; the preview of the extruded feature with the default values is displayed in the **Graphics** window. Also, a node for the extruded feature with the name **Extrude 1** is added below the three default planes in the Tree Outline.

   The default parameters used for generating the preview of the extruded feature are displayed in the **Details View** window. To get the required shape of the base feature, you need to edit the values in the **Details View** window.

   As per the requirement of this tutorial, the material should be added symmetrically on both sides of sketch in normal direction.

2. Select the **Both-Symmetric** option from the **Direction** drop-down list.

3. Enter **10** in the **FD1, Depth (>0)** edit box of the **Details View** window.

   The complete depth of material addition is 20mm, but the material will be added symmetrically by the same depth on both the sides of the sketch. Therefore, 10mm is specified as the depth value.

4. Choose the **Generate** tool from the **DesignModeler** toolbar; the base feature is created with the specified settings, refer to Figure 3-101.

   By default, the sketch is displayed only when the plane on which it is created is the active plane. Since the XY plane is the current active plane, the sketch and the dimensions of the *Sketch1* are still displayed in the **Graphics** window. You can hide the sketch as the sketch and its dimensions are not needed now.

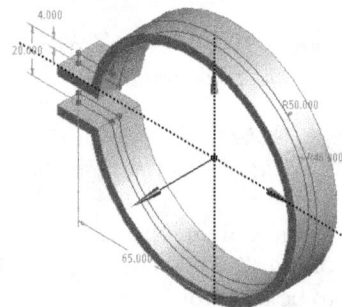

Figure 3-101 The base feature

5. Right-click on the **Sketch1** in the Tree Outline and choose the **Hide Sketch** option from the shortcut menu displayed.

> **Note**
> *If needed, you can again display the sketch and its dimensions. To do so, right-click on the **Sketch1** in the Tree Outline and choose the **Show Sketch** option from the shortcut menu displayed.*

Creating the Circular Cutout

Next, you need to create the circular cutout on the two rectangular flanges of the base feature. The sketch for this feature should be created on the rectangular flange. As the three default planes do not pass through the surface on which the cutout has to be created, these planes cannot be used for drawing the sketch. Therefore, you have to define a new plane on the top flat face of the rectangular flange and draw the sketch for the circular cutout.

1. Select the top face of the rectangular flange, refer to Figure 3-102.

Figure 3-102 *Selecting the flat face for defining the Sketching plane*

You can use the tools available in the **Select** toolbar to select an edge, face, vertex, and so on in a geometry, refer to Figure 3-103. For example, the **Edge** tool is used to select an edge, the **Face** tool is used to select a face, and so on. Alternatively, right-click in the **Graphics** window to display a shortcut menu and then choose the desired tool from the **Selection Filter** cascading menu, as shown in Figure 3-104.

2. Choose the **New Plane** tool from the **Active Plane/Sketch** toolbar; **Plane4** is added to the Tree Outline.

3. Choose the **Generate** tool available in the **Features** toolbar to generate the new plane.

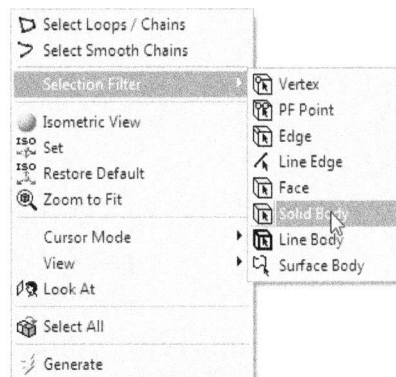

Figure 3-103 *The **Select** toolbar*

Figure 3-104 *Choosing a selection mode from the **Selection Filter** cascading menu*

4. Choose the **Sketching** tab available under the Tree Outline; the **Sketching** mode is invoked.

5. Choose the **Look At** tool from the **Graphics** toolbar.

The **Look At** tool is used to orient the view normal to the screen.

6. Choose the **Circle** tool from the **Draw** toolbox and draw a circle as shown in Figure 3-105.

7. Expand the **Dimensions** toolbox. The **General** tool is chosen by default in this toolbox.

8. Generate the dimensions of the circle and specify their values in the **Details View** window, refer to Figure 3-106. The sketch gets fully-defined and is ready to be used for feature creation.

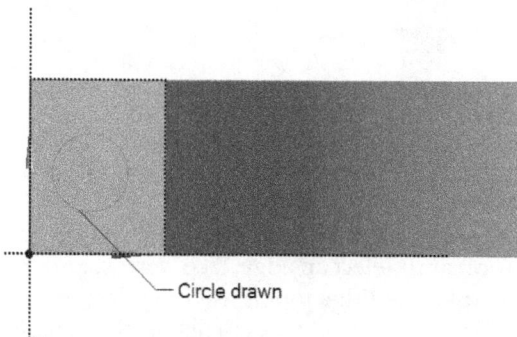

Figure 3-105 *Creating circle on the defined sketching plane*

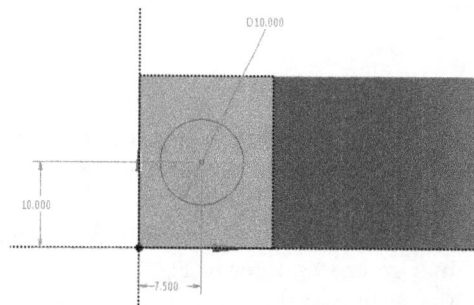

Figure 3-106 *Generating the dimensions of the sketch for cutout feature*

9. Invoke the **Modeling** mode by choosing the **Modeling** tab displayed below the **Sketching Toolboxes** window. The sketch of the cutout feature is created on Plane4 and is displayed as **Sketch2** in the Tree Outline, as shown in Figure 3-107.

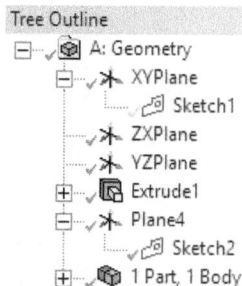

Figure 3-107 *The new plane and sketch added in the Tree Outline*

10. Change the view to isometric by clicking on the ISO ball of the Triad, displayed at the bottom right corner of the **Graphics** window.

After drawing the sketch, you need to remove the material from the base feature using the **Extrude** tool.

11. Choose the **Extrude** tool from the **Features** toolbar; the preview of the extruded feature with the default values is displayed in the **Graphics** window.

    [🔲 Extrude]

    Also, the Tree Outline is activated and **Extrude2** is added below **Extrude1** in the Tree Outline.

12. Select the **Cut Material** option from the **Operation** drop-down list.

    The **Cut Material** option is used to create cutouts, holes, and so on in an existing components. As per the design requirements, the material should be removed starting from the flat face on which the sketch is created and up to the bottom most face of the second rectangular flange.

13. Select the **To Surface** option from the **Extent Type** drop-down list, refer to Figure 3-108; the **Target Face** selection box is added in the **Details View** window.

    To extrude a sketch to a desired face of an existing model, choose the **To Surface** option.

14. Click on the **Target Face** selection box; the **Apply** and **Cancel** buttons are displayed in the **Target Face** selection box and you are prompted to select faces to create extrude.

15. Select the bottom face of the second flange, refer to Figure 3-109; the material is removed up to the specified surface.

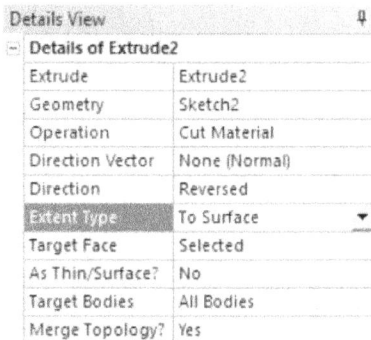

Details View	무
− **Details of Extrude2**	
Extrude	Extrude2
Geometry	Sketch2
Operation	Cut Material
Direction Vector	None (Normal)
Direction	Reversed
Extent Type	To Surface ▼
Target Face	Selected
As Thin/Surface?	No
Target Bodies	All Bodies
Merge Topology?	Yes

Figure 3-108 *Selecting the* ***To Surface*** *option from the* ***Extent Type*** *drop-down list in the* ***Details View*** *window*

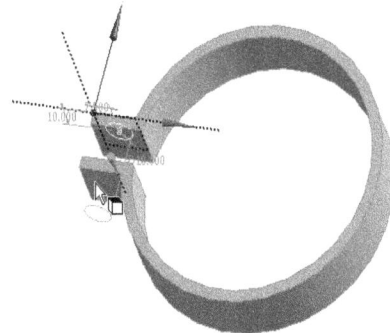

Figure 3-109 *Specifying the face up to which the material will be removed*

> **Note**
> *While selecting the target face, you need to rotate the view of the model. The process of dynamically rotating the model has been discussed in detail in the previous tutorial.*

16. Choose the **Apply** button from the **Target Face** selection box to accept the specified face.

17. Choose the **Generate** tool from the **Features** toolbar; the cutout feature is created, refer to Figure 3-110.

Creating the Blend Feature

To remove the sharp edges from the clamp, you need to create fillets of radius 5 mm at the vertical edges of the clamp. The fillet will be created on four vertical edges of the model using the **Fixed Radius** option of the **Blend** tool.

1. Choose the **Fixed Radius** tool from the **Blend** drop-down in the **Features** toolbar; you are prompted to select edges to blend.

2. Press the CTRL key and select the four vertical edges of the model, refer to Figure 3-111.

Figure 3-110 Model after creating the circular cutout

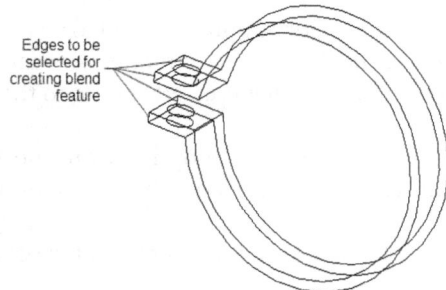

Figure 3-111 Edges to be selected for creating the blend feature

Note

*1. In Figures 3-110 and 3-111, the display of planes has been turned off for better visualization. As per the need, you can turn on or off the display of planes by choosing the **Display Plane** button from the **Graphics** toolbar.*

2. To facilitate the selection of edges without rotating the model and for generating the blend feature, change the display mode to wireframe. The procedure to change the display mode has been discussed in the previous tutorial.

3. Click on the **Geometry** selection box in the **Details View** window; the **Apply** and **Cancel** buttons are displayed. Next, choose the **Apply** button from the **Details View** window to accept the selected of edges to be blended.

4. Enter **5** in the **FD1, Radius (>0)** edit box as the radius of the edit box.

5. Choose the **Generate** tool from the **Features** toolbar; the blend feature is created. Figure 3-112 shows the final model.

Figure 3-112 *The final model*

Tip
*Even after creating a feature, you can modify some of its parameters. To do so, select the feature that you need to modify from the Tree Outline; the parameters of the selected feature will be displayed in the **Details View** window. The parameters that cannot be edited will be grayed out and the remaining parameters can be edited. Change the value of the required parameters and then choose the **Generate** button from the **Features** toolbar.*

6. Close the **DesignModeler** window; the **Workbench** window is displayed.

Saving the Project and Exiting ANSYS Workbench

After visualizing the model and restoring the default Isometric view, you need to save the project and exit ANSYS Workbench. This saved project will be used in later chapters for analysis.

1. In the **Workbench** window, choose the **Save Project** button from the **Standard** toolbar; the project is saved with the name *c03_ansWB_Tut03*.

2. Choose **File > Exit** from the **Workbench** window to exit the ANSYS Workbench session.

Self-Evaluation Test

Answer the following questions and then compare them to those given at the end of this chapter:

1. The **Arc by Tangent** tool can be invoked from the _____ toolbox.

2. The _____ tool is used to make two entities equal in length.

3. You can invoke the **Offset** tool from the _____ toolbox.

4. You can hide or show a sketch anytime by using the _____.

5. In the **DesignModeler** window, the P symbol represents the Coincident Point constraint. (T/F)

6. The **Extrude** tool can be invoked from the **Create** menu of the Menu bar. (T/F)

7. In the **DesignModeler** window, none of the constraints are automatically applied while drawing a sketch. (T/F)

8. The **Horizontal** tool in the **Constraints** toolbox can be used to make a linear entity horizontal. (T/F)

9. You can switch to the **Modeling** mode by choosing the **Modeling** tab available at the bottom of the **Sketching Toolboxes** window. (T/F)

Review Questions

Answer the following questions:

1. You can create line segments tangent to arcs by using the _____ tool from the **Draw** toolbox.

2. In the **DesignModeler** window, user actions are recorded in the _____ window.

3. In the **DesignModeler** window, you can change the dimension of the entities by specifying the new values in the _____ window.

4. In **DesignModeler**, the Vertical constraint is represented by _____ symbol.

5. The options in the **Details View** window are contextual in nature. (T/F)

6. You can create patterns of entities by using the **Replicate** tool available in the **Modify** toolbox. (T/F)

7. You can change the direction of extrusion by using the options in the **Direction** drop-down list in the **Details View** window. (T/F)

8. On choosing any tool from the **Draw** toolbox, the normal arrow cursor changes to Draw cursor. (T/F)

9. Like other tools in the **Graphics** toolbar, the **Rotate** tool is also a transparent tool. (T/F)

10. You can switch to the **Sketching** mode by choosing the **Sketching** tab available at the bottom of the **Tree Outline** window. (T/F)

EXERCISES

Exercise 1

Create the model shown in Figure 3-113. The dimensions of the model are shown in Figure 3-114. **(Expected time: 30 min)**

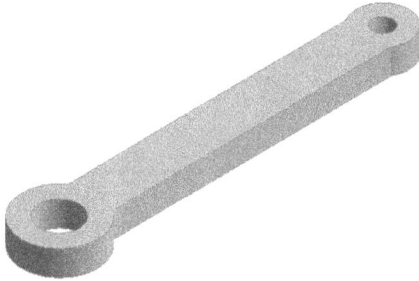

Figure 3-113 *Model for Exercise 1*

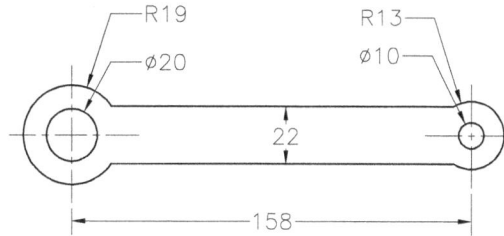

Figure 3-114 *Dimensions of the model for Exercise 1*

Exercise 2

Create the model shown in Figure 3-115. The dimensions of the model are shown in Figure 3-116. **(Expected time: 45 min)**

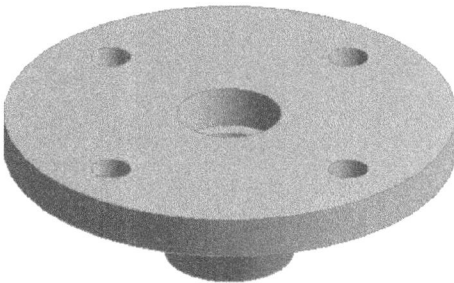

Figure 3-115 *Model for Exercise 2*

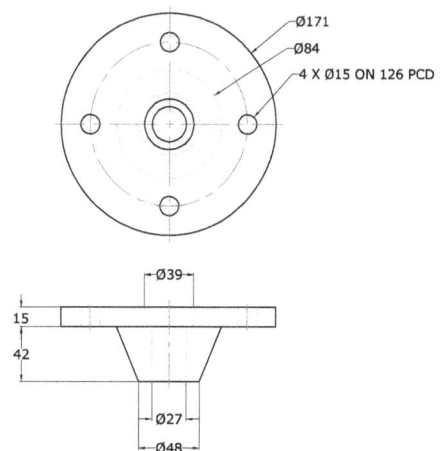

Figure 3-116 *Dimensions for Exercise 2*

Exercise 3

Create the model shown in Figure 3-117. The dimensions of the model are shown in Figure 3-118.   **(Expected time: 45 min)**

Figure 3-117 *Model for Exercise 3*

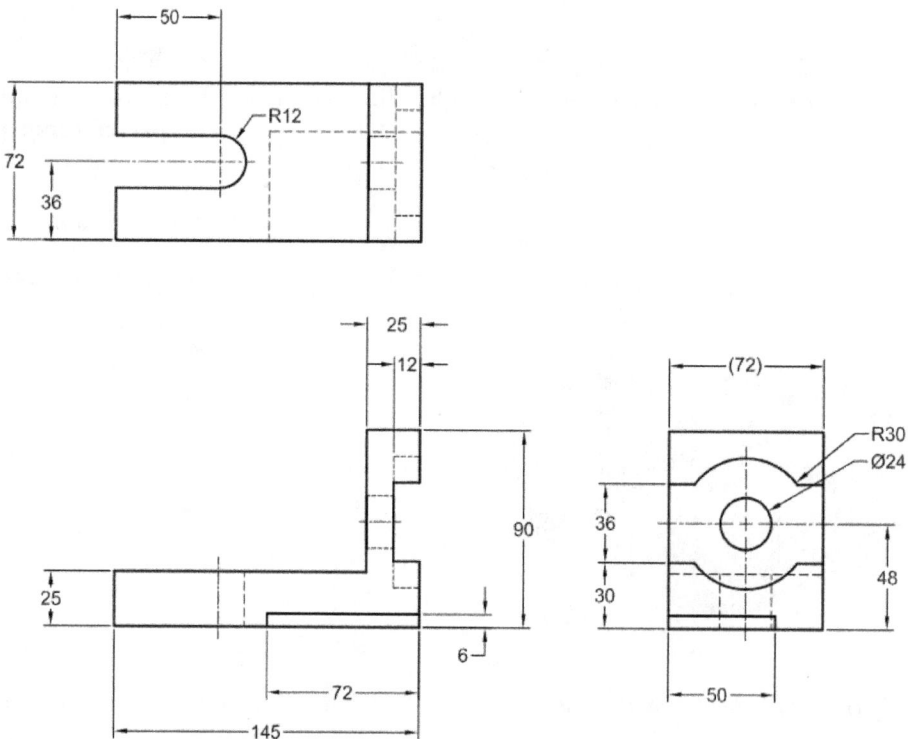

Figure 3-118 *Dimensions of the model for Exercise 3*

Answers to Self-Evaluation Test

1. Draw, 2. Equal Length, 3. Modify, 4. Tree Outline, **5.** T, **6.** F, **7.** F, **8.** T, **9.** T

Chapter 4

Part Modeling- II

Learning Objectives

After completing this chapter, you will be able to:

- *Understand line bodies and cross sections*
- *Apply cross-section to line bodies*
- *Create pattern features*
- *Create surfaces from sketches*

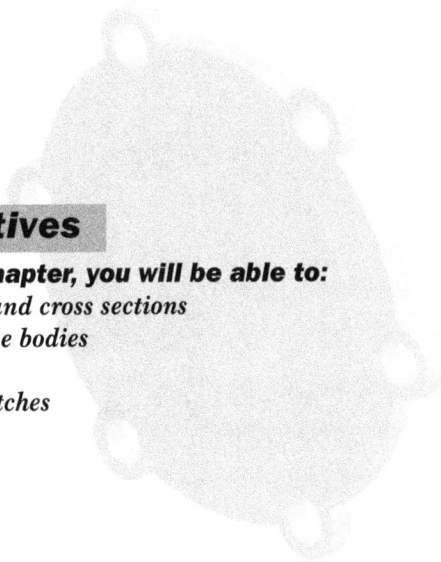

INTRODUCTION TO CONCEPT MENU

In the **DesignModeler** window of ANSYS Workbench, all the features and tools are available in the **Menu** bar. The **Menu** bar consists of the **File**, **Create**, **Concept**, **Tools**, **Units**, **View**, and **Help** menus. The **Concept** menu is discussed next.

Concept Menu

The **Concept** menu in the **Menu** bar contains verious options such as **Lines From Points**, **Lines From Sketches**, **Surfaces From Sketches**, **Cross Section**, and so on, as shown in Figure 4-1. These features are generally used for the creation of line bodies and surface bodies with the help of points, vertices, edges, sketches, faces, and so on.

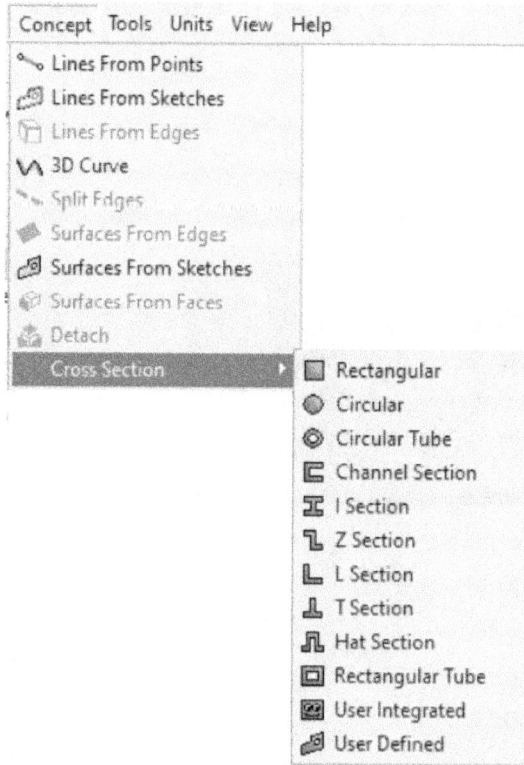

Figure 4-1 *The **Concept** menu displaying various cross sections available in it*

The **DesignModeler** window provides you a template of 12 cross-sections. These cross-sections assist you in assigning basic shapes such as rectangular, circular, I section, and so on, to the line bodies, refer to Figure 4-1. Some of the cross-sections are discussed next.

Rectangular

For creating the rectangular cross-section, you need to provide only two dimensions. These two dimensions are Width (B) and Height (H), as shown in Figure 4-2.

Circular Tube

For creating the circular tube cross-section, you need to provide inner and outer radius of the tube, as shown in Figure 4-2.

Here, Ri = Inner radius of the tube
Ro = Outer radius of the tube

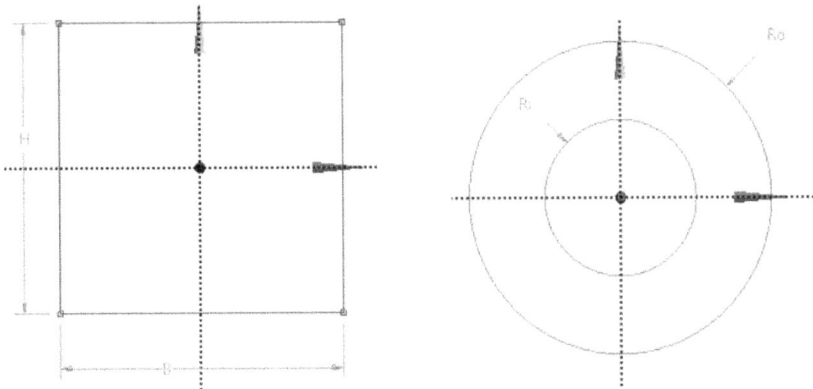

Figure 4-2 Displaying rectangular and circular tube cross-sections

In this chapter, some of the options under the **Concept** menu have been used to create the geometry.

TUTORIALS

Tutorial 1

In this tutorial, you will create the line body for a beam having the C cross-section, as shown in Figure 4-3. For dimensions, refer to Figure 4-4. **(Expected time: 45 min)**

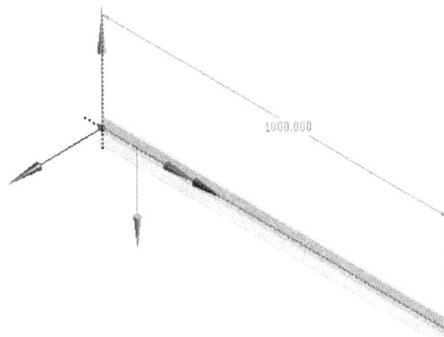

Figure 4-3 Model of the beam with C-section

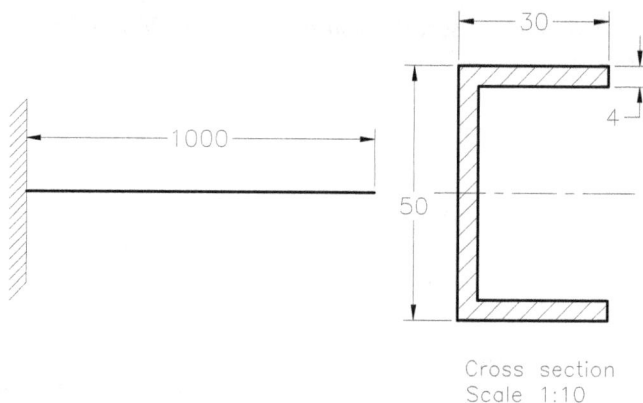

Figure 4-4 *Dimensions of the beam with C-section*

The following steps are required to complete this tutorial:

a. Start ANSYS Workbench and add the **Geometry** component system.
b. Draw the sketch for the beam.
c. Generate the dimension of the line and change its value to 1000 mm.
d. Generate the line body from the drawn sketch.
e. Define the cross-section of the line body.
f. View and rotate the cross-section as per your requirement.
g. Save the project and exit the ANSYS Workbench session.

Starting ANSYS Workbench and Adding the Geometry Component System

Before you start with the tutorial, you need to specify the project folder in which you need to save all the related files.

1. Create a folder with the name **c04** at the location *C:\ANSYS_WB*.

2. Create a folder with the name **Tut01** at the location *C:\ANSYS_WB\c04*.

    You will save all the work related to this tutorial in this folder.

    Now, you need to start ANSYS Workbench and then add a component system to the project.

3. Start ANSYS Workbench 2023 R2. The **Workbench** window is displayed.

4. In the **Workbench** window, choose the **Save** button from the **Main** toolbar; the **Save As** dialog box is displayed.

5. In this dialog box, browse to the location *C:\ANSYS_WB\c04\Tut01* and save the project with the name **c04_ansWB_tut01**.

After the project directory is specified, you need to add the **Geometry** component system to the **Project Schematic** window.

6.   Double-click on **Geometry** displayed under the **Component Systems** toolbox in the **Toolbox** window; the **Geometry** component system is added to the **Project Schematic** window.

7.   Once the **Geometry** component system is added to the **Project Schematic** window, its name is highlighted at the bottom of the component system in blue. If not, double-click on the default name and rename it to **C-Section Beam**.

Drawing the Sketch for the Beam

To create the C-section beam model, first you need to start the **DesignModeler** window and create the sketch for the beam.

1.   Right-click on the **Geometry** cell in the **C-Section Beam** component system; choose the **New DesignModeler Geometry**, the **DesignModeler** window is opened.

2.   Choose the **Millimeter** option from the **Units** menu of the Menu bar.

     Next, you need to create the sketch of the model. To do so, you need to specify the plane on which you want to create the sketch and then create the sketch on the selected plane.

3.   Select **XYPlane** from the **Tree Outline**; the selected plane is displayed in the graphics window.

4.   Choose the **Sketching** tab displayed below the **Tree Outline** to switch to the **Sketching** mode.

5.   Choose the **Look At Face/Plane/Sketch** tool from the **Graphics** toolbar; the plane is oriented normal to the viewing direction.

6.   Choose the **Line** tool from the **Draw** toolbox; the cursor changes to the Draw cursor.

7.   Move the cursor near the origin in the graphics window and click once the symbol of coincident point constraint (P) is displayed attached to the cursor, refer to Figure 4-5.

8.   Next, you need to specify the end point of the line. Move the cursor toward right in such a way that the symbol of horizontal constraint (H) is displayed over the line and symbol of coincident constraint (C) is displayed at the end point of the line, as shown in Figure 4-6.

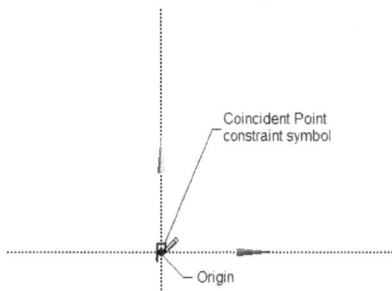

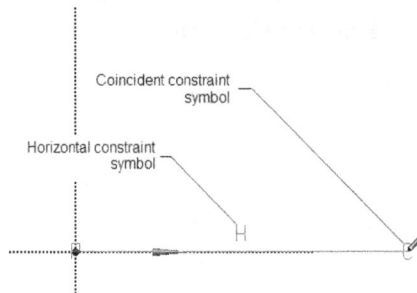

Figure 4-5 Specifying the origin as the start point of the line   *Figure 4-6 Specifying the end point of the line*

9. Click to specify the end point of the line; the line is created. Now, press the Esc key to exit the **Line** tool.

Generating Dimension

Now, you need to create the dimension of the line that was created in the previous step and edit its value to 1000 mm.

1. Click on the **Dimensions** toolbox in the **Sketching Toolboxes** window; the **Dimensions** toolbox is expanded.

2. Choose the **General** tool from the **Dimensions** toolbox, if it is not chosen by default.

3. Move the cursor over the line and select it; preview of the dimension is displayed attached to the cursor.

4. Move the cursor and click at the location where you want to place the dimension; the dimension is generated and its name is displayed on the dimension line.

5. In the **Details View** window, click in the edit box displayed on the right of the dimension name (**H1**) under the **Dimensions: 1** node and then enter **1000** in it; the length of line instantaneously changes to 1000 mm and is displayed in the graphics window, as shown in Figure 4-7.

Note
*To fit the drawn sketch in the graphics window, choose the **Zoom to Fit** tool from the **Graphics** toolbar.*

Generating the Line Body from the Drawn Sketch

Now, you need to convert the drawn sketch to a line body feature.

1. Choose the **Lines From Sketches** tool from the **Concept** menu of the Menu bar; you are prompted to select the base object to convert it into a line body feature. Also,

Line1 is added below the three default planes in the **Tree Outline**, as shown in Figure 4-8. A yellow thunderbolt is also displayed attached to **Line1** in the **Tree Outline** indicating that the feature needs to be generated.

Figure 4-7 *The dimension of the line*

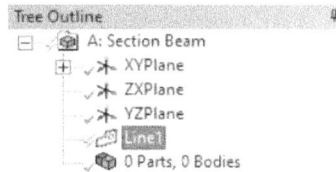

Figure 4-8 *Line 1 added to the Tree Outline representing the line body feature*

2. Select the line drawn earlier by selecting its end points and then choose the **Apply** button from the **Base Objects** selection box of the **Details View** window.

   The **Lines From Sketches** tool is used to create line bodies based on the object like faces and planes.

3. Choose the **Generate** tool from the **Features** toolbar; the yellow thunderbolt symbol is changed to a green check mark, indicating that the feature is updated.

4. Right-click in the graphics window and choose the **Isometric View** option from the shortcut menu displayed; the view is changed to Isometric, as shown in Figure 4-9.

Figure 4-9 *Isometric view of the generated line body feature*

Defining the Cross-section of the Line Body

After defining the line body, you need to define the cross-section of the line body.

1. Choose **Concept > Cross Section > Channel Section** from the **Menu** bar; the channel section is displayed in the graphics window along with its dimensions. Also, a new node with the name **1 Cross Section** is added in the **Tree Outline**.

The cross-sections are assigned to the line body to define the properties of the beam and are associated with some standard dimensions. The values of these dimensions can be changed to control the shape of the cross-sections.

2. Change the dimension values in the **Details View** window, refer to Figure 4-10. The final cross-section should be same as shown in Figure 4-11.

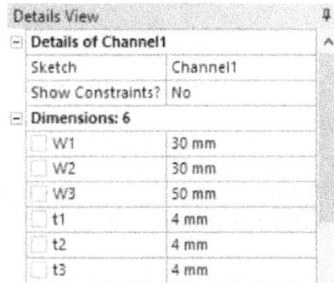

Figure 4-10 *The values to be entered in the* **Details View** *window*

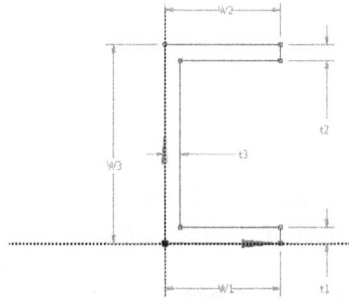

Figure 4-11 *The channel section after changing the values*

> **Tip**
> *You can change the locations of the dimensions displayed along with the cross-section. To do so, right-click in the graphics window and choose the* **Move Dimensions** *option from the shortcut menu displayed. Select the dimension to be moved and then move the cursor to the location where you want to place it. Click to place the selected dimension. After moving the selected dimensions, click on some other features in the Tree Outline or press the Esc key to exit the move dimension mode.*

3. Expand the **1 Part, 1 Body** node and select the **Line Body** displayed under it, as shown in Figure 4-12; the details of the selected line body are displayed in the **Details View** window. Note that the **Cross Section** property is highlighted in yellow indicating that you need to define the cross-section.

4. In the **Details View** window, select **Cross Section**; it turns yellow and a down arrow is displayed.

5. Click on the down-arrow and then select the **Channel1** option from the drop-down list displayed. The defined cross-section is assigned to the selected line body. Also, the **Offset Type** drop-down list is added in the **Details View** window.

6. Select the **Centroid** option from the **Offset Type** drop-down list, if it is not already selected.

    The options available in the **Offset Type** drop-down list are: **Centroid**, **Shear Center**, **Origin**, and **User Defined**.

    The **Centroid** option is selected by default and is used to center the cross-section on the edge according to the centroid of the cross-section.

Viewing and Rotating the Cross-section

Next, you need to view the cross-section assigned to the line body and change its orientation as per your requirement.

1. Choose the **Cross Section Solids** option from the **View** menu; the cross-section is displayed in the graphics window, as shown in Figure 4-13.

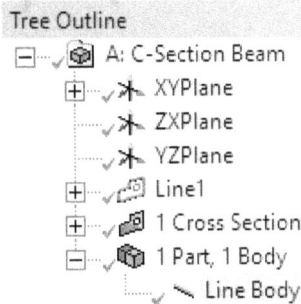

Figure 4-12 Selecting **Line Body** from the Tree Outline

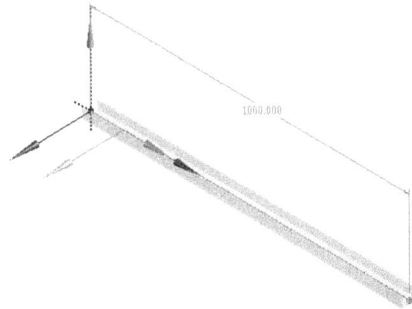

Figure 4-13 The cross-section displayed in the graphics window

> **Note**
> *Note that the coordinate system for the line body is different from that of **DesignModeler**. This difference in the coordinate system does not affect the final analysis results.*

2. Choose the **Edges** tool from the **Select** toolbar; the cursor changes to the Edge selection cursor.

   The tools available in the **Select** toolbar help you select points, edges, faces, bodies, and so on. There are various tools available in the **DesignModeler** window to facilitate selection of entities or parts. These tools can be accessed from the shortcut menu displayed, when you right-click in the graphics window.

3. Select the line body from the graphics window; the details of the selected line body are displayed in the **Details View** window.

4. Click on the **Rotate** edit box in the **Details View** window and enter **90** in the edit box; the model is rotated by 90 degrees, as shown in Figure 4-14. Figure 4-15 shows the zoomed partial view of the model.

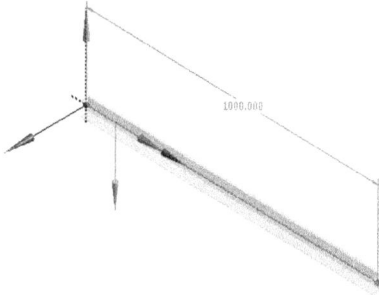

Figure 4-14 *Final rotated model of the beam with C-section*

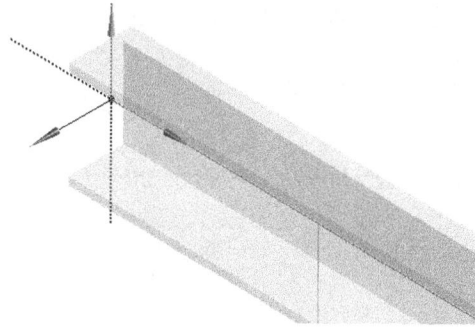

Figure 4-15 *Zoomed partial view of the beam with C-section*

5.   Close the **DesignModeler** window; the **Workbench** window is displayed.

Saving and Exiting ANSYS Workbench

Now, you need to save the project and exit ANSYS Workbench.

1.   In the **Workbench** window, choose the **Save Project** button from the **Main** toolbar; the project is saved with the name *c04_ansWB_tut01*.

2.   Choose **File > Exit** from the Menu bar to close the **Workbench** window and exit the ANSYS Workbench session.

Tutorial 2

In this tutorial, you will create the model of a car disk break rotor, as shown in Figure 4-16. The sketch of the model and its dimensions are shown in Figure 4-17.   **(Expected time: 45 min)**

Figure 4-16 *Model for Tutorial 2*

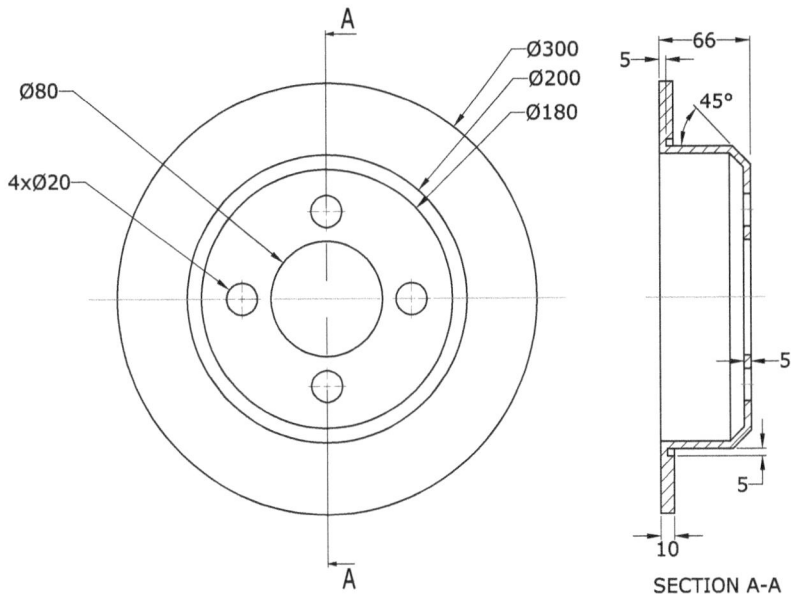

Figure 4-17 *Sketch and dimensions for the revolve feature*

The following steps are required to complete this tutorial:

a. Start ANSYS Workbench and add the **Geometry** component system.
b. Draw the sketch.
c. Apply constraints and generate dimensions.
d. Create the revolved feature.
e. Create the hole feature.
f. Create the pattern feature.
g. Save the model and exit Workbench.

Starting ANSYS Workbench and Adding the Geometry Component System

Before creating the model, you need to start the **Workbench** window and then add a component system to the **Project Schematic** window.

1. Start ANSYS Workbench 2023 R2. The **Workbench** window is displayed.

2. In the **Workbench** window, choose the **Save Project** button from the **Main** toolbar; the **Save As** dialog box is displayed.

3. In this dialog box, browse to the folder *C:\ANSYS_WB\c04* and create a sub folder **Tut02**.

4. Browse to the *Tut02* folder and then save the project with the name **c04_ansWB_tut02**.

Next, you need to add the **Geometry** component system to the **Project Schematic** window.

5. Double-click on the **Geometry** component system displayed under the **Component Systems** toolbox in the **Toolbox** window; the **Geometry** component system is added to the **Project Schematic** window.

Drawing the Sketch

To create the revolved feature, you need to start the **DesignModeler** window and create the sketch of the base feature.

1. Right-click on the **Geometry** cell in the component system and choose the **New DesignModeler Geometry**; the **DesignModeler** window is opened.

2. Choose the **Millimeter** option from the **Units** menu of the **Menu** bar.

3. Select the **XYPlane** from the **Tree Outline** to specify the plane on which you want to create the sketch for the model; the XY plane is displayed in the graphics window.

4. Choose the **Sketching** tab displayed below the **Tree Outline**; the **Sketching** mode is invoked.

5. Choose the **Look At Face/Plane/Sketch** tool from the **Graphics** toolbar; the sketching plane is oriented normal to the viewing direction.

6. Choose the **Line** tool from the **Draw** toolbox; the shape of the cursor changes to the draw cursor.

    Next, you need to create a line along the Y axis.

7. Move the cursor near the origin in the graphics window and then move it to some distance along the Y-axis and click when the symbol of coincident constraint (C) is still displayed, refer to Figure 4-18.

8. Move the cursor upward in such a way that the symbol of vertical constraint (V) is displayed over the line and the symbol of coincident constraint (C) is displayed at the end point of the line, as shown in Figure 4-19.

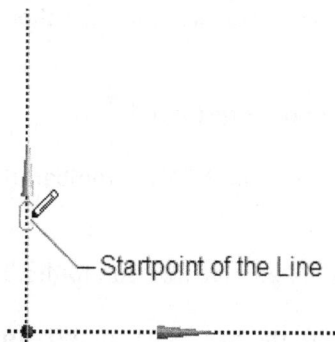

Figure 4-18 *Specifying the start point of the line*

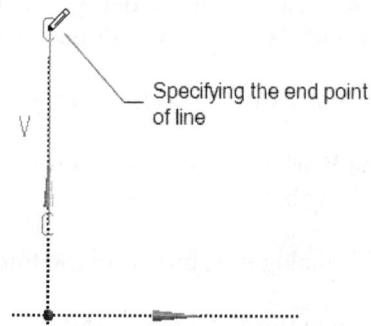

Figure 4-19 *Specifying the end point of the line*

9. Click to specify the end point of the line; the line is created. Now, press the Esc key to exit the line tool.

   Similarly, create all the remaining entities. The final sketch of the model after all entities are created is shown in Figure 4-20.

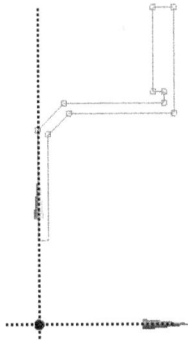

Figure 4-20 Final sketch of the model

Applying Constraints and Generating Dimensions

After the rough sketch is drawn, you need to apply required dimensions to the sketch.

1. Apply all the required constraints to the sketch.

2. Generate all the dimensions of the sketch, refer to Figure 4-21, and change the corresponding values in the **Details View** window. For values of the dimensions, refer to Figure 4-17.

Note
Note that you need to apply constraints to the sketch to fully define it.

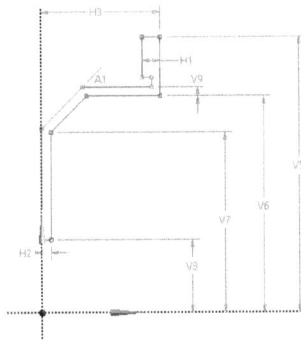

Figure 4-21 Sketch after the dimensions are generated

Note

*The names of the dimensions displayed in the graphics window may be different from the ones given in this textbook. If you want to change the names according to Figure 4-21, select a dimension in the graphics window and then right-click on it to display a shortcut menu. Choose the **Edit Name/Value** option from the shortcut menu; the **Geometry - DesignModeler** dialog box will be displayed. Enter the required name in the **Name** edit box of this dialog box and then choose the **OK** button to save the changes made and close the dialog box.*

Creating the Revolve Feature

Now, as the sketch is ready, you need to create the revolved feature, which is also the base feature.

1. Change the view to isometric by using the **ISO** tool from the **Graphics** toolbar.

2. Choose the **Revolve** tool in the **Features** toolbar; **Revolve1** along with a yellow thunderbolt symbol is added in the **Tree Outline**. Also the options corresponding to **Revolve1** are displayed in the **Details View** window.

3. Click on the **Geometry** selection box in the **Details View** window; the **Apply** and **Cancel** buttons are displayed. Click on **Apply** to assign the selected geometry for creating the revolve feature. Also, the color of the sketch in the graphics window turns green indicating that it is already selected.

Note

*If the desired sketch is selected by default in the graphics window, you need not select it to satisfy a parameter in the **Details View** window. If the sketch is not selected by default or there are multiple sketches in the graphics window, select the required sketch and then choose the **Apply** button in the corresponding selection box in the **Details View** window.*

4. Next, click on the **Axis** selection box; the **Apply** and **Cancel** buttons are displayed.

5. Select the X axis that is displayed on the model, as shown in Figure 4-22, and then choose the **Apply** button in the **Axis** selection box to specify the X axis as the axis of revolution; the preview of the feature is displayed in the graphics window.

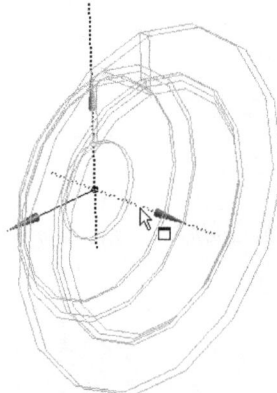

Figure 4-22 Selecting the X Axis that is displayed on the model

6. In the **Tree Outline**, right-click on **Revolve1** to display a shortcut menu, as shown in Figure 4-23. Choose the **Generate** option from this shortcut menu; the revolved feature is created, as shown in Figure 4-24.

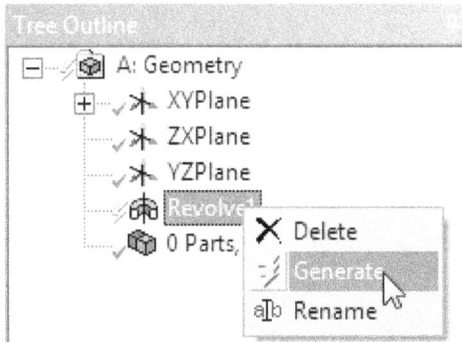

Figure 4-23 *The shortcut menu displayed*

Figure 4-24 *The revolved feature*

Creating the Hole Feature

After the revolve feature is created, the next task is to create the holes for accommodating the fasteners in the car brake disc rotor. To create the holes on the base feature, you first need to create the sketch for the hole feature on the inner face of the car brake disc rotor.

1. Choose the **Faces** tool from the **Select** toolbar.

2. Next, select the inner face of the base feature, as shown in Figure 4-25.

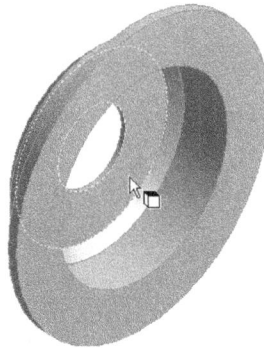

Figure 4-25 *The inner face of the base feature selected*

3. Choose the **New Plane** button available in the **Active Plane/Sketch** toolbar to create the sketch of the hole on the selected face; the preview of the new plane is displayed in the graphics window. Also, **Plane4** is added to the **Tree Outline**.

4. Choose the **Generate** tool; the new plane is created. Also, a green tick mark is placed before **Plane4** in the **Tree Outline** indicating that you can start creating the sketch for the hole feature.

    `⟲ Generate`

5. Next, choose the **Look At** tool in the **Graphics** toolbar; the model will be oriented normal to the new plane, refer to Figure 4-26.

6. Choose the **Sketching** tab available below the **Tree Outline** and then invoke the **Circle** tool from the **Draw** toolbox to create a circle.

    `◉ Circle`

7. Move the cursor to a point, as shown in Figure 4-26, and then click when the coincident point constraint symbol (P) is displayed to specify the center of the radius.

8. Next, move the cursor away from the center of the circle and then click at a random point to create a circle, as shown in Figure 4-27.

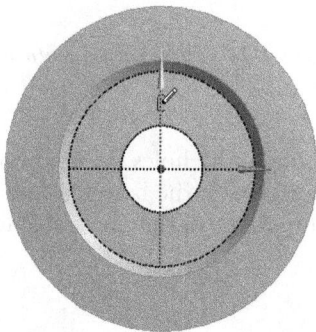

Figure 4-26 Specifying the center point of the circle

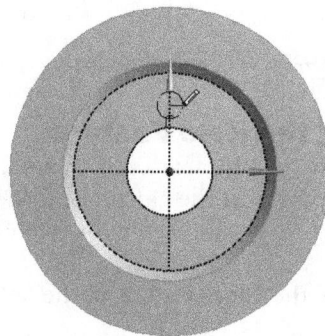

Figure 4-27 The circle created on the face of the model

9. Next, specify the diameter of the circle as **20**.

10. Change the view to isometric by using the **ISO** tool from the **Graphics** toolbar.

    `ISO`

11. Invoke the **Extrude** tool from the **Features** toolbar; **Extrude1** is added to the **Tree Outline**.

    `▣ Extrude`

12. Click on the **Geometry** selection box to display the **Apply** and **Cancel** buttons, if they are not already displayed. Choose the **Apply** button to specify the circle as the geometry for extrusion.

13. Next, in the **Details View** window, select the **Cut Material** option from the **Operation** drop-down list, as shown in Figure 4-28.

14. Next, select the **Reversed** option from the **Direction** drop-down list, as shown in Figure 4-29.

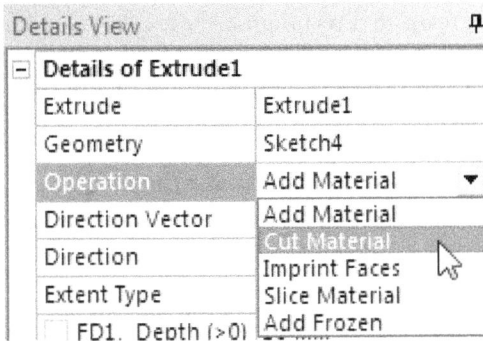

Figure 4-28 Selecting the **Cut Material** option from the **Operation** drop-down list

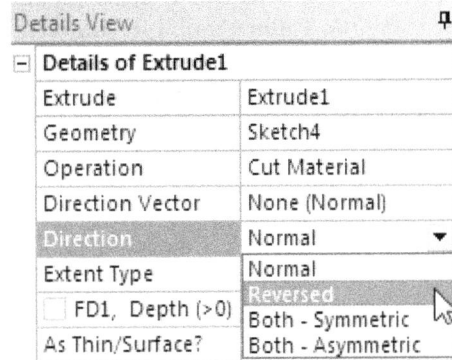

Figure 4-29 Selecting the **Reversed** option from the **Direction** drop-down list

15. Select the **Through All** option from the **Extent Type** drop-down list, as shown in Figure 4-30.

    The **Through All** option is selected when you need to create a cutout through the overall thickness of the feature.

16. Next, choose the **Generate** tool from the **Features** toolbar to generate the hole, as shown in Figure 4-31.

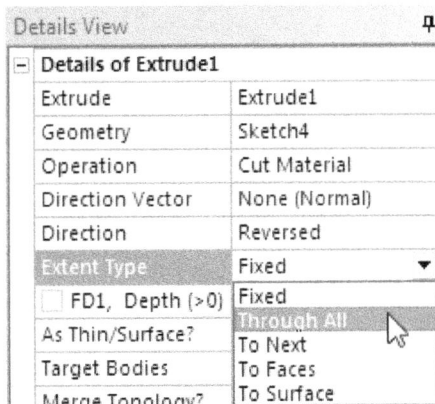

Figure 4-30 Selecting the **Through All** option from the **Extent Type** drop-down list

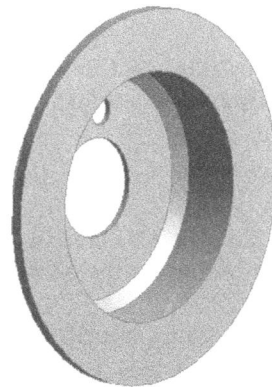

Figure 4-31 The hole created

Creating the Pattern of the Hole Feature

Now, after the hole feature is created, you need to pattern the hole to create three more instances of hole on the face of the rotor.

1. Invoke the **Pattern** tool from the **Create** menu in the Menu bar; **Pattern1** is attached to the **Tree Outline**.

The patterns are defined as the sequential arrangement of the copies of the selected entities. By using the **Pattern** tool, you can create the patterns in a rectangular or a circular mode.

2. Next, in the **Pattern Type** drop-down list, select the **Circular** option to specify the nature of the pattern as circular, as shown in Figure 4-32.

   Circular patterns are the patterns created around the circumference of a circle.

3. Choose the **Face** tool from the **Select** toolbar.

4. Next, click on the **Geometry** selection box to display the **Apply** and **Cancel** buttons.

5. Select the circular face of the hole created earlier in this tutorial, as shown in Figure 4-33, and then choose the **Apply** button in the **Geometry** selection box to confirm the selection.

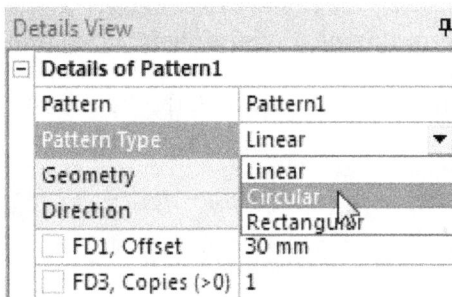

*Figure 4-32 Selecting the **Circular** option from the **Pattern Type** drop-down list*

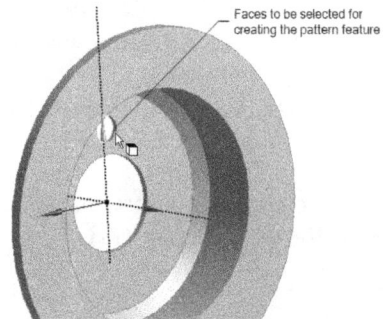

Figure 4-33 Partial view showing the selection of face of the circular hole

6. Next, click on the **Axis** selection box; the **Apply** and **Cancel** buttons are displayed.

7. Next, click on the **XYPlane** in the **Tree Outline** and then move the cursor to the X-axis in the graphics window; the X-axis gets highlighted, as shown in Figure 4-34.

8. Click on the X-axis; it turns yellow.

9. Next, move the cursor to the **Axis** selection box and then choose **Apply** to select the X-axis as the axis for creating the pattern.

10. Next, you need to specify the number of holes to be patterned. Enter **3** in the **FD3, Copies (>=0)** edit box in the **Details View** window.

11. Next, choose the **Generate** tool to generate the pattern. The model after creating the pattern feature is shown in Figure 4-35.

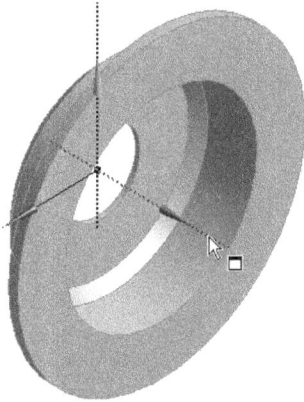

Figure 4-34 *Selecting the X-axis as the axis of circular pattern*

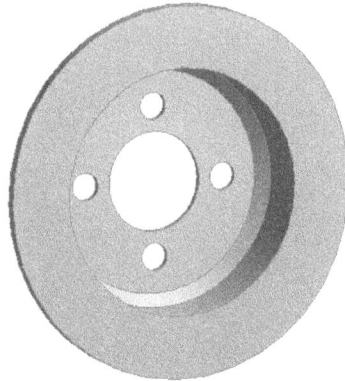

Figure 4-35 *Model after the hole feature is patterned*

12. Now, close the **DesignModeler** window by choosing the close ▬ button; the **Workbench** window is displayed.

Saving the Model and Exiting ANSYS Workbench

After the model is created, you need to save the project and exit the session.

1. Choose the **Save Project** button available in the **Main** toolbar; the project is saved with the name *c04_ansWB_tut02*.

2. Next, choose **File > Exit** from the Menu bar to close the **Workbench** window.

Tutorial 3

In this tutorial, you will create the revolved feature of the piston model, shown in Figure 4-36. The sketch of the model and its dimensions are shown in Figure 4-37.

(Expected time: 45 min)

Figure 4-36 *Model of the piston for Tutorial 3*

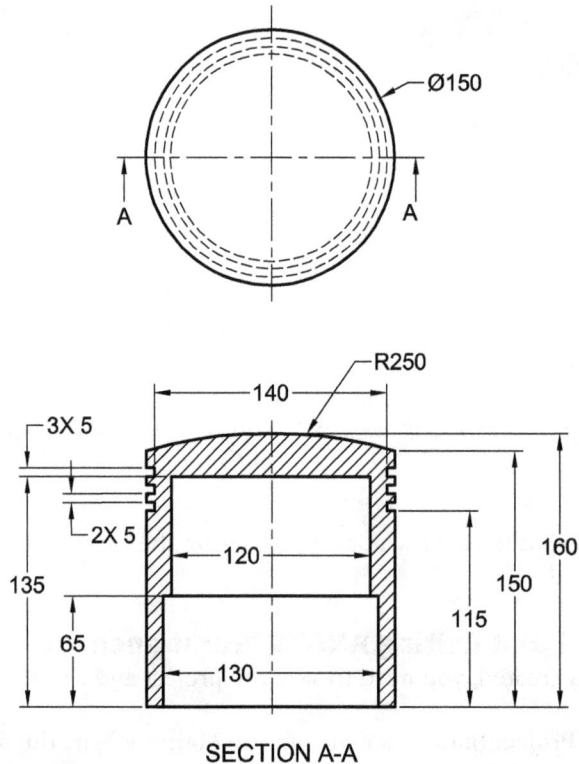

Figure 4-37 *Dimensions of the model for Tutorial 3*

The following steps are required to complete this tutorial:

a. Start ANSYS Workbench and add the **Geometry** component system.
b. Draw the sketch.
c. Create the revolve feature.
d. Save the model and exit ANSYS Workbench.

Starting ANSYS Workbench and Adding the Geometry Component System

Before starting the project, you need to start ANSYS Workbench and then add a component system to the project.

1. Start ANSYS Workbench 2023 R2. The **Workbench** window is displayed.

2. In the **Workbench** window, choose the **Save Project** button from the **Main** toolbar; the **Save As** dialog box is displayed.

3. In this dialog box, browse to the folder *C:\ANSYS_WB\c04* and create a subfolder **Tut03**.

4. Next, browse to the location *C:\ANSYS_WB\c04\Tut03* and save the project with the name **c04_ansWB_tut03**.

Next, you need to add the **Geometry** component system to the project.

5. Double-click on **Geometry** displayed under the **Component Systems** toolbox in the **Toolbox** window; the **Geometry** component system is added to the **Project Schematic** window.

6. Once the project is added to the **Project Schematic** window, its name gets highlighted in blue. If it is not highlighted, double-click on the default name and rename it to **Piston**.

Drawing the Sketch

To create the revolved feature, first you need to start the **DesignModeler** window. Here create the sketch of the base feature. Then, revolve the sketch to create the base feature.

1. Right-click on the **Geometry** cell in the **Piston** component system and choose the **New DesignModeler Geometry**; the **DesignModeler** window is opened.

2. Choose the **Millimeter** option from the **Units** menu of the **Menu** bar.

3. Select the **XYPlane** from the **Tree Outline** window; the XY plane is displayed in the graphics window. Now, all the sketches are created in the XY plane.

> **Note**
> *1. In the **DesignModeler** window, the XY plane is selected by default. To create a sketch on this plane, go to the **Sketching** mode and use the tools available in this mode.*
>
> *2. To create a sketch on a plane other than the XY plane, select the desired plane and then switch to the **Sketching** mode.*

4. Choose the **Sketching** tab displayed below the **Tree Outline**; the **Sketching Toolboxes** window is displayed, indicating that the **Sketching** mode is enabled.

5. Choose the **Look At Face/Plane/Sketch** tool from the **Graphics** toolbar to orient the plane normal to the viewing direction.

6. Create the sketch by using the tools available in the **Draw** toolbox, refer to Figure 4-38.

7. Add the required constraints by using the tools in the **Constraints** toolbox, refer to Figure 4-38.

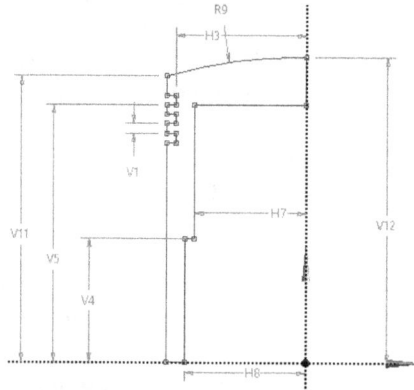

Figure 4-38 Sketch created on the XY plane

8. Similarly, add the required dimensions by using the tools available in the **Dimensions** toolbox, refer to Figure 4-38.

Creating the Revolved Feature

To create the revolved feature, you need to exit the **Sketching** mode and then switch to the **Modeling** mode.

1. Choose the **Revolve** tool in the **Features** toolbar; the **Modeling** mode is invoked and **Revolve1** is added to the **Tree Outline**, refer to Figure 4-39. Also, the options corresponding to the **Revolve1** node are displayed in the **Details View** window.

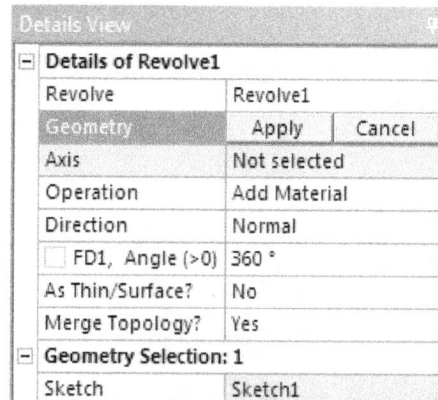

2. Click on the **Geometry** selection box in the **Details View** window; the **Apply** and **Cancel** buttons are displayed, as shown in Figure 4-40.

3. Select the **Sketch1** node under the **XYPlane** in the **Tree Outline**, as shown in Figure 4-39.

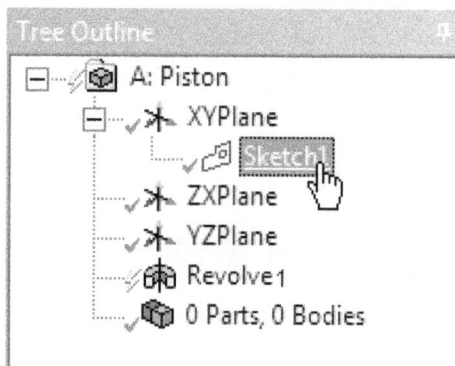

Figure 4-39 Selecting the sketch from the Tree Outline

Figure 4-40 The **Geometry** edit box with the **Apply** and **Cancel** buttons

4.   Next, choose the **Apply** button in the **Geometry** selection box in the **Details View** window; **Sketch1** is displayed in the **Geometry** selection box, indicating that the geometry is specified for the revolve operation.

5.   Next, click on the **Axis** selection box; the **Apply** and **Cancel** buttons are displayed.

6.   In the graphics window, select the Y axis, as shown in Figure 4-41; the preview of the revolve feature is displayed in the graphics window, as shown in Figure 4-42.

7.   Choose **Apply** from the **Axis** selection box; Y axis is specified as the axis of revolution.

     Note that in the **Details View** window, you need to specify the sketch to be revolved and the axis of revolution. You need not change other parameters in the **Details View** window.

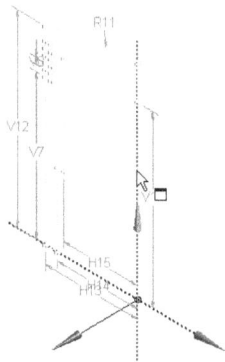

Figure 4-41 *Selecting the Y-axis as the axis of revolution*

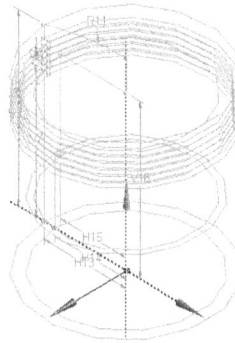

Figure 4-42 *Preview of the revolve feature*

8.   Choose the **Generate** tool from the **Features** toolbar; the revolved feature is created, as shown in Figure 4-43.

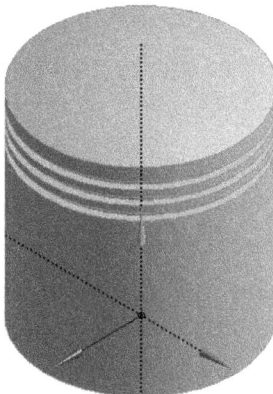

Figure 4-43 *The revolved feature created*

9. Exit the **DesignModeler** window; the **Workbench** window is displayed.

Saving the Model and Exiting ANSYS Workbench

After creating the model, you need to save the project and exit ANSYS Workbench session.

1. In the **Workbench** window, choose the **Save** button from the **Main** toolbar; the project is saved with the name *c04_ansWB_tut03*.

2. Next, choose **File > Exit** from the **Menu** bar to close the **Workbench** window.

Tutorial 4

In this tutorial, you will create a 5 mm thick plate, as shown in Figure 4-44. Also, you will create a sketch on the top surface of the base feature and then generate a surface on it. For dimensions, refer to Figure 4-45.

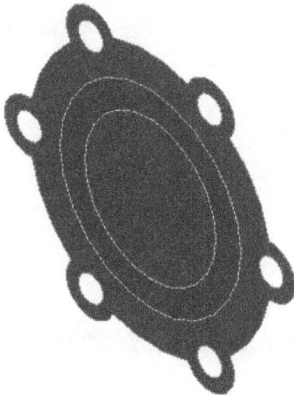

Figure 4-44 *Model of the surface body for Tutorial 4*

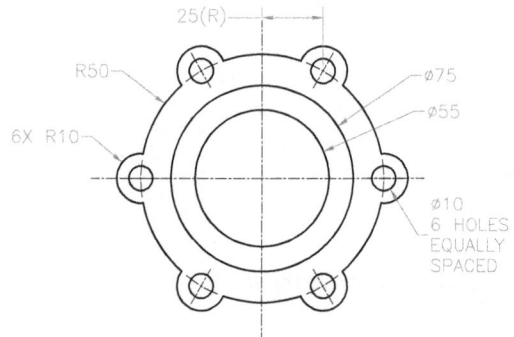

Figure 4-45 *Dimensions of the surface body for Tutorial 4*

The following steps are required to complete this tutorial:

a. Start ANSYS Workbench and add the **Geometry** component system.
b. Start the **DesignModeler** window and draw the sketch of the base feature.
c. Create the extrude feature.
d. Create a plane on the front face of the model.
e. Create a surface on the front face.
f. Save the project.

Starting ANSYS Workbench and Adding the Geometry Component System

First, you need to start ANSYS Workbench and then add a component system to the project.

1. Start ANSYS Workbench 2023 R2. The **Workbench** window is displayed.

2. In the **Workbench** window, choose the **Save** button from the **Main** toolbar; the **Save As** dialog box is displayed.

3. In this dialog box, browse to the folder *C:\ANSYS_WB\c04* and create a subfolder with the name **Tut04**.

4. Next, browse to the *Tut04* folder and save the project with the name **c04_ansWB_tut04**.

    After invoking the **Workbench** window, add an appropriate analysis or component system to the **Project Schematic** window. In this tutorial, you will create a surface body using the **Geometry** component system.

5. Expand the **Component Systems** toolbox if it is not already expanded. Double-click on the **Geometry** component system; the **Geometry** component system is added to the **Project Schematic** window.

6. Rename the added component system to **Surface Body**.

Creating the Sketch

Before you create the model, you need to start the **DesignModeler** window session and then specify the plane on which you want to create the sketch for the base feature.

1. Right-click on the **Geometry** cell in the **Surface Body** component system and choose the **New DesignModeler Geometry**; the **DesignModeler** window is opened.

2. Choose the **Millimeter** option from the **Units** menu of the **Menu** bar.

3. Choose the **Sketching** tab displayed below the **Tree Outline** of the **DesignModeler** window; the **Sketching** mode is invoked. Also, note that the default plane XY is displayed in the graphics window.

4. Choose the **Look At Face/Plane/Sketch** tool from the **Graphics** toolbar; the sketching plane is oriented normal to the viewing direction.

5. Choose the **Circle** tool from the **Draw** toolbox and draw a circle with its center at the origin.

6. Invoke the **General** tool from the **Dimensions** toolbox and generate the diameter dimension of the circle.

7. Edit the value of the dimension to **100** in the **Details View** window; the circle is modified according to the specified value and is displayed in the graphics window, as shown in Figure 4-46.

Note
*Change the display settings of dimension to **Value** as explained in previous tutorials. You may also need to zoom and pan the sketch to fit it into the graphics window.*

8. Invoke the **Circle** tool and draw two concentric circles such that their centers are coincident with the X-axis, refer to Figure 4-47.

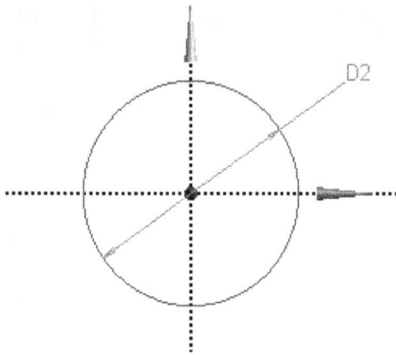

Figure 4-46 *The circle of diameter 100mm*

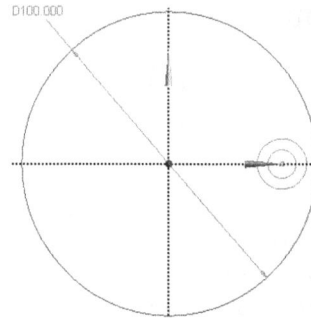

Figure 4-47 *Two concentric circles with their centers coincident with the X-axis*

After creating two concentric circles, you now need to make the center point of these circles lie on the circumference of the circle of diameter 100 mm.

9. Invoke the **Coincident** tool from the **Constraints** toolbox and then select ⌐ Coincident the center point of the concentric circles and the circumference of the circle of diameter 100 mm; the center of the concentric circles will becomes coincident with the circumference of the circle of diameter 100 mm, refer to Figure 4-48.

    The **Coincident** tool forces two points, or a point and line, to coincide.

10. Invoke the **Trim** tool from the **Modify** toolbox and trim the undesired portions ⊤ Trim of the sketch, refer to Figure 4-49.

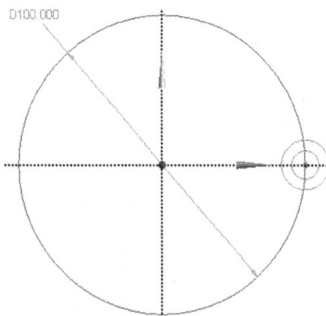

Figure 4-48 *The two concentric circles after applying the coincident constraint*

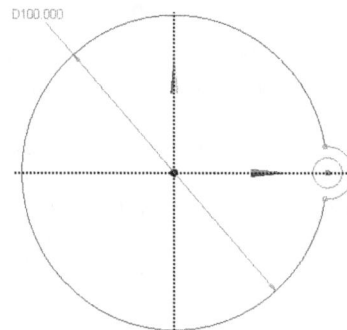

Figure 4-49 *Sketch after trimming the undesired portion*

11. Invoke the **General** tool from the **Dimensions** toolbox and generate the ⊘ Circle dimensions of the concentric arc and circle.

12. Edit the dimensions of the sketch, as shown in Figure 4-50.

Next, you need to create five more instances of the concentric arc and circle at an angle of 60 degrees to each other. These instances can be created by using the **Replicate** tool.

13. Invoke the **Replicate** tool from the **Modify** toolbox and select the concentric arc and circle as the entities to be replicated.

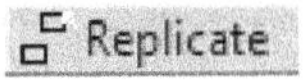

14. Right-click in the graphics window and choose the **End / Use Plane Origin as Handle** option from the shortcut menu displayed; the preview of the entities to be replicated is displayed attached to the paste handle, refer to Figure 4-51.

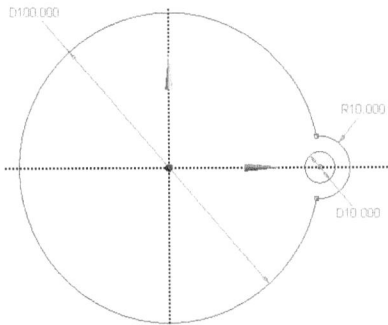

Figure 4-50 Sketch displaying various dimensions

Figure 4-51 Preview of the entities to be replicated

15. Enter **60** and **1** in the **r** and **f** edit boxes respectively, displayed on the right of the **Replicate** tool in the **Modify** toolbox, refer to Figure 4-52.

16. Right-click in the graphics window and choose the **Rotate by r** option from the shortcut menu displayed; the preview of the selected entities is rotated by 60 degrees, refer to Figure 4-53.

*Figure 4-52 The **r degree** edit box*

Figure 4-53 Preview of the entities to be replicated after rotating them by 60 degrees

17. Right-click in the graphics window and choose the **Paste at Plane Origin** option from the shortcut menu displayed; a similar instance of the selected entity is replicated at an angle of 60 degrees, refer to Figure 4-54.

18. Similarly, create four more instances of the selected entities. Figure 4-55 shows the sketch after creating all the instances.

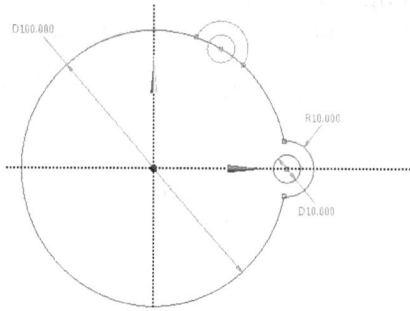

***Figure 4-54** Sketch after replicating the first instance of the selected entity*

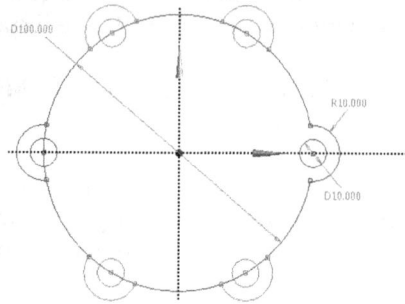

***Figure 4-55** Sketch after replicating all instances of the selected entity*

19. Invoke the **Trim** tool from the **Modify** toolbox and trim the undesired portions of the sketch, refer to Figure 4-56.   `⊤ Trim`

   Note
   There is a possibility that even after trimming all entities you may not get the desired sketch, as shown in Figure 4-56. This may happen because the extension lines of diameter dimension 100 mm are displayed on the locations where you have trimmed the entities. In such a case, you need to delete the diameter dimension of value 100 mm and create the radial dimension for any one of the arcs.

20. Invoke the **Symmetry** tool from the **Constraints** toolbox and make the center points, shown in Figure 4-57, symmetric about both X and Y axes.   `↑ Symmetry`

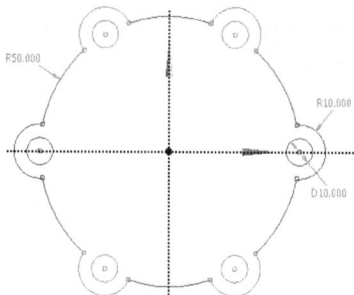

***Figure 4-56** Sketch after trimming the undesired portions of the sketch*

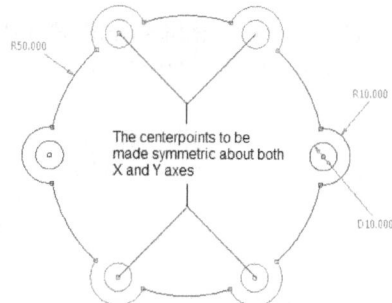

***Figure 4-57** The center points to be made symmetric about both X and Y axes*

The **Symmetry** tool is used to create symmetrical features about an axis. To create symmetries, invoke the tool and then select the axis about which you want to create the symmetry. Next, choose the required point or edge to apply symmetry constraint.

21. Invoke the **General** tool from the **Dimensions** toolbox and generate the dimension between the Y axis and the Center point A, refer to Figure 4-58. All sketched entities turn blue, indicating that the sketch is fully defined.

   As the sketch for the surface body is complete and fully defined, you need to exit the **Sketching** mode.

22. Choose the **Modeling** tab displayed below the **Sketching Toolboxes**; the **Modeling** mode is invoked.

Creating the Extrude Feature

After creating the sketch, you need to extrude the sketch using the **Extrude** tool.

1. Right-click on the graphics window and then choose the **Isometric View** option from the shortcut menu displayed; the view of the sketch becomes isometric.

   Note
   *You can also use the tools available in the **Graphics** toolbar to adjust the view of the model to isometric.*

2. Choose the **Extrude** tool from the **Features** toolbar; the contents of the **Details View** window are changed. Also, **Extrude1** is added to the **Tree Outline** and you are prompted to select the sketch to extrude.

3. Select **Sketch1** from the **Tree Outline**; the sketch in the graphics window is highlighted in yellow.

4. Choose the **Apply** button from the **Geometry** selection box in the **Details View** window; the sketch is now selected to be converted into a body.

5. Enter **5** in the **FD1, Depth (>=0)** edit box to specify the depth of extrusion.

6. Choose the **Generate** tool from the **Features** toolbar; the sketch is extruded, as shown in Figure 4-59.

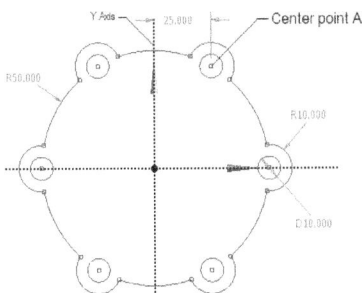

Figure 4-58 *Final sketch for generating the model*

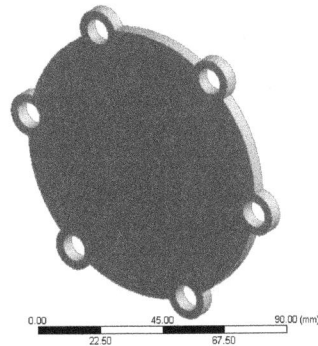

Figure 4-59 *The extruded feature*

Creating a Plane on the Front Face of the Model

The next step is to create a surface on the front face of the model.

1. Choose the **New Plane** tool in the **Active Plane/Sketch** toolbar; **Plane4** is attached to the **Tree Outline**. Also, the contents of the **Details View** window are changed.

2. In the **Details View** window, select **From Centroid** from the **Type** drop-down list; the **Base Face** selection box is displayed in the **Details View** window.

3. Click on the **Base Face** selection box to display the **Apply** and **Cancel** buttons.

4. Choose the **Face** tool from the **Select** toolbar and then select the front face of the model, as shown in Figure 4-60.

5. Choose the **Apply** button from the **Geometry** selection box; **Selected** is displayed in the **Geometry** selection box.

6. Choose the **Generate** tool from the **Features** toolbar; the new plane is created on the front face of the model, refer to Figure 4-61.

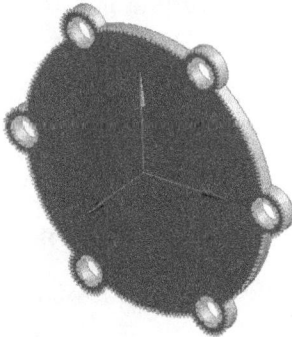

Figure 4-60 The front face of the model selected for creating the plane

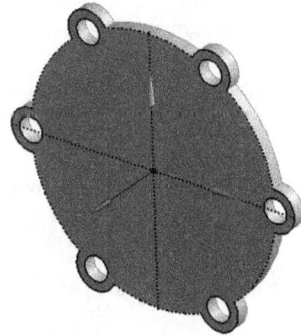

Figure 4-61 The new plane generated

Creating a Surface on the Solid Model

After the base feature is created, it is important to create a surface on the front face in such a manner that a different region can be created on it. To create a region on the model, follow the procedure given next:

1. Choose the **Look At** tool from the **Graphics** toolbar; the front view of the model is oriented, as shown in Figure 4-62.

2. Choose the **New Sketch** tool in the **Active Plane/Sketch** toolbar; **Sketch2** is added under the **Plane4** node in the **Tree Outline**, as shown in Figure 4-63.

3. Choose the **Sketching** tab; the **Sketching** mode is invoked.

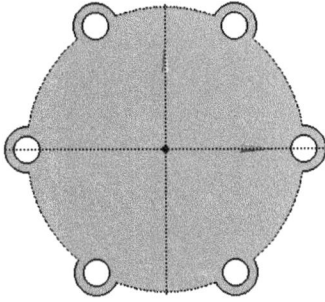

Figure 4-62 Front view of the model

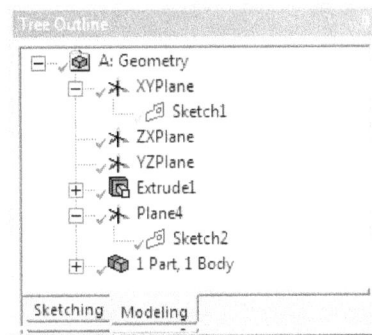

Figure 4-63 Sketch2 attached to the Plane4 node in the Tree Outline

Next, you need to create an intersecting surface from sketched entities. To do so, you need to create the sketch of two circles.

4. Invoke the **Circle** tool from the **Draw** toolbox in the **Sketching Toolboxes** window; the cursor changes to the Draw cursor. Also, you are prompted to specify the center of the circle.

5. Click on the origin to specify the center point of the circle.

6. Move the cursor away from the origin and click at a point such that the radius of the circle is smaller than the radius of the surface on which it is created, refer to Figure 4-64.

7. Similarly create another circle with a radius smaller than the radius of the circle created previously, refer to Figure 4-64.

8. Next, you need to specify the dimensions of the circles created. To do so, invoke the **General** tool from the **Dimension** toolbox in the **Sketching Toolboxes** window.

9. Next, click on the outer circle and then place the dimension such that it does not interfere with any other entity, refer to Figure 4-65.

10. Similarly, select the inner circle and place the dimension, refer to Figure 4-65.

11. In the **Details View** window, specify the outer and inner diameters as **75** and **55** respectively.

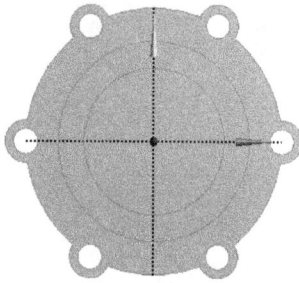

Figure 4-64 *Two circles created on the new plane*

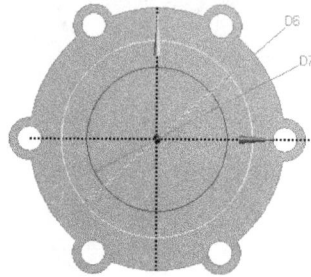

Figure 4-65 *Dimensioned sketches*

Next, you need to create a surface inside the sketch in such a way that a new region is created in the model.

12. Change the view of the model to isometric by using the **ISO** tool.

13. Invoke the **Surfaces From Sketches** tool from the **Concept** menu of the menu bar; **SurfaceSk1** is added in the **Tree Outline**.

    The **Surfaces From Sketches** tool is used to create surface bodies by using sketches. The sketches to be used for creating the surface should form a closed loop and the entities must not intersect each other at any point of time. The edges of the sketches are to be considered as the boundaries for the surface creation.

14. Click on the **Base Objects** selection box in the **Details View** window; the **Apply** and **Cancel** buttons are displayed in it.

15. Select **Sketch2** from the **Tree Outline**.

16. Choose the **Apply** button from the **Base Object** selection box; **Sketch2** is specified as the base object. Also, **1 Sketch** is displayed in the **Base Objects** selection box.

17. Choose the **Generate** tool from the **Features** toolbar; the surface is created on the model, as shown in Figure 4-66.

18. Exit the **DesignModeler** window; the **Workbench** window is displayed.

Saving the Project and Exiting ANSYS Workbench

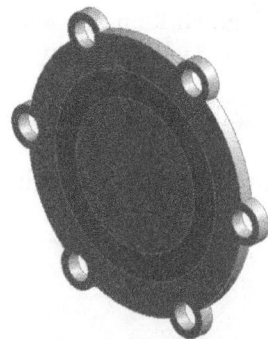

Figure 4-66 *Model with the newly created sketch*

After creating the model, you need to save the project and exit ANSYS Workbench. This saved project will be used in later chapters for analysis.

1. Choose the **Save** button from the **Main** toolbar; the project is saved with the name *c04_ansWB_tut04*.

2. Choose the **Exit** option from the **File** menu of the **Workbench** window to close the ANSYS Workbench session.

Self-Evaluation Test

Answer the following questions and then compare them to those given at the end of this chapter:

1. Which of the following Auto constraint symbols is used to make the end point of the current drawing entity coincident with a point?

    (a) C   (b) R
    (c) P   (d) T

2. You can create circular patterns using the _____ tool from the **Create** menu of the Menu bar.

3. The **DesignModeler** application is associated with the **Geometry** component cell. (T/F)

4. A system cannot be added to the **Project Schematic** window by dragging it from the **Toolbox** window. (T/F)

5. In the **DesignModeler** window, the XY, YZ, and ZX planes are displayed by default. (T/F)

6. In ANSYS Workbench, you can create L and I sections along with channel sections. (T/F)

7. You cannot cut material from an existing feature by using the **Revolve** tool. (T/F)

8. You can cut material from an existing feature by using the **Extrude** tool. (T/F)

9. You can use the plane origin as the Paste Handle while using the **Replicate** tool. (T/F)

10. In the **DesignModeler** window, you can save the sketch with a different name. (T/F)

Review Questions

Answer the following questions:

1. Which one of the following is displayed as dimension by default?

    (a) Dimension Value   (b) Dimension Name
    (c) Both   (d) None

2. Before extruding any sketch, you need to choose the _____ tool to create a feature.

3. If the sketched entity is displayed in blue, it represents that the sketch is _____ defined.

4. The _____ tool is used to orient the sketching plane perpendicular to the viewing direction.

5. The _____ option in the **Extent Type** edit box of the **Details View** window for the **Extrude** tool is used to add material to the specified surface.

6. In any system displayed in the **Project Schematic** window, you need to double-click on the desired cell to open the corresponding workspace. (T/F)

7. You can change the dimension of a sketch by using the options available on right-clicking on the particular dimension. (T/F)

8. When the **Zoom** tool is active, you can drag the cursor up and down to zoom in and out. (T/F)

9. You cannot create a pattern of a hole around a circular axis. (T/F)

10. The **Extrude** tool can also be used to remove material from the existing entity. (T/F)

11. You can also use the **Replicate** tool to scale and flip a sketched entity while replicating it. (T/F)

EXERCISE

Exercise 1

Create the model shown in Figure 4-67. The dimensions are given in Figures 4-68 through 4-70.

(Expected time: 45 min)

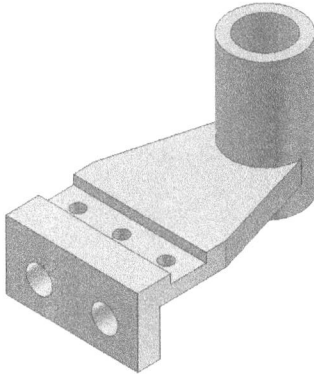

Figure 4-67 *Model for Exercise 1*

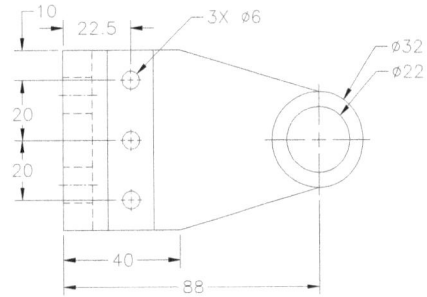

Figure 4-68 *Top view of the model*

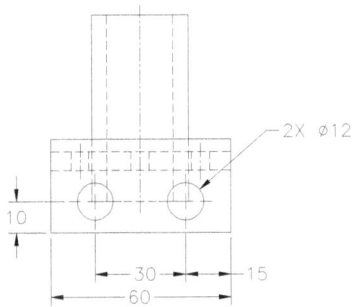

Figure 4-69 *Side view of the model*

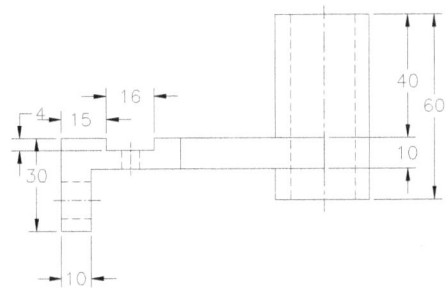

Figure 4-70 *Front view of the model*

Answers to Self-Evaluation Test

1. c, **2. Pattern**, **3.** T, **4.** F, **5.** T, **6.** T, **7.** F, **8.** T, **9.** T, **10.** T

Chapter 5

Part Modeling- III

Learning Objectives

After completing this chapter, you will be able to:

• *Create complex sketches*
• *Create cut features*
• *Create patterns*
• *Create sweep features*
• *Create planes*
• *Rotate and scale entities*
• *Create loft features*

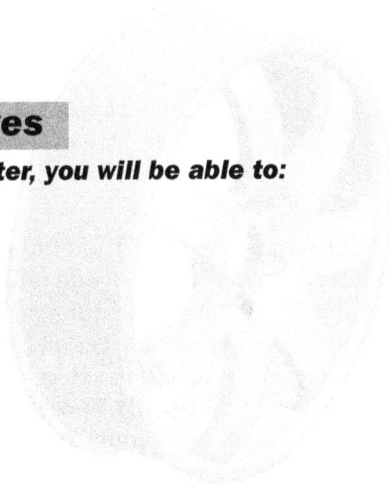

INTRODUCTION TO 3D FEATURES

In the **DesignModeler** window of ANSYS Workbench, various options such as **Share Topology**, **Extrude**, **Sweep**, **Blend**, and so on, are displayed in the **Feature** toolbar to create and revise models, as shown in Figure 5-1. These tools can also be invoked from the **Create** menu, as shown in Figure 5-2. Some of the features are discussed next.

Figure 5-1 The Feature toolbar displaying various options available

*Figure 5-2 The **Create** menu displaying various feature tools*

Extrude

The **Extrude** tool is used to add or remove material from the specified sketch along a straight line in the specified direction. You can invoke the **Extrude** tool from the **Feature** toolbar, refer to Figure 5-1. For the extrusion feature, geometric entities such as faces, edges, vertices, and points can also be selected as input. If you select any geometric entity other than sketch, then you need to define the direction vector. You can either select an open sketch or a closed sketch for extrusion but cannot be mixed type. The **Details View** window, as shown in Figure 5-3, is used to set the extrude parameters such as **Operation**, **Direction Vector**, **Direction**, **Extend Type**, and so on. Some of the nodes in the **Details View** window are discussed next.

Operation

In the **Operation** drop-down list, there are options that can be used to perform various modeling operations such as adding, cutting, slicing materials, and imprinting faces, and so on, as shown in Figure 5-4. The **Add Material** option is selected by default and it creates material model. The **Cut Material** option is used to create cutouts, holes, and so on in existing components. The **Slice Material** option slices bodies into multiple pieces. Similarly, the **Imprint Faces** option imprints curves onto the faces of active components in the model without splitting the bodies into multiple pieces. The **Add Frozen** option creates material without merging them with the active components.

Direction Vector

The direction vector is always normal to the plane on which the sketch is drawn. You can specify it by selecting direction references such as 2D lines, 3D edges, and so on.

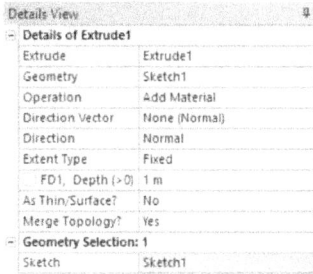

Figure 5-3 The **Details View** window

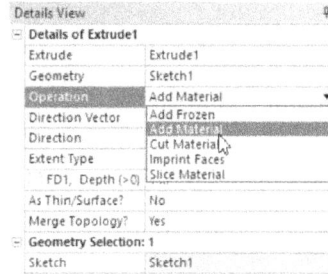

Figure 5-4 Displaying the **Operation** drop-down list

Direction

The options available in this drop-down list are shown in Figure 5-5. The **Normal** option extrudes the sketch in the positive Z direction. The **Reversed** option extrudes the sketch in the negative Z direction. The **Both - Symmetric** option extrudes the sketch in both directions symmetrically. The **Both - Asymmetric** option extrudes the sketch in both directions asymmetrically.

Extent Type

This drop-down list is used to define the extrusion in five different ways using the options **Fixed**, **Through All**, **To Next**, **To Faces**, and **To Surface**, refer to Figure 5-6. The **Fixed** option is used to specify the exact depth of material addition and it is selected by default. The **Through All** option is used to add material through the overall thickness of the feature. The **To Next** option is used to add material from one specified depth to the next available surface. The **To Faces** option is used to define the extent up to a boundary formed by one or more faces. Similarly, the **To Surface** option is used to define the extent surface upto which the extrusion will occur.

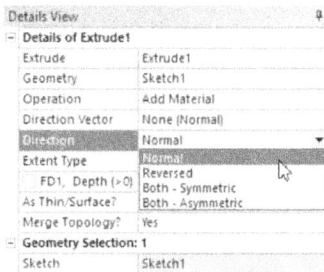

Figure 5-5 The **Direction** drop-down list

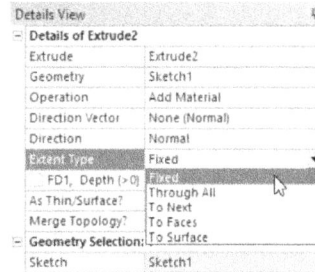

Figure 5-6 The **Extent Type** drop-down list

Revolve

The **Revolve** tool is used to create circular features like shafts, couplings, pulleys, and so on. You can also use this tool for creating cylindrical cut features. The default parameters for the this tool are displayed in the **Details View** window, as shown in Figure 5-7. You need to edit the values in the **Details View** window to get the required shape of the base feature. A revolved feature is created by revolving the sketch about an axis. To do this, you can use any straight 2D sketch edge, 3D model edge, or plane as the axis. In the **Direction** drop-down list of the **Details of Revolve1** window, as shown in Figure 5-8, the **Normal** option revolves the sketch in the positive Z direction.

The **Reversed** option revolves the sketch in a negative Z direction. The **Both - Symmetric** option revolves the sketch in both directions symmetrically. The **Both - Asymmetric** option revolves the sketch in both directions asymmetrically.

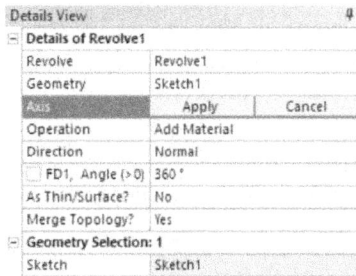

*Figure 5-7 The **Details View** window*

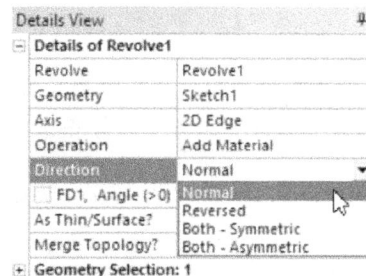

*Figure 5-8 The **Direction** drop-down list*

Sweep

This tool is used to create solid, surface, line bodies, and thin-walled features like gears, helical springs, bent pipes, and so on. For creating a swept surface, you need to create a sketch of the profile and path. The 2D cross-section profile is swept along a pre-set path to generate a 3D shape or surface. The default parameters for the sweep tool are displayed in the **Details View** window, refer to Figure 5-9. You can use either a sketch, plane, or geometry entries such as faces, edges, surface bodies, or line bodies as input for sweep profile or sweep path. Some of the parameters for the **Sweep** tool are discussed next.

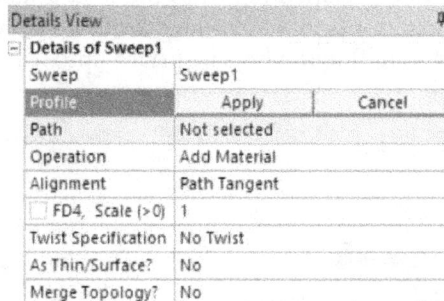

*Figure 5-9 The **Details View** window*

Alignment

In this drop-down list, two options are available: **Path Tangent** and **Global Axes**. In the case of the **Path Tangent** option, the orientation of the profile is remained consistent along the path, as shown in Figure 5-10. In case of the **Global Axes** option, the orientation of the profile remains constant regardless of the shape of the path, as shown in Figure 5-11. The **Path Tangent** option is selected by default.

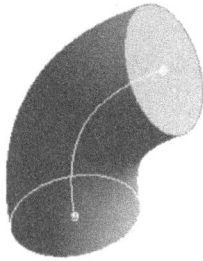

Figure 5-10 The model created using
the **Path Tangent** alignment option

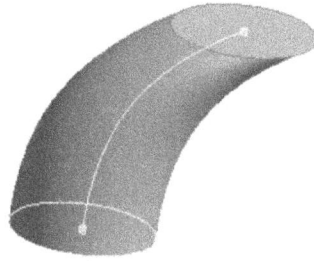

Figure 5-11 The model created using
the **Global Axes** alignment option

Twist Specification

In the **Twist Specification** drop-down list, there are options that can be used to create springs, threads, and circular stairways. The **No Twist** option is selected by default. To create helical sweeps, you need to change the option to **Pitch** or **Turns**. The **Pitch** option is used to create threads. The **Turns** option is used to create helical sweeps.

Skin/Loft

This tool is used to create solid or surface bodies using a series of profiles drawn on different planes. Thin-walled features can be created by using this tool. You can perform various operations such as adding material, cutting material, slicing material, imprinting faces or adding frozen material by using the options from the **Operation** drop-down list in the **Details View** window, as shown in Figure 5-12. You must select two or more profiles for creating a skin or loft feature. A profile is a sketch with one closed or open loop, but must be of the same type. All profiles must have the same number of edges. Additionally, open and closed profile combinations cannot be selected. In the **Details View** window, the **Profile Selection Method** node has two options, **Select All Profiles**, and **Select Individual Profile**. The **Select All Profiles** option is selected by default, refer to Figure 5-12, as a result, all profiles are selected. You can select the profiles from the graphics window using the **Profiles** node.

Using the second option, **Select Individual Profile,** you can select individual profiles and can edit them. You can add or insert new profile groups and delete existing profile groups. To do this, right-click on **Profiles 1** node and choose the required option from the shortcut menu displayed, as shown in Figure 5-13. The entities present in that selected group will be shown in cyan and entities of other groups will be shown in magenta.

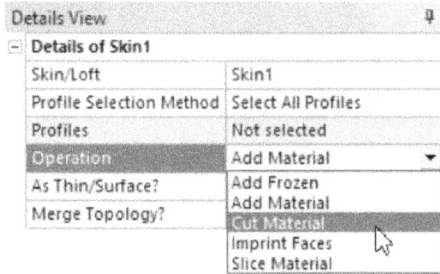

*Figure 5-12 The **Details View** window*

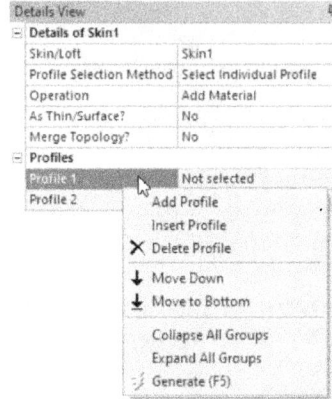

Figure 5-13 Shortcut menu displaying various options

Thin/Surface

The **Thin/Surface** tool assists you to convert solids into thin solids or surfaces. You can create the thin surfaces by using the options corresponding to the **Selection Type** field in the **Details View** window, as shown in Figure 5-14. These selection types are discussed next.

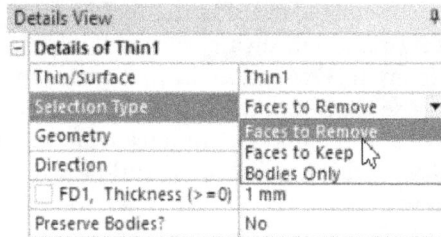

Figure 5-14 Displaying selection types drop-down list

Faces To Remove

This option is used to remove the selected faces from the body, refer to Figure 5-15 and Figure 5-16.

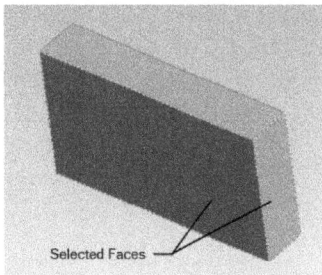

Figure 5-15 Selected faces

Figure 5-16 Thin surfaces generated

Faces To Keep

Use this option to retain the selected faces and remove the unselected faces from the body.

Bodies Only

Select this option to create a closed shell body. However, in this process you will select the body not the face of the body.

You can provide the thickness to the generated model in three different ways as inward, outward, and midplane by using the options **Inward**, **Outward** and **Mid-Plane**.

In the previous chapter, you learnt to work with various part modeling tools. In this chapter, you will learn to work with some more tools used in part modeling.

TUTORIALS

Tutorial 1

In this tutorial, you will create the solid model of the Rim shown in Figure 5-17. The dimensions of the model are shown in Figure 5-18. Save the project with the name *c05_ansWB_tut01* at the location *C:\ANSYS_2023_R2\c05\Tut01*.

(Expected time: 40 min)

Figure 5-17 Model for Tutorial 1

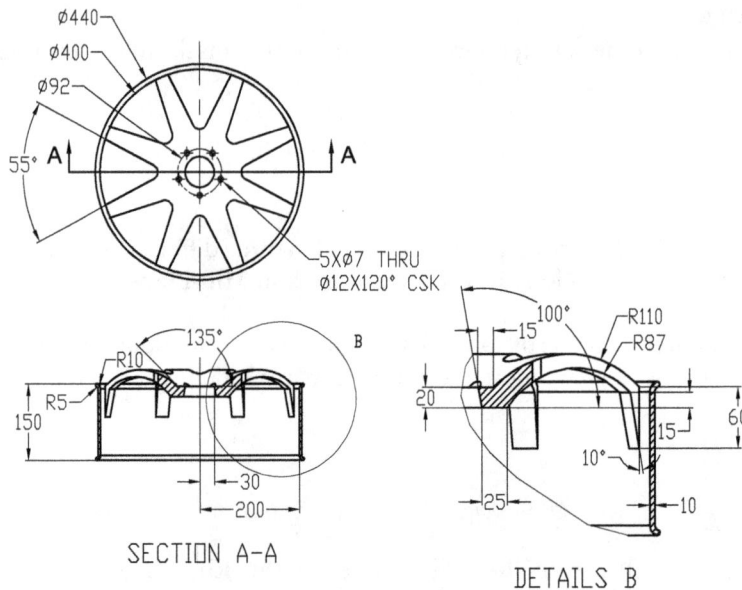

Figure 5-18 *Dimensions for the model in Tutorial 1*

The following steps are required to complete this tutorial:

a. Start ANSYS Workbench.
b. Add the **Geometry** component system to the **Project Schematic** window.
c. Start **DesignModeler** window and specify unit system.
d. Draw the sketch for the base feature on the XY plane.
e. Create the base feature.
f. Create the cut feature.
g. Create the pattern of the cut feature.
h. Create revolved cut feature for the nut hole.
i. Create pattern of the nut hole.
j. Create revolved feature for the rim.
k. Create the blend.
l. Save the project and exit the ANSYS Workbench session.

Starting ANSYS Workbench and Adding the Geometry Component System

To create the model, you need to start ANSYS Workbench and then add a component system to the project.

1. Start ANSYS Workbench 2023 R2. The **Workbench** window is displayed.

After invoking the **Workbench** window, you have to add appropriate analysis system or component system to the **Project Schematic** window. In this tutorial, you will create a solid model using the **Geometry** component system.

2.   In the **Workbench** window, expand the **Component Systems** toolbox in the **Toolbox** window and drag the **Geometry** component system to the **Project Schematic** window; a green colored rectangular boundary will be displayed, refer to Figure 5-19.

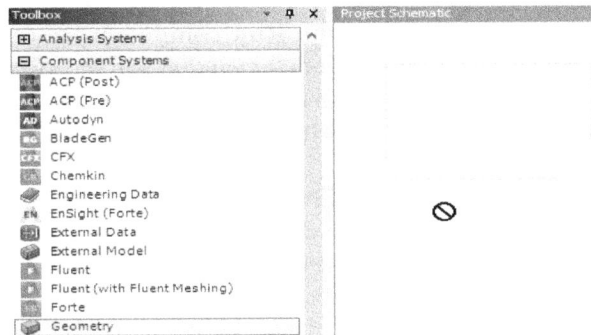

Figure 5-19 *Prospective location for adding component system*

The green colored rectangular boundary indicates the prospective location where the analysis or the component system can be added.

3.   Drag the cursor over this rectangular boundary; the boundary will turn red, refer to Figure 5-20. Next, drop the component system over this rectangular boundary; the **Geometry** component system is added to the **Project Schematic** window.

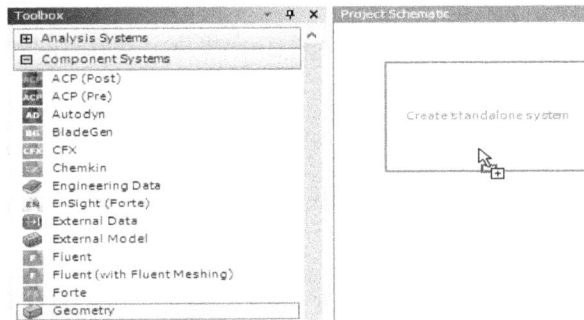

Figure 5-20 *Adding the component system to the* ***Project Schematic*** *window*

4.   Once the component system is added to the **Project Schematic** window, rename it as **Rim**.

5.   Choose the **Save Project** button from the **Main** toolbar; the **Save As** dialog box is displayed.

6.   In this dialog box, browse to the location *C:\ANSYS_WB* and then create a folder with the name **c05**.

7.   Browse to the *c05* folder and then create a sub folder with the name **Tut01**.

8.   In this folder, save the project with the name **c05_ansWB_tut01**.

Starting DesignModeler Window and Specifying Unit System

After the **Geometry** component system is added to the **Project Schematic** window and the project is saved, you now need to open **DesignModeler** to create the model.

1. Right-click on the **Geometry** cell in the **Rim** component system; choose the **New DesignModeler Geometry**, the **DesignModeler** window will be opened.

2. Choose the **Millimeter** option from the **Units** menu of the **Menu** bar.

Drawing the Sketch for the Base Feature

Now, you have to specify a plane on which you want to create sketch for base feature. In this tutorial, the sketch for the base feature will be created on the XY plane which is the default plane in **DesignModeler**. Therefore, you do not need to specify the plane for sketching. The sketch will now be drawn on the XY plane.

1. Choose the **Sketching** tab displayed at the lower left corner of the Tree Outline to invoke the **Sketching** mode.

   Now, you need to orient the sketching plane normal to the viewing direction so that you can easily draw the sketch on the specified plane.

2. Choose the **Look At** tool from the **Graphics** toolbar; the plane is oriented normal to the viewing direction.

3. Choose the **Line** tool from the **Draw** toolbox; you will be prompted to specify the start point of the line.

4. Draw an inclined line in the fourth quadrant, refer to Figure 5-21.

5. Choose the **General** tool from the **Dimensions** toolbox and generate the dimensions of the inclined line, as shown in Figure 5-21.

To generate the angular dimension using the **General** tool, select the inclined line and right-click in the graphics window to display a shortcut menu. Next, choose the **Angle** option from the shortcut menu, refer to Figure 5-22. Now, select the Y axis; the angular dimension will be displayed attached to the cursor. If the displayed angle is not the one that is required, right-click in the graphics window and choose the **Alternate Angle** option from the shortcut menu displayed; the alternate angle will be displayed. Keep on choosing the **Alternate Angle** option from the shortcut menu until you get the angle of the desired quadrant. Now, place the dimension at the desired location. Alternatively, you can use the **Angle** tool from the **Dimensions** toolbox and then select the two lines between which you want to measure the dimension.

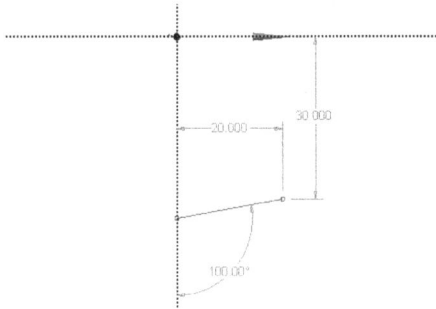

Figure 5-21 *The inclined line along with its dimensions*

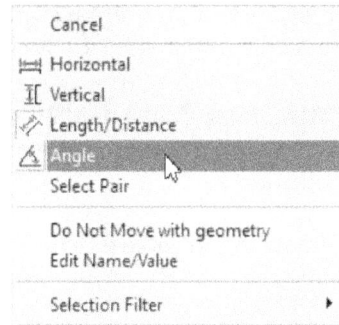

Figure 5-22 *Choosing the* **Angle** *option from the shortcut menu*

6. Use the **Line** and **Arc by 3 Points** tools to create the remaining sketch for the base feature and then generate its dimensions, refer to Figure 5-23.

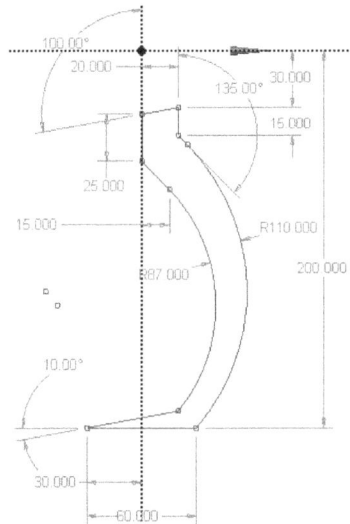

Figure 5-23 *Complete sketch for the base feature*

The **Arc by 3 Points** tool is used to create arcs by specifying three points in the graphics window. The first two points of the arc specify the start and end points of the arc whereas the last point specifies the radius of the arc.

7. Choose the **Modeling** tab displayed at the bottom of the **Sketching Toolboxes** window; the **Sketching** mode is exited and the **Modeling** mode is invoked. Also, **Sketch 1** is displayed under the **XYPlane** node.

After exiting the **Sketching** mode, the sketching plane will still be normal to the viewing direction. Therefore, to proceed further with the feature creation operation, it is advised to change the view of the sketching plane to Isometric view.

8.   Right-click in the graphics window and choose the **Isometric View** option from the shortcut menu displayed; the view is changed to Isometric.

Creating the Base Feature

After creating the sketch, revolve the drawn sketch about the X axis to create the base feature of the model.

1.   Choose the **Revolve** tool from the **Features** toolbar; you will be prompted to select the base object. Also, **Revolve1** is added below the three default planes in the Tree Outline.

2.   Select **Sketch1** from the Tree Outline and click on the **Apply** button in the **Geometry** selection box of the **Details View** window; the sketch to be revolved is now selected.

3.   For this model, the material should be added by revolving the sketch about the X axis. Select the X axis from the graphics window, refer to Figure 5-24. Figure 5-25 shows the preview of the revolved feature after the axis is specified.

> **Note**
> *If there is a disjoint line in the sketch of the revolve feature, it will be selected as the default axis of revolution.*

4.   Choose the **Apply** button from the **Axis** selection box in the **Details View** window; **2D Edge** is displayed in the **Axis** selection box indicating that the axis for revolution is specified.

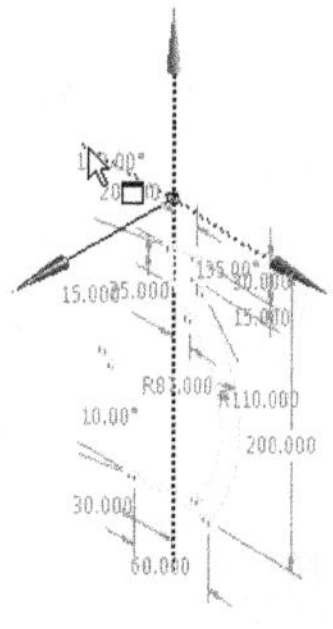

Figure 5-24 *Selecting the X axis from the Graphics window*

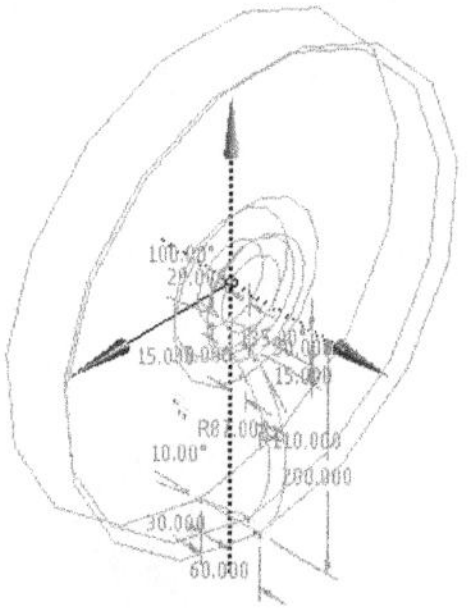

Figure 5-25 *Preview of the revolved feature after specifying the axis of revolution*

5.   Next, we need to specify the angle of revolution. Enter **360** in the **FD1, Angle (>0)** edit box of the **Details View** window, if it is not already specified; the sketch will be revolved by the angle specified in this edit box.

6.   Choose the **Generate** tool from the **Features** toolbar; the base feature is created by revolving the sketch about the X axis by 360 degrees, refer to Figure 5-26.

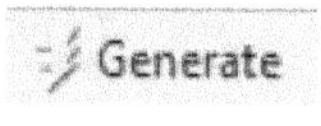

7. Right-click on the **Sketch1** in the Tree Outline and choose the **Hide Sketch** option from the shortcut menu displayed.

Figure 5-26 *The base feature created*

Note

*In Figure 5-26, the display of planes has been turned off for better visibility of the model. You can turn off the display of the planes by choosing the **Display Plane** tool from the **Graphics** toolbar.*

Creating the Cut Feature

After creating the base feature, you need to remove material from it to generate the spoke of the rim. The cut feature will be created using the **Extrude** tool. The sketch for the extrude feature will be created on the YZ plane.

1. Select **YZPlane** from the Tree Outline.

2. Choose the **Sketching** tab displayed below the Tree Outline; the **Sketching** mode is activated.

3. Choose the **Look At** button from the **Graphics** toolbar; the sketching plane is oriented normal to the viewing direction.

4. Choose the **Arc by Center** tool from the **Draw** toolbox. Next, click to specify the center of arc at the origin, and draw the arc, as shown in Figure 5-27.

    The **Arc by Center** tool is used to create arcs by specifying the center of the arc. After this tool is invoked, click to specify the center of the arc. Next, move the cursor to specify the radius of the arc. The point specified for the circle also acts as the start point of the arc. Move the cursor to specify the end point of the arc.

5. Choose the **Polyline** tool from the **Draw** toolbox and draw the two line segments as shown in Figure 5-28.

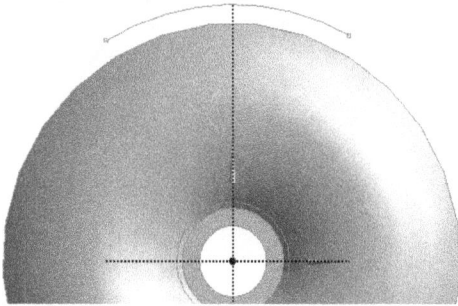

Figure 5-27 *The arc created using the Arc by Center tool*

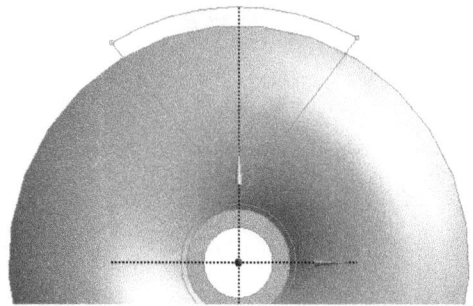

Figure 5-28 *The lines created using the Polyline tool*

6. Expand the **Constraints** toolbox and choose the **Symmetry** tool. Next,   ⫩ Symmetry
    make the two inclined lines symmetric about the Y axis.

7. Expand the **Modify** toolbox and choose the **Fillet** tool.   ⌐ Fillet

    Filleting is the process of rounding the sharp corners of a sketch. This is done to reduce the
    stress concentration in the model. Using the **Fillet** tool, you can round the corners of the
    sketch by creating an arc tangent to both the selected entities.

8. Enter **15** as fillet radius in the **Radius** edit box displayed on the right of the **Fillet** tool.

9. Select the intersection point of the two inclined line segments of the polyline; the fillet is
    created, as shown in Figure 5-29. Next, exit the **Fillet** tool.

10. Expand the **Dimensions** toolbox and choose the **General** tool, if it is not   ◈ General
    chosen by default.

11. Apply the dimensions to the sketch and edit their values, as shown in Figure 5-30.

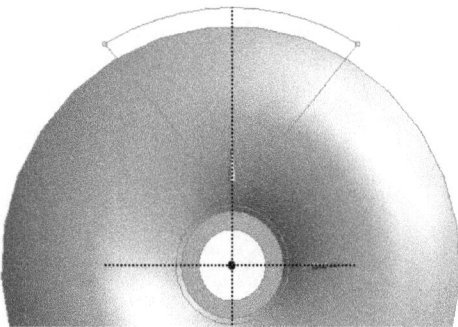

Figure 5-29 *The fillet created at the intersection of two line segments*

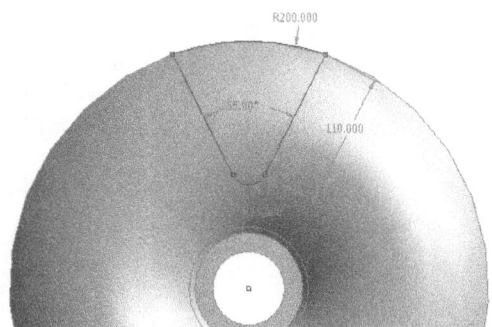

Figure 5-30 *The final sketch after the dimensions are applied*

Now, the sketch is ready to be used for creating the cut feature.

12. Exit the **Sketching** mode by selecting the **Modeling** tab displayed below the **Sketching Toolboxes** window.

13. Choose the **ISO** tool from the **Graphics** toolbar; the view is changed to isometric.

After drawing the sketch, you need to remove the material from the base feature using the **Extrude** tool to create the spoke of the rim.

14. Choose the **Extrude** tool from the **Features** toolbar; the preview of the extruded feature with the default values is displayed in the graphics window. Also, **Extrude1** is added below **Revolve1** in the Tree Outline.

The default parameters used for generating the preview of the extruded feature are displayed in the **Details View** window. You need to edit the values in the **Details View** window to get the cutout.

15. Select the **Cut Material** option from the **Operation** drop-down list.

The **Cut Material** option is used to remove material from an existing feature.

To create the model for this tutorial, the material should be removed from both the sides of the sketch and through all the features that are normal to the sketch.

16. Select the **Both - Symmetric** option from the **Direction** drop-down list in the **Details View** window, refer to Figure 5-31.

The **Both - Symmetric** option is used to perform an extrusion operation on both sides of the sketch, equally.

17. Select the **Through All** option from the **Extent Type** drop-down list in the **Details View** window, refer to Figure 5-31.

The **Through All** option is used to perform the extrusion operation through the total thickness of the existing feature.

18. Choose the **Generate** tool from the **Features** toolbar; the cut feature is created, refer to Figure 5-32.

Figure 5-31 *The **Details View** window with the options selected for extrude operation*

Figure 5-32 *The cut feature created using the **Extrude** tool*

Note
For better visibility, the display of the planes in this figure has been turned off.

19. Right-click on **Sketch1** in the Tree Outline and choose the **Hide Sketch** option from the shortcut menu displayed.

The **Hide Sketch** option is used to make any sketch temporarily invisible in the graphics window. To hide a sketch, right-click on the sketch instance in the Tree Outline and then choose **Hide Sketch** from the shortcut menu displayed.

Creating the Pattern of the Cut Feature

Next, you need to create seven more similar instances of the cut feature on the base feature to get the final shape of the spoke.

1. Choose the **Pattern** tool from the **Create** menu, refer to Figure 5-33; you are prompted to select a geometry to create the pattern. Also, **Pattern 1** is added to the Tree Outline.

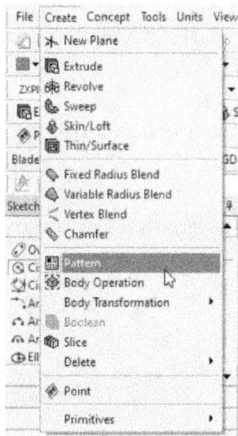

Figure 5-33 *Partial view of the **Create** menu*

The **Pattern** tool is used to create multiple instances of the selected faces along a linear direction or along a circular path. The patterns are defined as the sequential arrangement of the copies of the selected entities. You can create the patterns in rectangular or circular fashion.

You have to select the faces of the cut feature that need to be patterned.

2. Choose the **Face** tool from the **Select** toolbar; the cursor will get modified.

The tools in the **Select** toolbar are used to apply filters while selecting entities from the graphics window, refer to Figure 5-34. For example, if you want to select only the edges of the model, you can use the **Edges** tool to filter the edges. It will enable you to select the edges only, and will disallow selection of any other entity of the model. These tools are also available in the shortcut menu that is displayed by right-clicking in the graphics window.

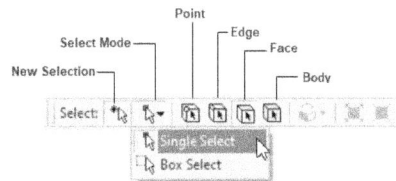

Figure 5-34 The tools in the Select toolbar

3. Press the Ctrl key and select the three faces of the cut feature to be patterned, refer to Figure 5-35.

Note
While selecting the faces, you have to rotate the view of the model. The process of dynamically rotating the model has been discussed in detail in the previous tutorial.

4. In the **Details View** window, choose the **Apply** button from the **Geometry** selection box to accept the specified faces; **3 Faces** is displayed in the **Geometry** selection box in the **Details View** window.

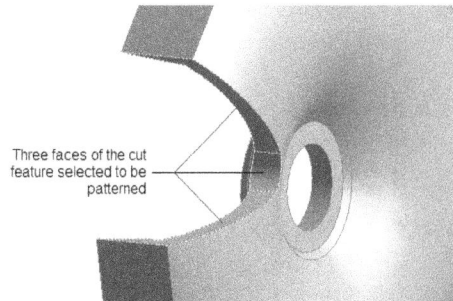

Figure 5-35 Faces of the cut feature selected to be patterned

5. Select the **Circular** option from the **Pattern Type** drop-down list in the **Details View** window.

The **Circular** option will enable you to create a pattern of the cut feature in a circular manner about the specified axis. The **Linear** option in the **Pattern Type** drop-down list is used to create a pattern of the selected feature along the specified linear direction. The **Rectangular** option in this list is used to create the pattern of the selected feature along two specified linear directions so that the final pattern results in a rectangular form.

Notice that the **Axis** selection box in the **Details View** window is highlighted in yellow indicating that you have to still specify the axis about which the pattern will be created.

6. Click on the **Axis** selection box; you are prompted to select an axis to create the pattern.

7. Select **XYPlane** from the Tree Outline; the XY plane is displayed in the graphics window.

8. Click on the X axis, refer to Figure 5-36; the selected axis is highlighted in yellow.

9. Choose the **Apply** button from the **Axis** selection box; the selected axis is specified as the axis for creating the circular pattern.

   After specifying the axis of the circular pattern, you need to specify the angular value between two consecutive instances of the pattern feature. In this tutorial, all instances of the circular pattern should be equally spaced and arranged in 360 degrees.

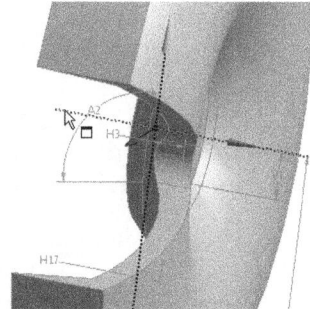

Figure 5-36 Selecting the axis about which the circular pattern will be created

10. Make sure **Evenly Spaced** is displayed in the **FD2, Angle** edit box in the **Details View** window. However, if this option is not displayed in this edit box, enter **0** in the **FD2, Angle** edit box; **Evenly Spaced** will be displayed in it.

    The **Evenly Spaced** option is used to specify same distance between the instances of a pattern feature.

    After specifying the angle between two consecutive instances of the pattern feature, you need to specify the number of instances in the pattern. In this tutorial, you need to create seven more instances of the feature to be patterned.

11. Enter **7** in the **FD3, Copies (>=0)** edit box. Figure 5-37 shows the **Details View** window with all parameters that you have specified.

    The **FD3, Copies (>=0)** edit box is used to specify the number of instances required in a pattern.

12. Choose the **Generate** tool from the **Features** toolbar; the circular pattern is generated by creating seven more instances of the cut feature about the X axis, refer to Figure 5-38.

Creating Revolved Cut Feature for the Nut Hole

Next, you need to create a nut hole on the rim. This nut hole is used to assemble the rim with the driving shaft. The nut hole can be created by removing material from the base feature. You will use the **Revolve** tool to remove the material. The sketch for the revolve feature will be created on the XY plane.

1. Select **XYPlane** from the Tree Outline and choose the **New Sketch** tool from the **Active Plane/Sketch** toolbar; **Sketch3** is added under the **XYPlane**.

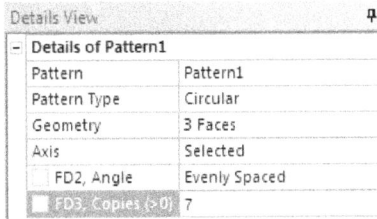

Details View	무
− **Details of Pattern1**	
Pattern	Pattern1
Pattern Type	Circular
Geometry	3 Faces
Axis	Selected
FD2, Angle	Evenly Spaced
FD3, Copies (>0)	7

Figure 5-37 The Details View window with parameters for creating the circular pattern

Figure 5-38 Model after patterning the cut feature

2. Select **Sketch3** from the Tree Outline and invoke the **Sketching** mode. Also, orient the view of the sketching plane parallel to the screen.

3. Draw the sketch with the dimensions shown in Figure 5-39.

4. Choose the **Modeling** tab to switch to the **Modeling** mode and then change the view to Isometric.

5. Choose the **Revolve** tool from the **Features** toolbar; you are prompted to select the axis of revolution. Also, **Revolve2** is added to the Tree Outline.

Note
*Since there is only one sketch in the **Sketching** mode, it is automatically highlighted in the graphics window. In case it is not highlighted, you may need to select **Sketch3** from the Tree Outline and then choose the **Apply** button from the **Geometry** selection box in the **Details View** window.*

6. Select the longest horizontal line of the sketch as the axis for the revolved feature, refer to Figure 5-40.

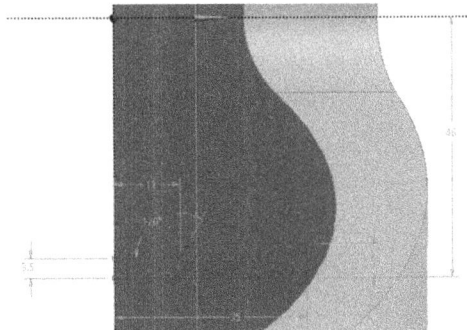

Figure 5-39 Sketch for the revolved nut hole feature

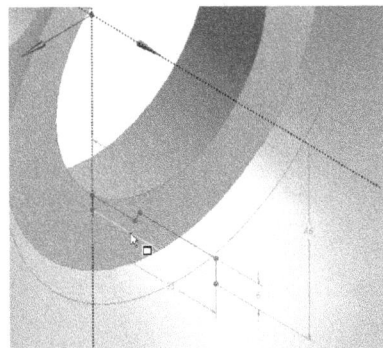

Figure 5-40 The line to be selected as axis for the revolve feature

7. Choose the **Apply** button from the **Axis** selection box in the **Details View** window; the line is specified as the axis for the revolve feature.

8. To remove material from the base feature, select the **Cut Material** option from the **Operation** drop-down list in the **Details View** window.

9. Enter **360** in the **FD1, Angle (>0)** edit box, if not specified by default.

10. Choose the **Generate** tool from the **Features** toolbar; the revolved cut feature is created by revolving the sketch about the specified line by 360 degrees, refer to Figure 5-41.

11. Right-click on **Sketch3** in the Tree Outline and choose the **Hide Sketch** option from the shortcut menu displayed; the sketch is not displayed in the graphics window.

Creating the Pattern of the Hole

Next, you need to create four more instances of the hole feature on the base feature. This can be achieved by creating circular pattern of the revolved cut feature that was created earlier.

Figure 5-41   The model after creating the revolved cut feature

1. Choose the **Pattern** option from the **Create** menu; you are prompted to select a geometry to pattern. Also, **Pattern2** is added to the Tree Outline.

    Now, you need to select the faces of the revolved cut feature that need to be patterned.

2. Choose the **Face** tool from the **Select** toolbar; the cursor is modified.

3. Press the Ctrl key and select the faces of the cut feature to be patterned, refer to Figure 5-42.

4. Choose the **Apply** button from the **Geometry** selection box to accept the specified faces; **3 Faces** is displayed in the **Geometry** selection box.

5. In the **Details View** window, select the **Circular** option from the **Pattern Type** drop-down list.

    Notice that the **Axis** selection box in the **Details View** window is highlighted in yellow color indicating that you still need to specify the axis about which the pattern will be created.

6. Click on the **Axis** selection box; the **Apply** and **Cancel** buttons are displayed. Also, you are prompted to select an axis to create the circular pattern.

7. Select **XYPlane** from the Tree Outline; the XY plane is displayed in the graphics window.

8. Click on the X axis, refer to Figure 5-43; the selected axis is highlighted in yellow indicating that the X axis can now be specified as the axis of the circular pattern.

9. Choose the **Apply** button from the **Axis** selection box; the axis is specified.

10. By default, **Evenly Spaced** is displayed in the **FD2, Angle** edit box. However, if this option is not displayed in the **FD2, Angle** edit box, then enter **0** in it.

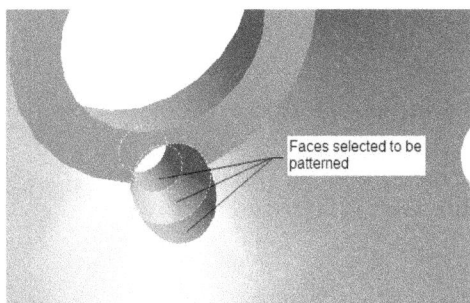

Figure 5-42 Faces of the revolved cut feature selected to be patterned

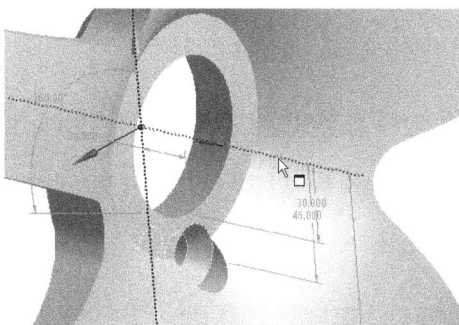

Figure 5-43 Selecting the axis about which the circular pattern will be created

11. Enter **4** in the **FD3, Copies (>=0)** edit box. Figure 5-44 shows the **Details View** window with all parameters that you have specified.

12. Choose the **Generate** tool from the **Features** toolbar; the circular pattern is generated by creating four more instances of the cut feature about the X axis, refer to Figure 5-45.

*Figure 5-44 The **Details View** window displaying parameters for creating the circular pattern*

Figure 5-45 Model after patterning the cut feature

Creating Revolved Feature for the Rim

Next, you need to create rim around the spokes. The **Revolve** tool can be used to create the rim. The sketch for the revolve feature will be created on the XY plane.

1. Select the **XYPlane** from the Tree Outline and choose the **New Sketch** button from the **Active Plane/Sketch** toolbar; **Sketch4** is added under the **XYPlane** node.

2. Select **Sketch4** from the Tree Outline and invoke the **Sketching** mode. Next, orient the view of the sketching plane parallel to the screen using the **Look At** tool.

3. Draw the sketch for the revolved feature using the dimensions given in Figure 5-46.

    Note
    You need to apply geometric constraints, such as horizontal, vertical, equal length, and so on to fully define the sketch.

4. Choose the **Modeling** tab and change the view to Isometric.

5. Choose the **Revolve** tool from the **Features** toolbar; you are prompted to select the axis for the revolve feature. Also, **Revolve3** is added to the Tree Outline.

6. Click on the **Geometry** selection box in the **Details View** window to display the **Apply** and **Cancel** buttons.

    As there is only one sketch in the graphics window, it is automatically highlighted for the revolve operation.

7. Choose the **Apply** button from the **Geometry** selection box to specify the recently created sketch as the sketch for the revolve feature.

8. Click on the **Axis** selection box and then select the X axis from the graphics window; the selected axis is highlighted in yellow color and the preview of the revolved feature is displayed.

9. Choose the **Apply** button from the **Axis** selection box.

10. Make sure that the **Add Material** option is selected in the **Operation** drop-down list. Also, make sure that the angle of revolution is specified as **360** degrees in the **FD1 Angle, (>0)** edit box.

11. Choose the **Generate** tool from the **Features** toolbar; the revolved feature is created by revolving the sketch about the X axis by 360 degrees, refer to Figure 5-47.

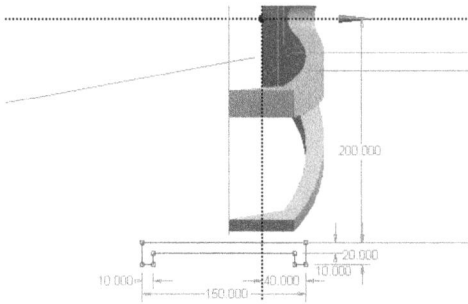

Figure 5-46 *Sketch for the revolved feature of the rim*

Figure 5-47 *Model after creating the revolved feature for the rim*

Creating the Blend Feature

Now, to remove the sharp edges from the rim, you need to create the blend feature (fillet) with radius of 10 mm, 5 mm, and 3 mm at the vertical edges of the base feature.

1. Choose the **Fixed Radius** tool from the **Blend** drop-down in the **Features** toolbar; you are prompted to select 3D edges, faces, or edit an existing blend.

2. Press the Ctrl key and select the two edges shown in Figure 5-48 for creating a blend of radius 10 mm.

3. Choose the **Apply** button from the **Details View** window to accept the selection of edges to be blended, .

4. Enter **10** in the **FD1 Radius (>0)** edit box to specify the radius of the blend feature.

5. Choose the **Generate** tool from the **Features** toolbar; the blend feature is created.

6. Similarly, create two blend features of radius 5 mm and 3 mm respectively. Refer to Figure 5-48 for the edges to be selected.

The final model of the Rim is shown in Figure 5-49.

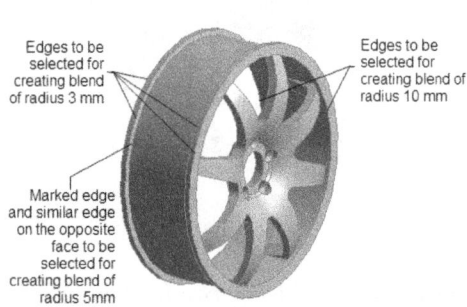

Edges to be selected for creating blend of radius 3 mm

Edges to be selected for creating blend of radius 10 mm

Marked edge and similar edge on the opposite face to be selected for creating blend of radius 5mm

Figure 5-48 Edges to be selected for creating the blend feature

Figure 5-49 The final model of the rim

7.   Close the **DesignModeler** window; the **Workbench** window is displayed.

Saving the Project and Exiting ANSYS Workbench

After the **DesignModeler** window is closed, you need to save the project and exit ANSYS Workbench.

1.   Choose the **Save Project** button from the **Main** toolbar; the project is saved with the name *c05_ansWB_tut01*.

2.   Choose the **Exit** option from the **File** menu to exit the current ANSYS Workbench session.

Tutorial 2

In this tutorial, you will create model of a Basket Ball Hoop by using the **Sweep** tool. You will also use the **Extrude** and **Fillet** tools to add required support to the hoop. Figure 5-50 shows the model of the hoop and Figure 5-51 shows the major dimensions for creating the model.   **(Expected time: 40 min)**

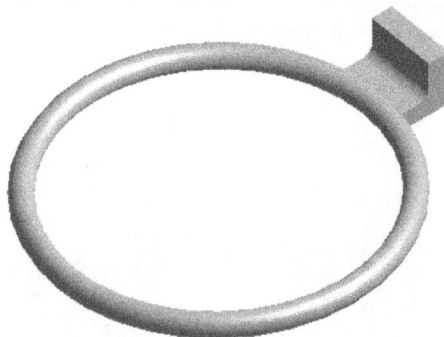

Figure 5-50 Model of the Basket Ball Hoop

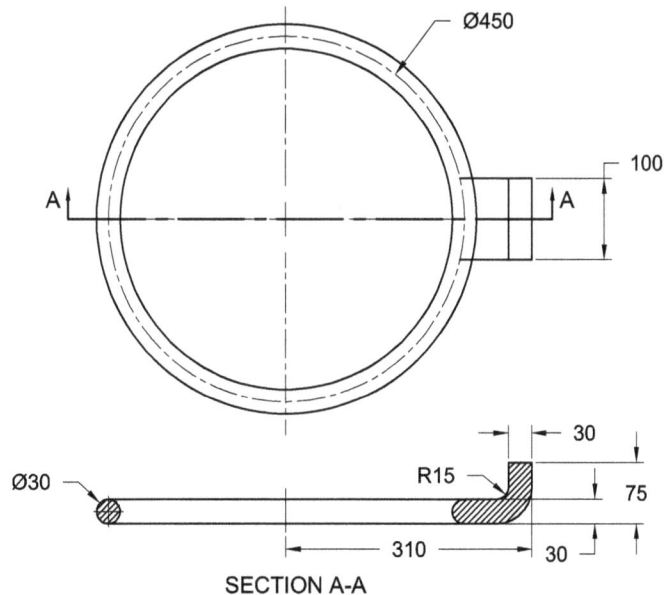

Figure 5-51 *Dimensions of the hoop*

The following steps are required to complete this tutorial:

a. Start ANSYS Workbench.
b. Add the **Geometry** component system to the **Project Schematic** window.
c. Draw the sketch.
d. Create the sweep feature.
e. Create the clamp feature as the second feature.
f. Create the blend feature.
g. Save the project and exit the ANSYS Workbench session.

Starting ANSYS Workbench and Adding the Geometry Component System

To start the tutorial, you first need to start ANSYS Workbench and then add a component system to the project.

1. Start ANSYS Workbench 2023 R2. The **Workbench** window is displayed.

    Next, you need to add the **Geometry** component system to the **Project Schematic** window.

2. Double-click on the **Geometry** option displayed under the **Component Systems** node in the **Toolbox** window; the **Geometry** component system is added to the **Project Schematic** window.

3. Once the system is added to the **Project Schematic** window, its name gets highlighted at the bottom of the component system in blue. If it is not highlighted, double-click on the default name and rename it to **Basket Ball Hoop**.

4. Choose the **Save Project** button from the **Main** toolbar; the **Save As** dialog box is displayed.

5. Browse to the location *C:\ANSYS_WB\c06* and then create a folder with the name **Tut02**.

6. Browse to the *Tut02* folder and then save the project with the name **c06_ansWB_tut02**.

Drawing the Sketch

After adding the component system and saving the project, you need to create the profile of the circular feature of the Hoop in **DesignModeler**.

1. Right-click on the **Geometry** cell in the **Basket Ball Hoop** component system. Choose **New DesignModeler Geometry**; the **DesignModeler** window gets opened.

2. Choose the **Millimeter** option from the **Units** menu of the **Menu** bar.

    Next, you need to create the sketch for the sweep feature. To do so, first you need to specify the plane on which you want to create the sketch.

3. Select the **XYPlane** from the Tree Outline; the XY plane becomes the active plane. Now, you can create sketch on this plane.

4. Choose the **Sketching** tab displayed at the bottom of the Tree Outline; the **Sketching** mode is invoked.

5. Choose the **Look At** tool from the **Select** toolbar to orient the plane normal to screen.

6. Invoke the **Circle** tool from the **Draw** toolbox; the cursor is modified.

7. Move the cursor close to X axis; the symbol of Coincident Constraint (C) is displayed.

8. Move the cursor toward your left to some distance and click while the Coincident Constraint symbol is still displayed; the center of the circle is specified, as shown in Figure 5-52.

9. Next, move the circle away from the center point of the circle and click again to specify the radius and then create the circle, as shown in Figure 5-53.

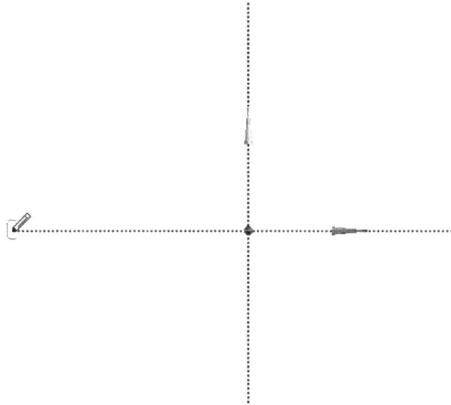

Figure 5-52 *Specifying the center for the circular profile*

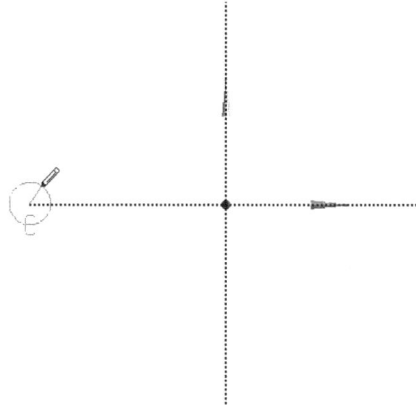

Figure 5-53 *Creating the circular profile*

10. Expand the **Dimensions** toolbox to display the tools available in it.

11. Choose the **General** tool and then place the dimension of the circular profile, as shown in Figure 5-54.

12. In the **Details View** window, click on the **D1** edit box and change the dimension of the circular profile to 30.

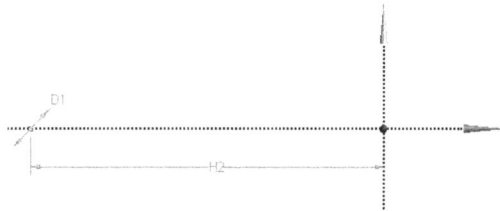

Figure 5-54 *Placing the dimension*

13. Invoke the **Horizontal** tool from the **Dimensions** toolbox; you are prompted to specify the start point or edge for placing the horizontal dimension.

14. Select the center point of the circular profile; you are prompted to select the second point or edge for horizontal dimensioning.

15. Next, click on the Y axis to specify the horizontal dimension and then place the dimension anywhere on the screen, refer to Figure 5-54.

16. In the **H2** edit box in the **Details View** window, specify **225** as the distance between the origin and the center of the circular profile.

17. Choose the **Zoom to Fit** button in the **Graphics** toolbar to fit the complete sketch in the graphics window.

18. Next, click on the **Modeling** tab at the bottom of the **Sketching Toolboxes** window to switch to the **Modeling** mode.

19. Change the view to Isometric by using the **ISO** tool.

20. Select **ZXPlane** then choose the **New Plane** tool from the **Active Plane/Sketch** toolbar.

21. **Sketch2** is added under the **ZXPlane** node, refer to Figure 5-55.

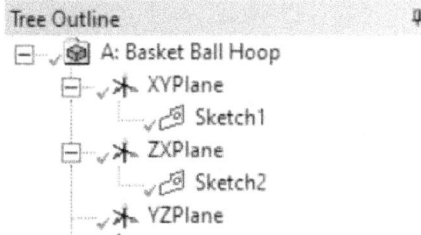

Figure 5-55 Sketch2 is added under ZXPlane

22. Choose the **Sketching** tab displayed under the **Sketching Toolboxes** window to switch to the **Sketching** mode.

23. Choose the **Look At** tool to orient the sketching plane normal to the viewing direction.

24. Choose the **Circle** tool from the **Draw** toolbox and create a circle with origin as the center of the circle, as shown in Figure 5-56.

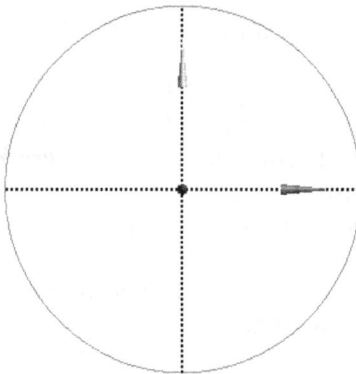

Figure 5-56 Circle created on the ZX plane

25. Choose the **General** tool from the **Dimensions** toolbox.

26. Next, generate and place the diametric dimension of the circle in the graphics screen.

27. Enter **450** as the diameter of the circle in the **D1** edit box in the **Details View** window; the size of the circle changes.

28. Choose the **Zoom to Fit** tool in the **Graphics** toolbar to fit the complete sketch in the graphics window.

29. Choose the **Modeling** tab at the bottom of the **Sketching Toolboxes** window to switch to the **Modeling** mode.

30. Change the view to Isometric by using the **ISO** tool.

Creating the Sweep Feature

After the sketch of the profile and the path are drawn, it is now required to sweep the profile along the path.

1. Choose the **Sweep** tool available in the **Features** toolbar; **Sweep1** is attached to the Tree Outline with a yellow thunderbolt symbol attached indicating that immediate action needs to be taken to create the sweep feature.

2. In the **Details View** window, click on the **Profile** selection box; the **Apply** and **Cancel** button are displayed.

3. Click on **Sketch1** under the **XYPlane** node in the Tree Outline; the sketch of the circular profile is selected in the graphics window. Next, choose the **Apply** button displayed in the **Profile** selection box; **Sketch1** is displayed in the **Profile** selection box.

4. Next, click on the **Path** selection box in the **Details View** window; the **Apply** and **Cancel** buttons are displayed.

5. Select **Sketch2** under the **ZXPlane** node in the Tree Outline and then choose the **Apply** button in the **Path** selection box.

   After the profile and the path are specified for the sweep feature, you now need to generate the sweep feature.

6. Choose the **Generate** tool from the **Features** toolbar; the sweep feature is created, as shown in Figure 5-57.

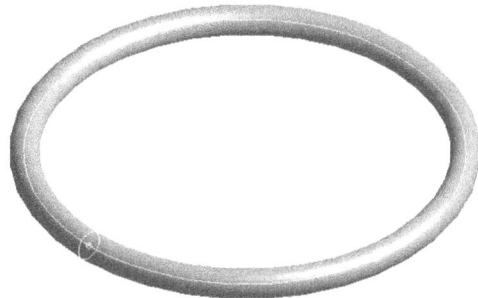

Figure 5-57 The sweep feature

Creating the Second Feature

Next, you need to create second feature of the model, which is the clamp.

1. Choose the **New Plane** button available in the **Active Plane/Sketch** toolbar; **Plane4** gets attached to the Tree Outline. Also, the corresponding options are displayed in the **Details View** window.

2. In the **Type** drop-down list of the **Details View** window, select the **From Plane** option.

3. Next, click on the **Base Plane** selection box in the **Details View** window; the **Apply** and **Cancel** buttons are displayed.

4. Select **YZPlane** in the Tree Outline and then choose the **Apply** button in the **Base Plane** selection box to specify YZ plane as the base plane.

5. In the **Transform 1 (RMB)** drop-down list, choose the **Offset Z** option; the **FD1, Value1** edit box is displayed.

6. Enter **250 FDI, Value 1** in this edit box to specify the offset distance of the new plane from the default YZ plane.

7. Choose the **Generate** tool to create the new plane.   Generate

8. Switch to the **Sketching** mode by choosing the **Sketching** tab at the bottom of the Tree Outline.

9. Choose the **Look At** tool to orient the new plane normal to the viewing direction, refer to Figure 5-58.

Note
*When you choose the **Look At** tool, the orientation of the plane may be different from the one shown in Figure 5-58. You can use the options available in the **Graphics** toolbar to orient the plane normal to the graphics window with the Y axis upward.*

10. Choose the **Rectangle** tool from the **Draw** toolbox; the cursor is modified.   Rectangle

11. Draw a rectangle such that the horizontal axis lies in the middle, as shown in Figure 5-59.

12. Invoke the **General** tool from the **Dimensions** toolbox and place the horizontal and vertical dimensions for the rectangle, refer to Figure 5-60.

13. In the **Details View** window, click on the **H5** edit box and enter **30** as the width of the rectangle.

14. Similarly, click on the **V6** edit box and then enter **100** as the length of the rectangle.

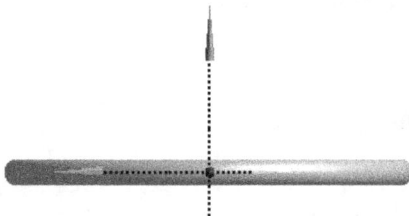

Figure 5-58 *The new plane oriented normal to the viewing direction*

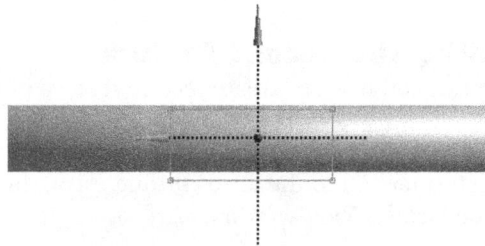

Figure 5-59 *The rectangular sketch*

15. Invoke the **Symmetry** tool from the **Constraints** toolbox; you are prompted to select a line to specify as the axis of symmetry.

16. Select the vertical axis to specify the axis of symmetry, as shown in Figure 5-60; you are prompted to select the first point or 2D edge to apply the Symmetric constraint.

17. Select the left vertical line of the rectangle as the first edge to apply Symmetric constraint; you are prompted to select the second point or 2D edge to apply Symmetric constraint.

18. Select the right vertical line as the second edge; the sketch is adjusted symmetrically about the vertical axis, as shown in Figure 5-61.

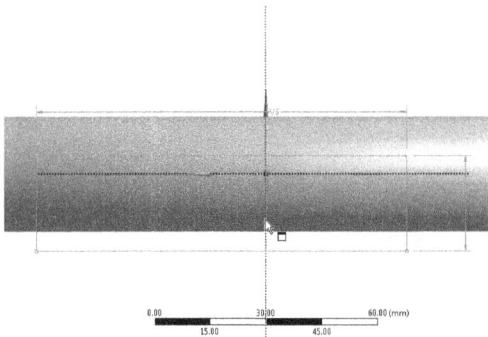

Figure 5-60 *Selecting the axis of symmetry*

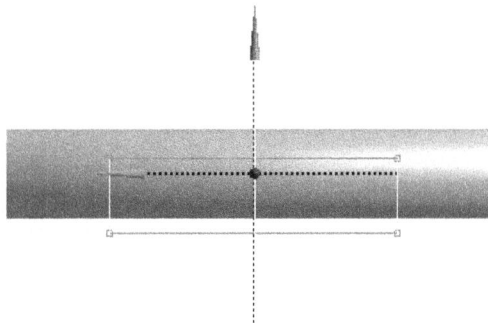

Figure 5-61 *The sketch after the vertical lines are made symmetric*

Now, you need to apply symmetric constraint between the top horizontal line and the bottom horizontal line of the rectangle.

19. Choose the **Symmetry** tool from the **Constraints** toolbox; you are prompted to select the axis of symmetry.

20. Click on the horizontal axis, as shown in Figure 5-62, as the axis of symmetry; you are prompted to select a point or a 2D edge to apply Symmetric constraint.

21. Select the top horizontal line of the rectangle as the first line to apply symmetry; you are prompted to select the second point or 2D edge to apply Symmetric constraint.

22. Select the bottom horizontal line of the rectangle as the second line; the horizontal lines of the rectangular sketch are now symmetrical about the horizontal axis, as shown in Figure 5-63.

After the dimensional and symmetric constraints are applied to the sketch, the rectangular sketch becomes fully constrained, refer to Figure 5-63.

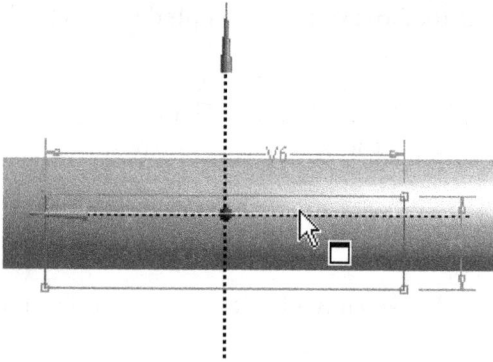

Figure 5-62 *Selecting the horizontal axis for applying Symmetric constraint*

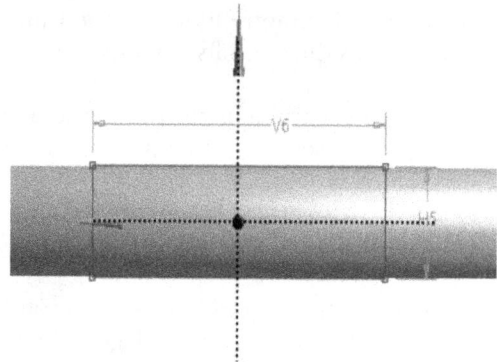

Figure 5-63 *The sketch after the horizontal lines are made symmetric*

23. Next, choose the **ISO** button from the **Graphics** toolbar; the view is changed to Isometric.

24. Choose the **Extrude** tool from the **Features** toolbar; **Extrude1** is attached to Tree Outline. Also, the options in the **Details View** window are changed.

25. In the **Details View** window, select the **Geometry** selection box to display the **Apply** and **Cancel** buttons if not already displayed.

26. Select **Sketch3**, available under the **Plane4** node in the Tree Outline and then click on the **Apply** button in the **Geometry** selection box in the **Details View** window; preview of extrusion is displayed in the graphics window.

27. Select the **Add Material** option from the **Operation** drop-down list if not already selected, as shown in Figure 5-64.

28. Next, from the **Direction** drop-down list, select the **Both-Asymmetric** option, as shown in Figure 5-65; the **Extent Type 2** drop-down list and the **FD4, Depth 2 (>0)** edit box are added to the **Details View** window.

29. In the **FD1, Depth (>0)** edit box, enter **30** and then press Enter; the changes will be displayed in the preview of the extruded feature.

30. Select **To Next** option from the **Extent Type 2** drop-down list in the **Details View** window.

31. Choose the **Generate** tool from the **Features** toolbar; the extruded feature is created

Details of Extrude1

Extrude	Extrude1
Geometry	Sketch3
Operation	Add Material
Direction Vector	Add Frozen
Direction	Add Material
Extent Type	Cut Material
FD1, Depth (>0)	Imprint Faces / Slice Material
Extent Type 2	Fixed
FD4, Depth 2 (>0)	30 mm
As Thin/Surface?	No
Merge Topology?	Yes

Geometry Selection: 1

| Sketch | Sketch3 |

Details of Extrude1

Extrude	Extrude1
Geometry	Sketch3
Operation	Add Material
Direction Vector	None (Normal)
Direction	Both - Asymmetric
Extent Type	Normal / Reversed
FD1, Depth (>0)	Both - Symmetric
Extent Type 2	Both - Asymmetric
FD4, Depth 2 (>0)	30 mm
As Thin/Surface?	No
Merge Topology?	Yes

Geometry Selection: 1

| Sketch | Sketch3 |

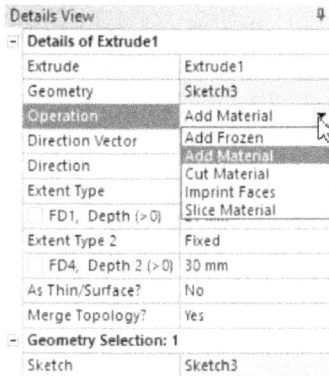

*Figure 5-64 Selecting the **Add Material** option from the **Operation** drop-down list*

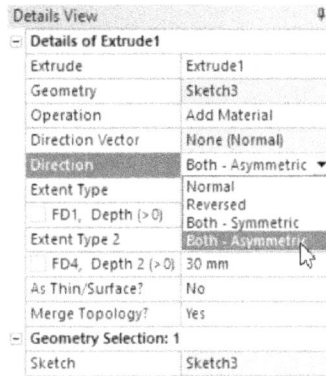

*Figure 5-65 Selecting the **Both-Assymetric** option from the **Direction** drop-down list*

32. Click on the **New Plane** button available in the **Active Plane/Sketch** toolbar; **Plane5** is attached to the Tree Outline. Also, the options in the **Details View** window are modified.

33. In the **Details View** window, select the **From Face** option from the **Type** drop-down list; the **Base Face** selection box is displayed.

34. Click on the **Base Face** selection box; the **Apply** and **Cancel** buttons are displayed. Also, you are prompted to select the base face required for the plane creation.

35. Choose the **Face** button from the **Select** toolbar.

36. Select the top face of the first extruded feature, as shown in Figure 5-66. Next, click on the **Apply** button available in the **Base Face** selection box in the **Details View** window.

37. Choose the **Generate** tool from the **Features** toolbar; the **Plane2** is generated.

After creating the new plane on the required face, you are required to create the sketch for the second extruded feature.

Figure 5-66 Selecting the top face of the first extruded feature

38. Choose the **Look At** tool to orient the new plane normal to the viewing direction.

Note that Plane5 is the current plane and you need to create a sketch on this plane.

39. Choose the **New Sketch** button from the **Active Plane/Sketch** toolbar; a new sketch with the name **Sketch5** is added under the **Plane5** node.

Note
*In your case, a different name may be displayed under the **Plane5** node in the Tree Outline. To edit this name, specify a new name in the **Sketch** edit box in the **Details View** window. Note that this window will be displayed when the sketch is selected from the Tree Outline.*

40. Next, click on the **Sketching** tab to switch to the **Sketching** mode.

41. Create a sketch for the second extruded feature by using the **Polyline** tool, as shown in Figure 5-67.

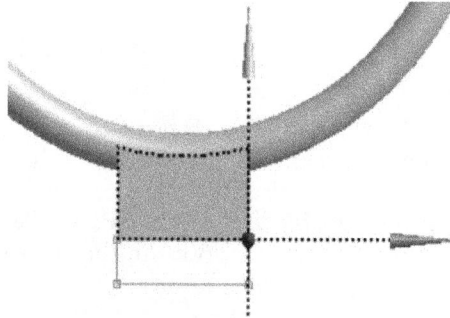

*Figure 5-67 Partial view of the model with the newly created sketch on **Plane5***

42. Next, click on the **Extrude** tool from the **Features** toolbar; **Extrude2** is attached to the Tree Outline. Also, the **Details View** window is displayed with various options.

43. Click on the **ISO** button available in the **Graphics** toolbar; the view is changed to isometric.

44. Next, click on the **Geometry** selection box in the **Details View** window; the **Apply** and **Cancel** buttons are displayed.

45. Expand the **Plane5** node in the Tree Outline and then select **Sketch5** in it.

46. Click on the **Apply** button in the **Geometry** selection box to specify Sketch5 as the sketch to be extruded.

47. Select the **Both - Asymmetric** option from the **Direction** drop-down list, as shown in Figure 5-68; the **Extent Type 2** drop-down list along with the **FD4, Depth 2 (>0)** edit box is added to the **Details View** window.

48. Click on the **FD1, Depth (>0)** edit box and enter **50** as the distance for extrusion.

49. In the **Extent Type 2** drop-down list, select the **To Surface** option, as shown in Figure 5-69; the **Target Face 2** selection box is displayed under the **Extent Type 2** drop-down list and is highlighted in yellow.

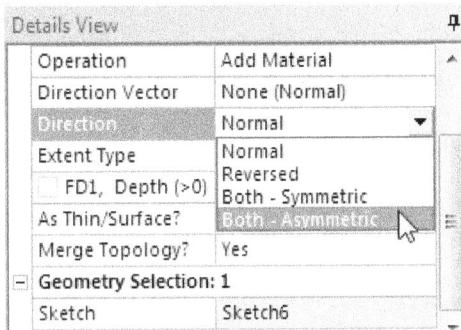

Figure 5-68 *Selecting the **Both-Asymmetric** option from the **Direction** drop-down list*

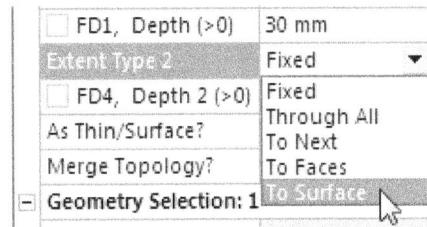

Figure 5-69 *Selecting the **To Surface** option from the **Extent Type 2** drop-down list*

50. Click on the **Target Face 2** selection box; the **Apply** and **Cancel** buttons are displayed in it. Also, you are prompted to select the surface to extrude the sketch to.

51. Next, select the bottom face of the first extruded feature, as shown in Figure 5-70.

52. Next, click on the **Apply** button displayed in the **Target Face 2** selection box in the **Details View** window; **Selected** is displayed in the **Target Face 2** selection box indicating that selected face is specified as the extent of extrusion in that direction.

53. After all the parameters are specified, choose the **Generate** tool; the extruded feature is generated, as shown in Figure 5-71.

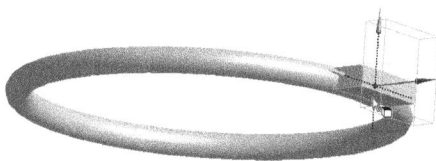

Figure 5-70 *Selecting the bottom face of the first extruded feature*

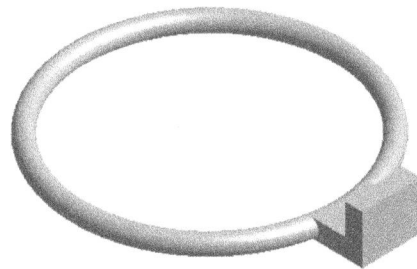

Figure 5-71 *Model with the extruded feature*

Creating the Blend Feature

After the second feature is created, you need to remove the sharp edges.

1. Choose the **Fixed Radius** tool from the **Blend** drop-down in the **Features** toolbar; **FBlend1** is attached to the Tree Outline. Also, the options in the **Details View** window are changed.

2. Choose **Wireframe** from the **View** menu in the Menu bar; the view is changed to Wireframe and only the second feature is visible.

3. Select the **Geometry** selection box; the **Apply** and **Cancel** buttons are displayed in it.

4. Next, select the first edge to apply blend, refer to Figure 5-72.

5. Choose **Apply** to specify the selected edge for blending; **1 Edge** is displayed in the **Geometry** selection box.

6. In the **FD1, Radius (>0)** edit box, enter **15** to specify the radius of the blend and then press Enter.

7. Choose the **Generate** tool; the blend feature is created, as shown in Figure 5-73.

8. Again, choose the **Fixed Radius** tool from the **Blend** drop-down in the **Features** toolbar; **FBlend2** is attached to the Tree Outline. Next, select the second edge to apply blend, refer to Figure 5-72.

9. Choose **Apply** to specify the selected edge for blending; **1 Edge** is displayed in the **Geometry** selection box. Next, In the **FD1, Radius (>0)** edit box, enter **45** to specify the radius of the blend and then press Enter.

10. Choose **Shaded Exterior and Edges** from the **View** menu to change the view from Wireframe to Shaded.

11. Choose the **Generate** tool; the blend feature is created, as shown in Figure 5-73.

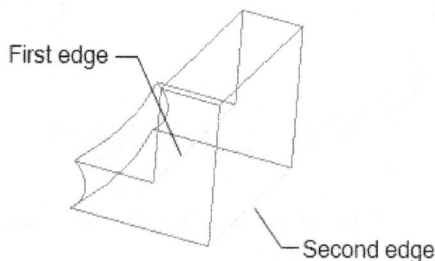

Figure 5-72 Selecting the edges to apply blend

Figure 5-73 Model with the blend feature

12. Close the **DesignModeler** window.

Saving the Project and Exiting ANSYS Workbench

After the blend feature is created, you need to save the project and exit ANSYS Workbench.

1. Choose the **Save Project** button from the **Main** toolbar; project is saved with the name *c05_ansWB_tut02*.

2. Choose the **Exit** option from the **File** menu to exit the current ANSYS Workbench session.

Tutorial 3

In this tutorial, you will create the model of a Helical Gear, as shown in Figure 5-74. For dimensions of the model, refer to Figure 5-75. You will use the **Revolve** and **Loft** tools for creating this model. **(Expected time: 45 min)**

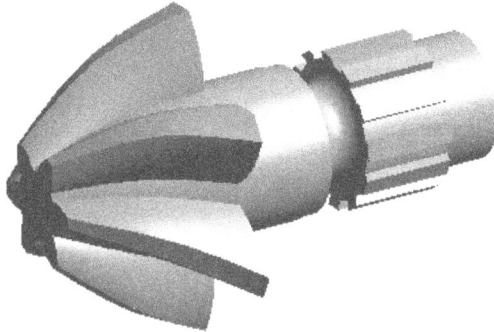

Figure 5-74 *Model for Tutorial 3*

NOTE:
1.   CREATE A BLEND FEATURE WITH 3 SKETCHES. SIZE OF THE SECOND SKETCH IS 75%
     OF THE FIRST SKETCH AND, SIZE OF THE THIRD SKETCH IS 50% OF THE FIRST
     SKETCH.

2.   THE PLANE FOR CREATING THE SECOND SKETCH IS 45 UNITS FROM THE FIRST
     SKETCH. THE PLANE FOR CREATING THE THIRD SKETCH IS 95 UNITS FROM THE
     FIRST SKETCH.

3.   ROTATION ANGLE FOR SECOND AND THIRD SECTION IS 15° AND 30° FROM THE FIRST
     SECTION RESPECTIVELY.

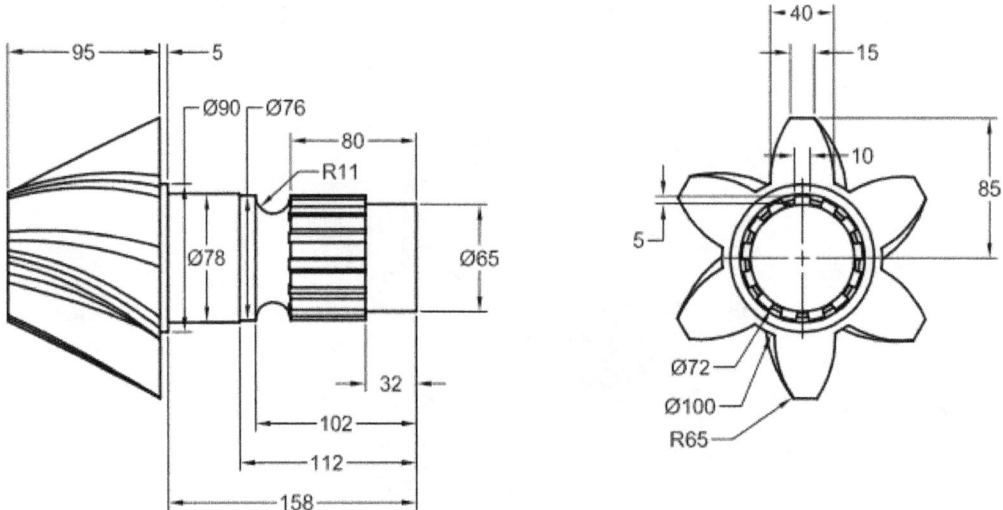

Figure 5-75 *Dimensions for the model*

The following steps are required to complete this tutorial:

a.   Start ANSYS Workbench.
b.   Add the **Geometry** component system to the **Project Schematic** window.
c.   Draw the sketch.
d.   Create the revolved feature.
e.   Create the extruded feature.
f.   Create the sketches for the loft feature.
g.   Create the loft feature.
h.   Save the project.

Starting ANSYS Workbench and Adding the Geometry Component System

Before you create the model, you first need to start ANSYS Workbench and then add a
component system to the project.

1.   Start ANSYS Workbench 2023 R2. The **Workbench** window is displayed.

2. Double-click on the **Geometry** component system displayed under the **Component Systems** toolbox in the **Toolbox** window; the **Geometry** component system is added to the **Project Schematic** window.

3. Once the project is added to the **Project Schematic** window, its name gets highlighted at the bottom of the component system in blue. If it is not already highlighted, double-click on the default name and rename it as **Lofted Feature**.
   Now, you need to save the project.

4. Choose the **Save Project** button from the **Main** toolbar; the **Save As** dialog box is displayed.

5. Browse to *C:\ANSYS_WB\c05* folder

6. Create another subfolder with the name **Tut03** under the *c05* folder and then choose the **Open** button from the **Save As** dialog box.

7. Enter **c05_ansWB_tut03** in the **File name** edit box and choose the **Save** button from the **Save As** dialog box; the project is saved.

Drawing the Sketch

You now need to start the **DesignModeler** window and then create the sketch for the revolve feature.

1. Right-click on the **Geometry** cell in the **Lofted Feature** component system. Choose the **New DesignModeler Geometry**; the **DesignModeler** window gets opened.

2. Choose the **Millimeter** option from the **Units** menu of the **Menu** bar.

3. Select the **XYPlane** from the Tree Outline; the XY plane becomes the active plane. Now, you need to create the sketch for the base feature.

4. Choose the **Sketching** tab displayed at the bottom of the Tree Outline to switch to the **Sketching** mode; the **Sketching Toolboxes** window is displayed.

5. Choose the **Look At** tool to orient the plane normal to the viewing direction.

6. Use the **Polyline** and **Arc** tools from the **Draw** toolbox to draw the sketch for the revolved feature, as shown in Figure 5-76.

7. Apply required constraints and dimensions to the sketch to make it fully constrained. For dimensions, refer to Figure 5-75.

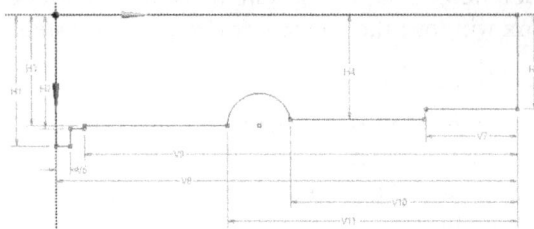

Figure 5-76 *Sketch for the revolved feature*

Creating the Revolved Feature

1. Choose the **Revolve** tool available in the **Features** toolbar; **Revolve1** is added to the Tree Outline. Also, the respective options in the **Details View** window are displayed. By default, the **Geometry** selection box displays the **Apply** and **Cancel** buttons.

2. Click on the **Apply** button to confirm the sketch as the geometry for the revolved feature. Figure 5-77 shows the geometry and the axis to be selected for the revolve operation.

3. Next, click on the **Axis** selection box; the **Apply** and **Cancel** buttons are displayed.

4. Next, click on the X axis; the preview of the revolved feature is displayed in the graphics window, as shown in Figure 5-78.

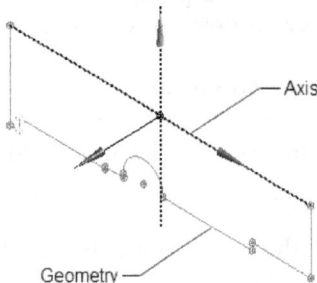

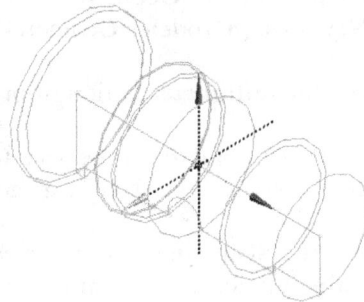

Figure 5-77 *The geometry and the axis to be selected*

Figure 5-78 *Preview of the revolved feature*

5. Choose the **Generate** tool in the **Features** toolbar; the revolved feature is created, as shown in Figure 5-79.

Creating the Extrude Feature

1. Choose the **Faces** tool in the **Display** toolbar and then click on the flat face of the revolve feature, as shown in Figure 5-80.

2. Choose the **New Plane** tool available in the **Active Plane/Sketch** toolbar; **Plane4** is added to the Tree Outline.

3. Choose the **Generate** button in the **Features** toolbar; the new plane is generate. Also, the new plane is displayed on the selected face in the graphics window.

    Generate

4. Choose the **Look At** tool in the **Graphics** toolbar; the plane is oriented normal to the screen.

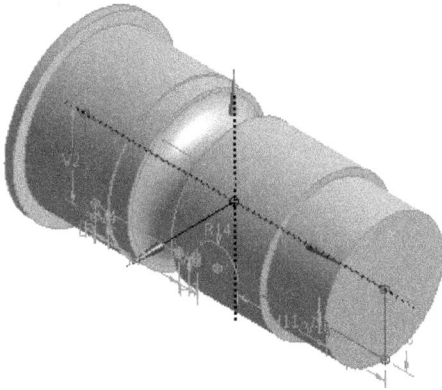

Figure 5-79 Revolved feature created

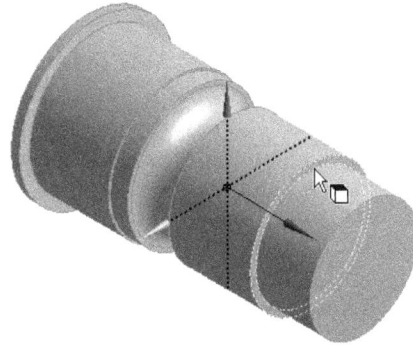

Figure 5-80 Selecting the flat face of the revolved feature

5. Draw a sketch for the extruded feature on the new plane, as shown in Figure 5-81.

6. Apply required constraints and dimensions to the sketch, as shown in Figure 5-82.

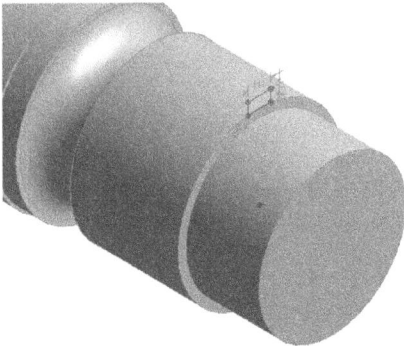

Figure 5-81 The sketch drawn on the new plane

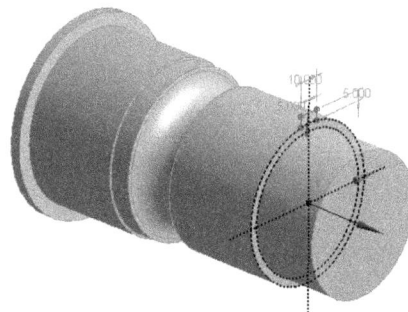

Figure 5-82 Sketch of the extruded feature along with the dimensions

After the sketch is drawn, it is now important to create a pattern around the circular face. To do so, use the **Replicate** tool.

7. Choose the **Replicate** tool available in the **Modify** toolbox of the **Sketching Toolboxes** window; the **r** and **f** edit boxes are displayed next to the **Replicate** tool. Also, you are prompted to select the edges to replicate.

8. Enter **30** in the **r** edit box and **1** in the **f** edit box.

Note

The ***r*** *edit box in the* ***Replicate*** *tool is used to rotate the object about certain angle whereas the* ***f*** *edit box is used to scale the selected object by a certain fraction.*

9. Next, select the sketch drawn for the extruded feature, as shown in Figure 5-83.

10. Right-click in the graphics window to display a shortcut menu. Next, choose the **End / Use Plane Origin as Handle** option from the shortcut menu.

11. Orient the view of the plane normal to the screen. Next, right-click and choose **Rotate by r Degrees** option from the shortcut menu displayed.

12. Right-click again and choose **Paste at Plane Origin** to create an instance of the pattern, as shown in Figure 5-84.

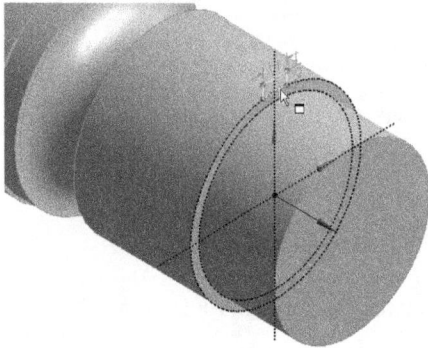

Figure 5-83 The sketch selected for creating a pattern around the circular edge

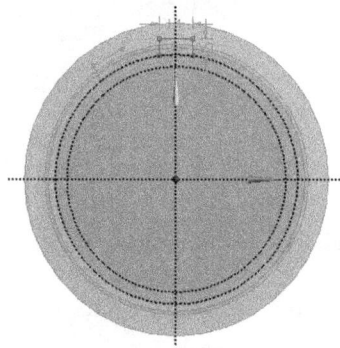

Figure 5-84 Creating instance of the sketch

13. Similarly, create 10 more instances of the sketch, as shown in Figure 5-85.

14. Next, click on the **Extrude** tool in the **Features** toolbar; **Extrude1** is added to the Tree Outline. Also, preview of extrusion is displayed on the model.

15. Change the view to isometric.

16. Select the **Reversed** option from the **Direction** drop-down list in the **Details View** window, as shown in Figure 5-86.

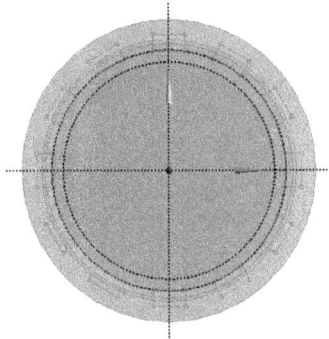

Figure 5-85 *Pattern created using the* **Replicate** *tool*

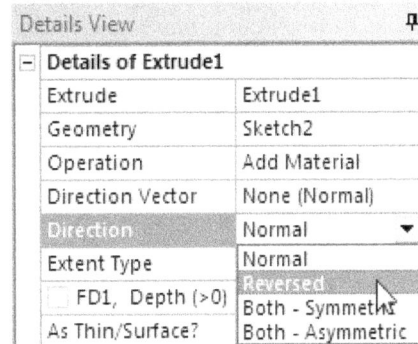

Figure 5-86 *Selecting the* **Reversed** *option from the* **Direction** *drop-down list*

17. In the **Extent Type** drop-down list, select the **Fixed** option if not already selected, as shown in Figure 5-87.

18. In the **FD1, Depth (>0)** edit box, enter **48** and then press Enter; the changes will be displayed in the preview of the extruded feature, as shown in Figure 5-88.

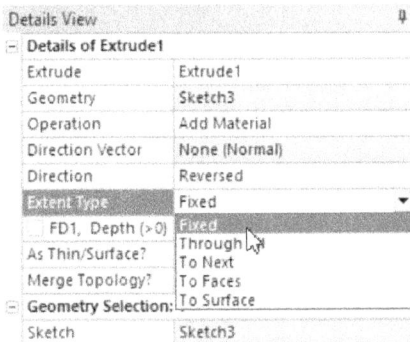

Figure 5-87 *Selecting the* **Fixed** *option from the* **Extent Type** *drop-down list*

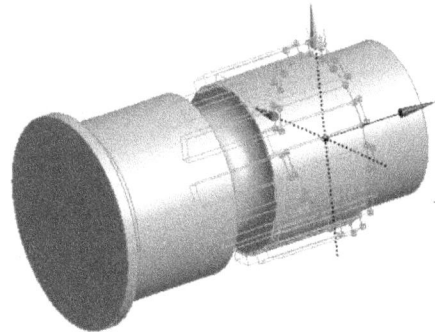

Figure 5-88 *Displaying preview of the extruded part*

19. Choose the **Generate** tool from the **Features** toolbar; the extrude feature is created, as shown in Figure 5-89.

Creating the Sections for the Loft Feature

After the extrude feature is created, it is now required to create the three sections for the loft feature.

1. Select the front face of the model and then choose the **New Plane** tool in the **Active Plane / Sketch** toolbar; **Plane5** is added to the Tree Outline.

2. Choose the **Generate** tool in the **Features** toolbar; a new plane is created and is displayed on the selected face, as shown in Figure 5-90.

> Generate

3. Choose the **Look At** tool in the **Graphics** toolbar to orient the model normal to the viewing direction, refer to Figure 5-90.

4. Switch to the **Sketching** mode.

5. Choose the **Circle** tool from the **Draw** toolbox.

> Circle

Figure 5-89 *Extruded feature created*

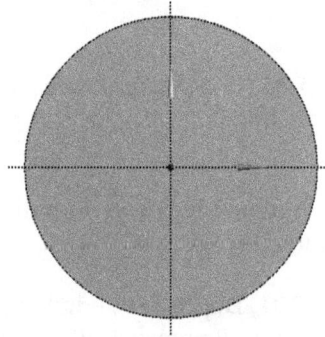

Figure 5-90 *Model after being oriented*

6. Draw a circle with the origin of the new plane as the center. Change the diameter of the circle to 100.

7. Create a line at an offset from the horizontal axis, as shown in Figure 5-91. For dimensions, refer to Figure 5-75.

8. Choose the **Arc by 3 Points** tool from the **Draw** toolbox and then draw an arc, as shown in Figure 5-92. For dimensions and placement of the arc, refer to Figure 5-75.

9. Apply dimensions and constraints to the arc created, refer to Figure 5-92.

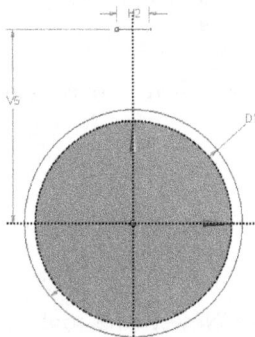

Figure 5-91 *A line created at an offset*

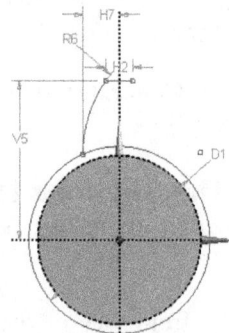

Figure 5-92 *The arc created*

10. Click on the **Modify** toolbox in the **Sketching Toolboxes** window to display the tools in it.

11. Invoke the **Replicate** tool from the **Modify** toolbox and then select the arc, as shown in Figure 5-93.

12. Right-click in the graphics window to display the shortcut menu and then choose **End / Use Plane Origin as Handle** from the menu displayed.

13. Right-click in the graphics window again and choose the **Flip Horizontal** option from the shortcut menu displayed.

14. Right-click and then choose **Paste at Plane Origin** to create a mirror of the entity across the Y axis, refer to Figure 5-94.

15. Right-click again and then choose **End** from the shortcut menu displayed to end the operation.

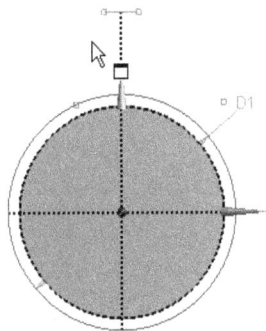

Figure 5-93 The selected arc

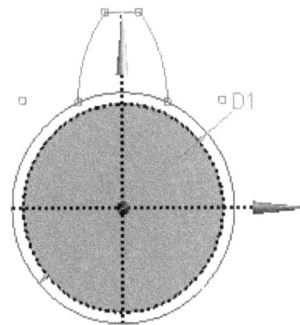

Figure 5-94 The complete sketch

16. Invoke the **Replicate** tool again and then select the two arcs and the line created at an offset, refer to Figure 5-95.

17. Enter **60** in the **r** edit box corresponding to the **Replicate** tool.

18. Right-click in the graphics window and then choose **End / Use Plane Origin as Handle** from the shortcut menu displayed.

19. Right-click and then choose **Rotate by r Degrees** option from the shortcut menu displayed.

20. Right-click again and then choose the **Paste at Plane Origin** option from the shortcut menu displayed; one instance of the sketch is created at an angle of 60 degrees.

21. Similarly, create other instances by using the Steps 16 through 20. The instances of the sketch created are shown in Figure 5-96.

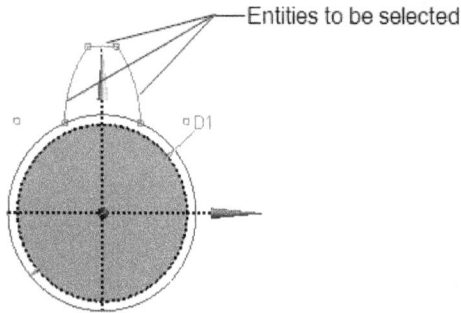

Figure 5-95 Selecting entities of the sketch

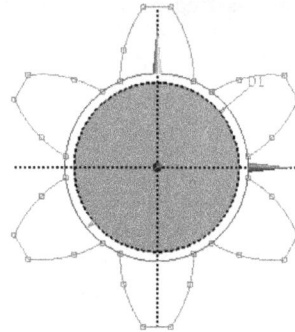

Figure 5-96 Instances of the sketch after being replicated

22. After all the instances are created, you need to trim the unwanted entities by using the **Trim** tool. The sketch after the unwanted entities are removed will appear as shown in Figure 5-97.

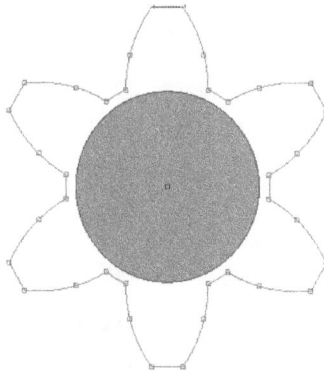

Figure 5-97 Sketch after trimming unwanted entities

23. Switch to the **Modeling** mode and then choose the **New Plane** tool from the **Active Plane /Sketch** toolbar; **Plane6** is attached to the Tree Outline. Also options related to this tool are displayed in the **Details View** window.

24. In the **Transform1 (RMB)** drop-down list, select the **Offset Z** option; the **FD1, Value1** edit box is activated.

25. Enter **45** in the **FD1, Value1** edit box.

26. Next, choose the **Generate** tool; the plane is created.

27. Expand the **Plane5** node in the Tree Outline and then select **Sketch3**; Sketch3 becomes the active sketch.

28. Invoke the **Sketching** mode by choosing the **Sketching** tab.

29. Select the **Box Select** option from the **Select Mode** drop-down list in the   ⬚ Box Select
Graphics toolbar.

30. Select the sketch on plane 5; the sketch turns yellow.

31. Next, expand the **Modify** toolbox in the **Sketching Toolboxes** window; the tools in this toolbox are displayed.

32. Choose the **Copy** tool; you are prompted to select the entities to copy. Next, right-click in the graphics window; a shortcut menu is displayed.

33. Choose the **Use Plane Origin as Paste Handle** option from the shortcut menu displayed; the center of the sketch gets selected as the reference point for the copy object.

34. Choose the **Modeling** tab displayed at the bottom of the **Sketching Toolboxes** window; the **Modeling** mode is activated.

35. Select **Plane6** in the Tree Outline; Plane6 becomes the active plane.

36. Invoke the **Sketching** mode by choosing the **Sketching** tab from the bottom of the Tree Outline.

37. Expand the **Modify** toolbox; the tools in it are displayed.

38. Choose the **Paste** tool available in the **Modify** toolbox; the **r** and **f** edit boxes are displayed next to the **Paste** tool. Also, you are prompted to paste the sketch copied from Plane5 on the graphics window.

39. Enter **15** in the **r** edit box and **0.75** in the **f** edit box.

40. Right-click in the graphics window to display a shortcut menu. Now, choose **Rotate by r Degrees** from the shortcut menu.

41. Right-click again and then choose **Scale by factor f** from the shortcut menu.

On choosing the **Rotate by r Degrees** option from the shortcut menu, the sketch rotates by an angle specified in the **r** edit box. Similarly, the **Scale by factor f** option scales the object by a factor specified in the **f** edit box.

42. After the copied sketch is rotated and scaled, paste it on the origin. To do so, right-click in the graphics window and then choose **Paste at Plane Origin** from the shortcut menu displayed; the modified sketch section is pasted on Plane6, as shown in Figure 5-98.

Note that the new sketch is rotated and scaled as it is pasted on the origin of the new plane.

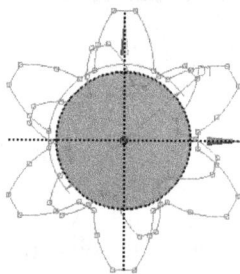

Figure 5-98 *Sketch copied on Plane6*

43. Create another plane at an offset of **90** from the **Plane6**.

44. Next, copy the sketch from Plane6 and then rotate and scale it by following the procedure described in steps 34 through 40. Figure 5-99 shows the copied sketch which has been rotated about 30 degrees, and scaled to a fraction of 0.5.

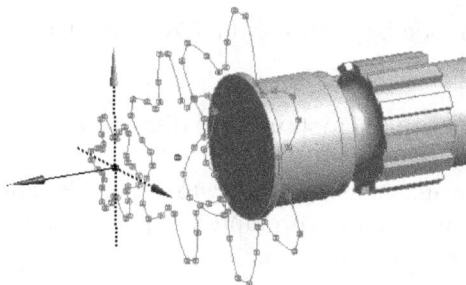

Figure 5-99 *Three sections created for the loft feature*

Note
The view of the model has been adjusted for a better visibility of the sections.

Creating the Loft Feature

After creating the three sketches, you are required to create the loft feature to complete the model.

1. Choose the **Skin/Loft** tool in the **Features** toolbar; **Skin1** with a yellow thunderbolt symbol is added to the Tree Outline. Also, the corresponding options of the **Skin/Loft** tool are displayed in the **Details View** window.

2. The **Select All Profiles** option is selected by default in the **Profile Selection Method** drop-down list in the **Details View** window. Select the **Select Individual Profile** option from this list, as shown in Figure 5-100; **Profiles** is attached below the **Details of Skin1** node in the **Details View** window.

Details View	⊓
− **Details of Skin1**	
Skin/Loft	Skin1
Profile Selection Method	Select All Profiles ▼
Profiles	Select All Profiles
Operation	Select Individual Profile
	Add Material
As Thin/Surface?	No
Merge Topology?	No

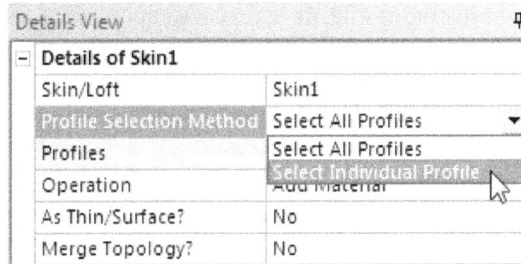

Figure 5-100 *Selecting the* **Select Individual** **Profile** *option from the drop-down list*`

3. Click on the **Profile 1** selection box highlighted in yellow; the **Apply** and **Cancel** buttons are displayed in it.

4. In the Tree Outline, expand the **Plane5** node and then select **Sketch3** under the **XYPlane** node.

5. Next, choose the **Apply** button displayed in the **Profile 1** selection box under the **Profiles** node in the **Details View** window.

6. Click on the **Profile 2** selection box available below the **Profile 1** selection box; the **Apply** and **Cancel** buttons are displayed.

7. Next, in the Tree Outline, expand the **Plane6** node and then select **Sketch4** displayed under this node.

8. Choose the **Apply** button in the **Profile 2** selection box to select this sketch as the second section for the loft feature.

9. Right-click on the **Profile 2** selection box and then choose the **Add Profile** option from the shortcut menu displayed; the **Profile 3** selection box is added under the **Profiles** node in the **Details View** window.

10. Click on the **Profile 3** selection box; the **Apply** and **Cancel** buttons are displayed.

11. In the Tree Outline, expand the **Plane7** node and then select **Sketch5** from it.

12. Choose the **Apply** button from the **Profile 3** selection box to specify the sketch for the loft feature; a guide line connecting the vertices of the sketches is displayed in the graphics window.

Note
After the selection of profiles is completed, ANSYS generates a guide line automatically. If the generated guide line (path) is not the one that is required, you can change it manually.

A guide line is the curve that connects corresponding vertices of the sections being used for the **Skin/Loft** operation. As soon as the selection of profiles for feature creation is done, a guide line is displayed in the model. It is very easy to create a guide line in a simple loft feature. However, it is quite challenging to create a guide line in a twisted loft, where it passes through various points in the profiles under consideration. In case, the default path viewed in ANSYS is not the required one and you need to change it, you can do so by using the **Fix Guide Line** option from the shortcut menu, which is displayed on right-clicking in the graphics window, as shown in Figure 5-101. Figure 5-102 shows three profiles (sections) selected for creating a loft feature. The corresponding vertices through which the guide line passes is shown in Figure 5-102. In this figure, three rectangles are used to create a loft feature. Figure 5-103 shows the corresponding loft created using the guide line.

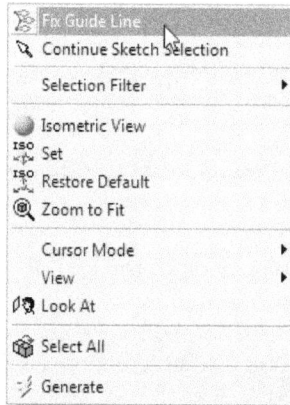

*Figure 5-101 Choosing the **Fix Guide Line** option from the shortcut menu*

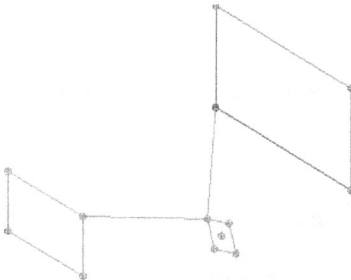

Figure 5-102 Guide line for a loft feature

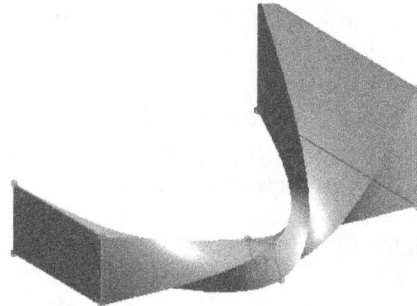

Figure 5-103 Loft feature created using the guide line

You can also modify a guide line. To do so, you need to change the vertices that define it. To do so, right-click after you specify the sketches for the loft feature in the **Details View** window; a shortcut menu will be displayed. Choose the **Fix Guide Line** option from it; you will be prompted to select a line or a vertex to change the loft path.

Based on the requirement, you may need to select lines, points, edges, and so on to define the required guide line. To select points on the profiles, right-click again in the graphics window to display a shortcut menu and then choose **Selection Filter > Point**. Next, you will select a particular vertex or point through which the guide line would pass. In this way you can alter the path of the loft feature. By comparing Figures 5-102 and 5-104, you can observe that guide line has been modified using the **Fix Guide Line** option. The corresponding loft feature created after the guide line is modified is shown in Figure 5-105.

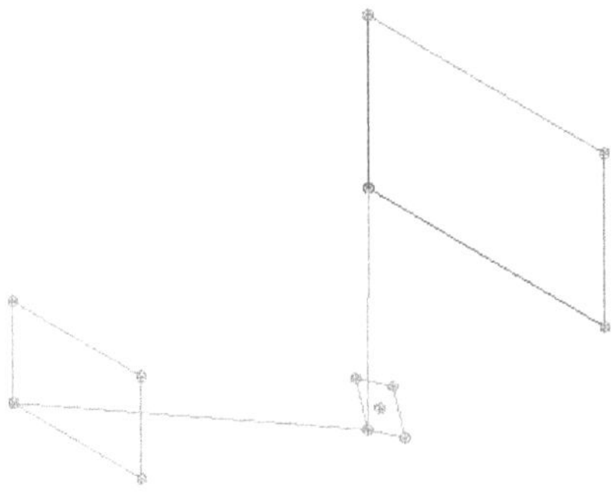

Figure 5-104 Modified guide line

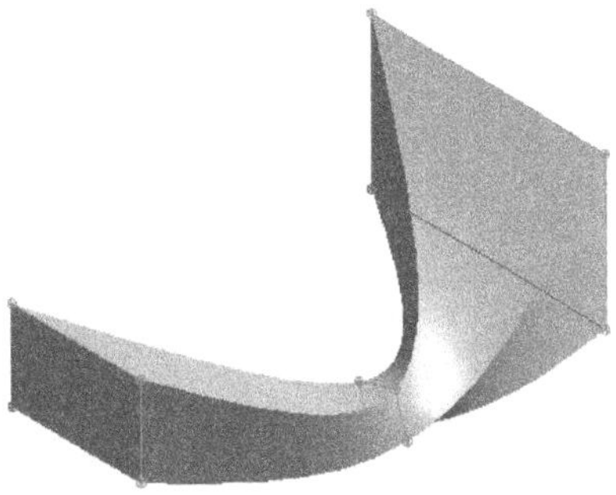

Figure 5-105 Loft feature created using the modified guide line

Note
*As in this case the guide line generated does not connect to the corresponding points of different profiles, you need to modify the path of the guide line by reallocating the points using the **Fix Guide Line** option. Figure 5-106 shows the sketch created for the loft feature with various points annotated.*

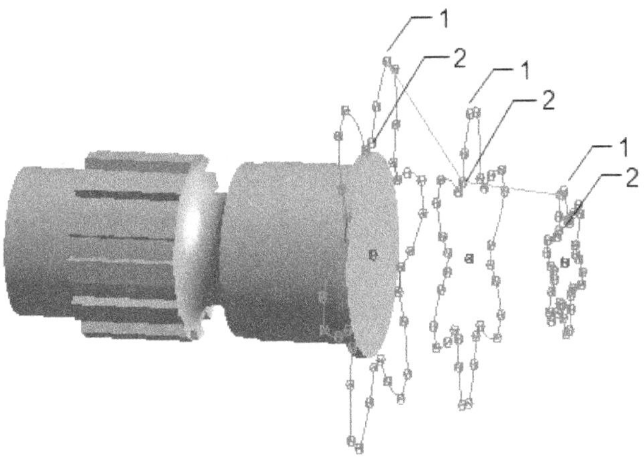

Figure 5-106 Sketch created for the loft feature with various points annotated

13. Right-click in the graphics window to alter the guide line; a shortcut menu is displayed. Choose the **Fix Guide Line** option from the shortcut menu displayed; you are prompted to select lines, edges, or vertices to alter the guide line.

14. Since the guide line passes through points in different profiles, you need to select points to alter the guide line. To do so, right-click again in the graphics window; a shortcut menu is displayed. Next, choose **Selection Filter > Point** from the shortcut menu displayed.

15. Select point 1 in the first profile.

16. Similarly, select point 1 of the second and third profiles; the path of the guide line changes, as shown in Figure 5-107.

17. Next, choose the **Generate** tool in the **Features** toolbar; the changes are applied and the loft feature created, as shown in Figure 5-108.

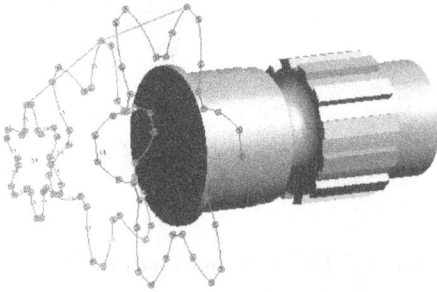

Figure 5-107 Modified guide line

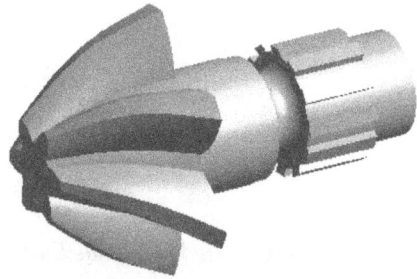

Figure 5-108 Loft feature created using the modified guide line

18. Exit the **DesignModeler** window; the **Workbench** window is displayed.

Saving the Project and Exiting ANSYS Workbench

1. In the **Workbench** window, choose the **Save Project** button from the **Main** toolbar; the project is saved with the name *c05_ansWB_tut03*.

2. Next, choose the **Exit** option from the **File** menu to exit the current ANSYS Workbench session.

Self-Evaluation Test

Answer the following questions and then compare them to those given at the end of this chapter:

1. Which of the following tools is used to create a pattern of a sketch?

    (a) **Replicate**   (b) **Generate**
    (c) **Body Operation**   (d) **New Plane**

2. Which of the following tools is used to view the sketching plane at right angle?

    (a) **New Plane**   (b) **Look At**
    (c) **Imprint**   (d) **Display Plane**

3. Which of the following tools is invoked to perform a mirror operation?

    (a) **Imprint**   (b) **Boolean**
    (c) **Body Transformation**   (d) **Freeze**

4. While using the **Extrude** tool, _____ extent type will extrude the profiles to the exact distance.

5. You cannot create circular pattern of features by using the **Pattern** tool. (T/F)

6. To create a revolved feature, choose the **Revolve** tool from the **Create** menu of the Menu bar. (T/F)

7. The **Revolve** tool is used to revolve a sketch only around the horizontal axis. (T/F)

8. While creating a sweep feature, you cannot confirm the selection of the profile in the **Details of "Sweep"** window. (T/F)

9. By using the **Revolve** tool, you can add as well as cut material from a feature. (T/F)

10. You cannot create a loft feature between sketches in two different plane. (T/F)

Review Questions

Answer the following questions:

1. Which of the following options is chosen by default in the **Operation** drop-down list in the **Details View** window.

    (a) **Add Material**   (b) **Cut Material**
    (c) **Add Frozen**   (d) **Imprint Faces**

2. You can share a geometry of the **Geometry** component system with the _____ cell of any other component or analysis system.

3. To create the projection of a sketch on a model, you need to use the _____ tool.

4. You can use the tools available in the **Modify** toolbox of the **Sketch** tab to modify sketches. (T/F)

5. You can turn on the display of sketching planes by using the **Display Plane** tool. (T/F)

6. You can turn off the display of the model in the graphics window by choosing the **Display Model** tool. (T/F)

7. To delete an unwanted face, you need to choose the **Face Delete** tool. (T/F)

8.   The **Generate** tool is used after most of the operations are performed. (T/F)

9.   To create a new sketch, you first need to specify the sketching plane. (T/F)

10. You can use the middle mouse button to rotate the model freely in the graphics window. (T/F)

EXERCISE

Exercise 1

Create the model shown in Figure 5-109. The dimensions of the model are shown in the same figure.   **(Expected time: 45 min)**

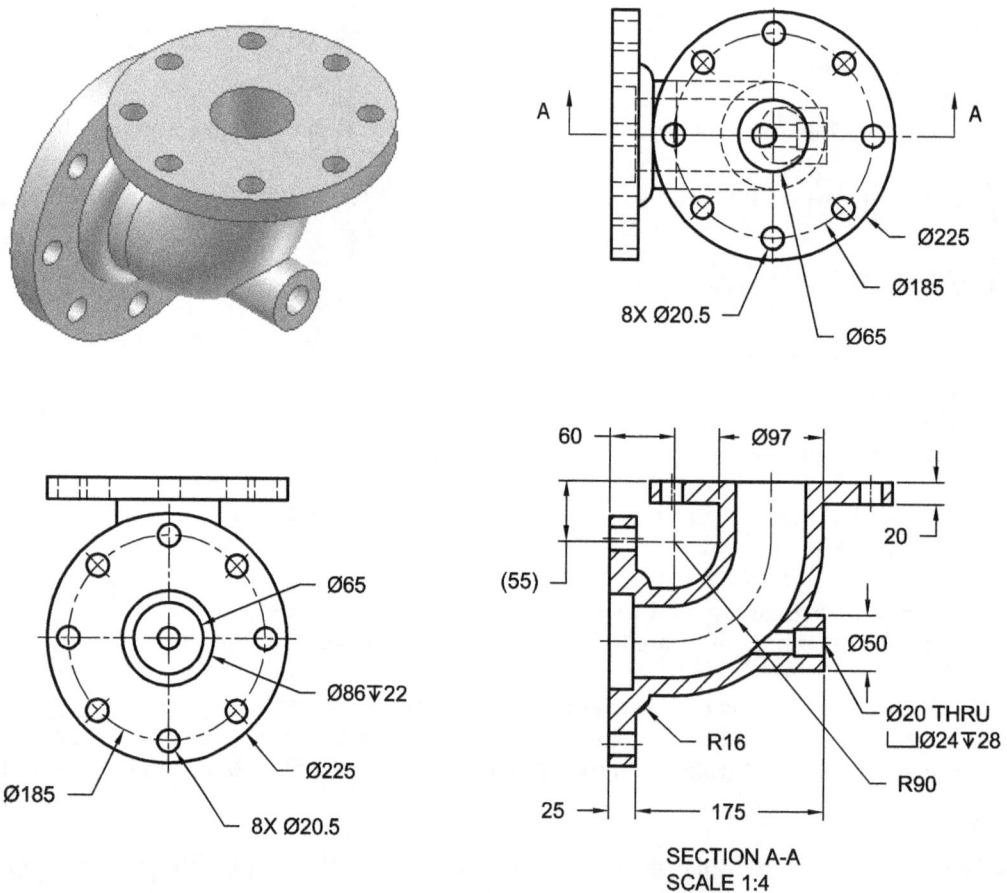

Figure 5-109 *Isometric and Orthographic views of the model for Exercise 1*

Answers to Self-Evaluation Test
1. a, 2. b, 3. c, 4. Fixed, 5. F, 6. T, 7. T, 8. F, 9. T, 10. F

Chapter 6

Defining Material Properties

INTRODUCTION TO ENGINEERING DATA WORKSPACE

For performing an analysis, you need to define the material properties of a model. You can do so by using the Engineering Data workspace of ANSYS Workbench. The Engineering Data workspace can be invoked by using the **Engineering Data** cell of an analysis system. The **Engineering Data** cell is added to almost all the analysis systems where material properties are required to be defined. To define material properties, right-click on the **Engineering Data** cell in the analysis or component system and then choose the **Edit** option from the shortcut menu displayed, as shown in Figure 6-1. On doing so, the **Project Schematic** window will be replaced with four default windows, as shown in Figure 6-2. These windows are collectively known as the Engineering Data workspace.

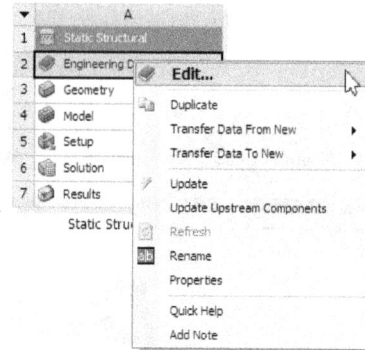

Figure 6-1 Choosing the **Edit** option from the flyout

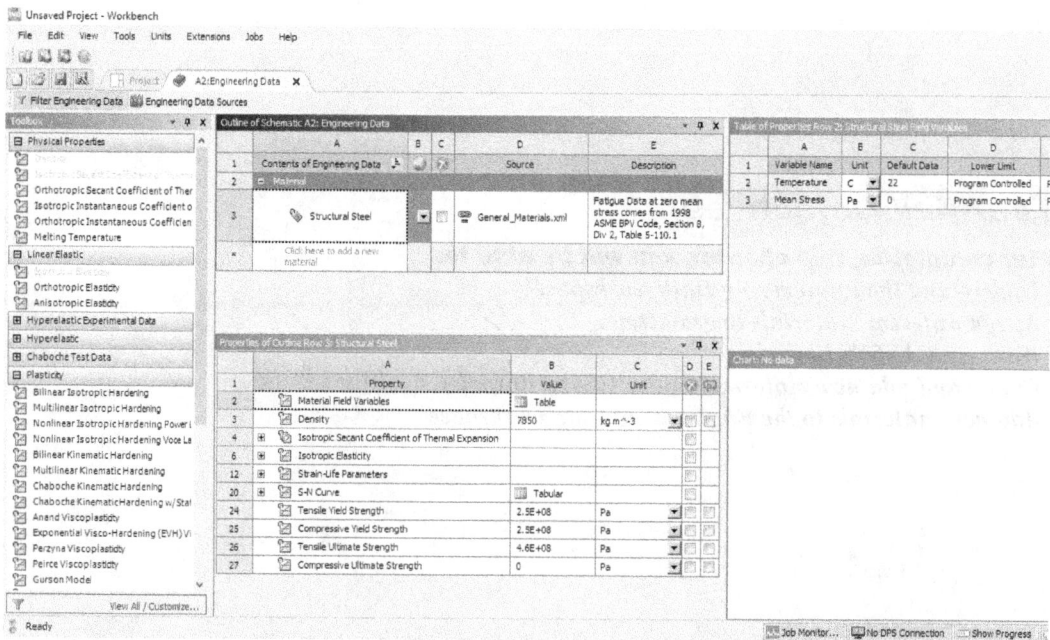

Figure 6-2 Initial screen of the Engineering Data workspace

The default windows displayed in the Engineering Data workspace are: **Outline of Schematic A2: Engineering Data**, **Properties of Outline Row 3: Structural Steel**, **Table of Properties Row 2: Structural Steel Field Variables**, and **Chart: No data**. When the Engineering Data workspace is invoked, the contents of the **Toolbox** window get changed. When you expand a toolbox displayed in the **Toolbox** window, various material properties are displayed in it. You can use the options available in the toolboxes to assign properties to any newly created material.

Note

1. The name Outline of Schematic A2: Engineering Data window changes according to the analysis system which was used for invoking it. Hereafter, it is called Outline window.

2. Similarly, the name of Properties of Outline Row 3: Structural Steel window (will be called Properties of Outline window) may vary depending on the material selected in the Outline window.

3. Depending on the property selected in the Properties of Outline window, the name and contents of the Table of Properties Row 2: Density (hereafter, called Table of Properties window) and the Chart of Properties Row 2: Density (hereafter, called Chart of Properties window) will change.

By default, when you invoke the Engineering Data workspace, the **Structural Steel** material will be available in this workspace. Figure 6-3 shows the **Structural Steel** material selected in the **Outline** window and Figure 6-4 shows the corresponding **Properties of Outline** window, displaying the properties of the selected Structural Steel material.

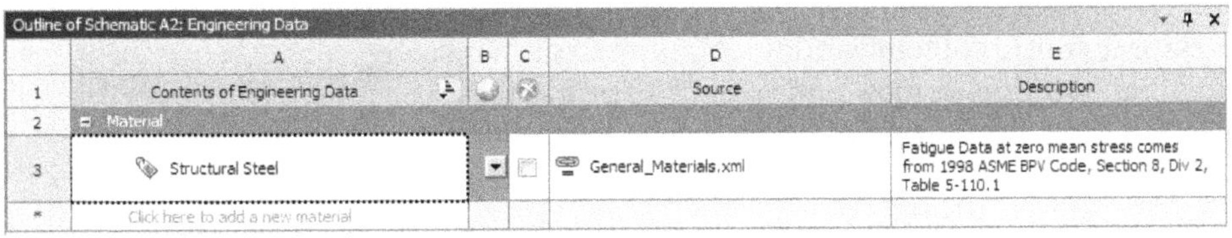

Figure 6-3 The Structural Steel material selected in the Outline window

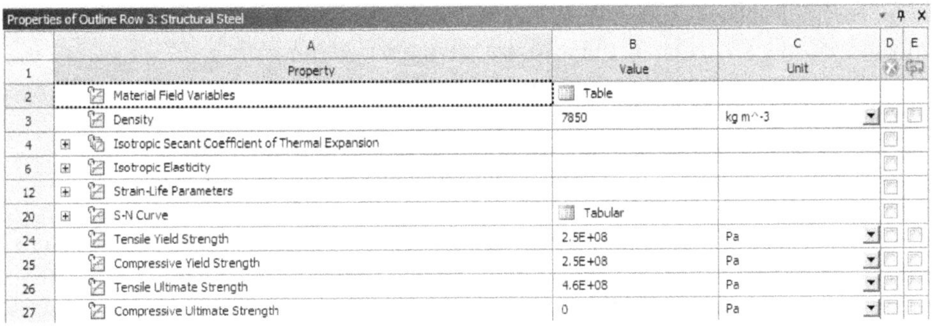

Figure 6-4 The Properties of Outline window displaying properties of the selected Structural Steel material

As discussed earlier, when the Engineering Data workspace is invoked, the contents of the **Toolbox** window get changed and display various material properties that can be assigned to a material, refer to Figure 6-5. In the **Toolbox** window, all the material properties are grouped together as per their category into various toolboxes. For example, the **Strength** toolbox contains properties pertaining to the strength of a material such as Tensile Yield Strength, Compressive Yield Strength, and so on. Similarly, the **Physical Properties** toolbox contains Density, Isotropic Secant Coefficient of Thermal Expansion, and so on.

CREATING AND ADDING MATERIALS

ANSYS Workbench 2023 R2 contains almost all the standard materials in its libraries. You can select the required material from the libraries and assign it to your project. Apart from using the materials from the libraries, you can create a new material as per the requirement and use it. In ANSYS Workbench, you can create a new material either in the **Outline** or **Engineering Data Sources** window. The procedure of creating new material in both these windows is discussed next.

Creating a New Material in the Outline Window

As discussed earlier, you can invoke the Engineering Data workspace by double-clicking on the **Engineering Data** cell of a component or analysis system. By default, the Engineering Data workspace consists of four windows. You can create a new material in the **Outline** window and specify its properties in the corresponding **Properties of Outline** window, refer to Figures 6-3 and 6-4.

By default, **Structural Steel** is displayed in the **Outline** window. If you do not add any other material to the **Outline** window of the Engineering Data workspace, then only Structural Steel material will be available to be applied to any geometry. To create a new material, specify its name in the **Click here to add a new material** edit box in the **Outline** window, refer to Figure 6-3. The material is added to the **Outline** window with a question symbol displayed in front of the name. Also, the **Properties of Outline** window is displayed with no material

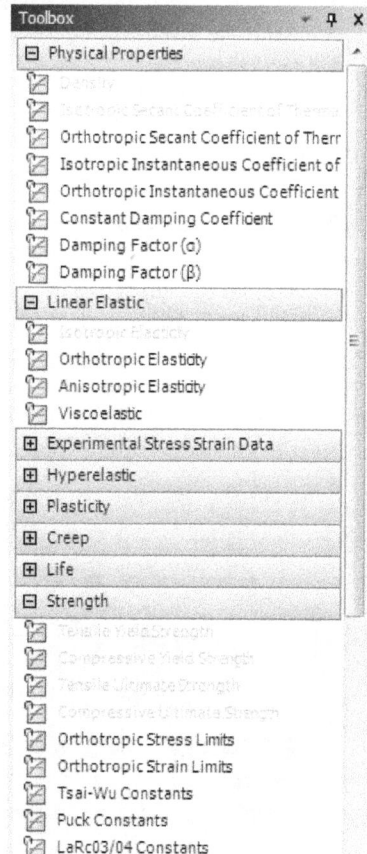

Figure 6-5   Partial view of the Toolbox window

properties. After specifying the name of the material in the **Outline** window, you need to add its properties in the **Properties of Outline** window. To add a property to the newly created material, expand the corresponding toolbox in the **Toolbox** window and double-click on the property to be added to the material; it will be added under the **Property** column in the **Properties of Outline** window. You can specify a value for the property in the corresponding **Value** column. Similarly, to specify a desired unit for the property, select it from the drop-down list available in the **Unit** column in the **Properties of Outline** window.

TUTORIALS

Tutorial 1

In this tutorial, you will open the file *c03_ansWB_tut03* from *C:\ANSYS_WB\c03\Tut03* and then save it with the name *c06_ansWB_tut01*. Figure 6-6 shows the model for the tutorial. Next, you will add the **Static Structural** analysis system to the **Project Schematic** window and then assign Steel material to the model. The properties of the Steel material are given next.

(Expected time: 30 min)

Properties of Steel:

Density:	8100 Kg/m3
Young's Modulus:	1.9E+11 Pa
Poisson's ratio:	0.27
Tensile strength:	2.1E+8 Pa

Figure 6-6 *Model for Tutorial 1*

The following steps are required to complete this tutorial:

a. Open an existing project and save it.
b. Add an analysis system and share geometry from the existing project.
c. Create a new material.
d. Apply the material to the model.
e. Save the project.

Opening an Existing Project and Saving it

Before starting the tutorial, you need to open an existing project and save it with a new name. Next, you need to add the **Static Structural** analysis system to the **Project Schematic** window.

1. Start ANSYS Workbench to display the **Workbench** window.

2. In the **Workbench** window, choose the **Open Project, Archive or Script** button from the **Main** toolbar; the **Open** dialog box is displayed.

3. Browse to the folder *C:\ANSYS_WB\c03\Tut03* and then double-click on **c03_ansWB_tut03**; the *c03_ansWB_tut03* is opened in the **Workbench** window.

4. In the **Workbench** window, choose the **File > Save As** from the Menu bar; the **Save As** dialog box is displayed.

5. Browse to *C:\ANSYS_WB* folder and then create a new folder with the name **c06**.

6. Browse to *C:\ANSYS_WB\c06* and then create a folder with the name **Tut01**.

7.   Next, browse to the *Tut01* folder and save the project with the name **c06_ansWB_tut01**.

Adding the Static Structural Analysis System and Sharing Geometry

Notice that in the **Workbench** window, the **Clamp** component system is already displayed. You need to share the geometry created in the **Clamp** component system with an analysis system to assign material to it.

1.   Drag and drop the **Static Structural** analysis system to the right of **Clamp** component system from the **Analysis Systems** toolbox in the **Toolbox** window; the **Static Structural** analysis system is added to the **Project Schematic** window. Note that a green tick mark is displayed in the **Engineering Data** cell in the **Static Structural** analysis system, indicating that this cell is satisfied.

2.   Next, click and drag the **Geometry** cell of the **Clamp** component system; two cells: **Geometry** and **Model** are highlighted with a green outline, as shown in Figure 6-7, indicating that the existing geometry can be shared with the **Geometry** cell or with the **Model** cell of the **Static Structural** analysis system.

3.   Drag and drop the **Geometry** cell from the **Clamp** component system to the **Geometry** cell of the **Static Structural** analysis system, refer to Figure 6-8; the geometry is shared.

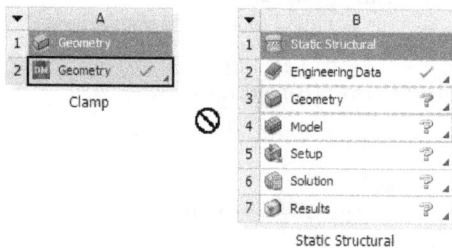

*Figure 6-7 The **Static Structural** analysis system with the **Geometry** and **Model** cells highlighted*

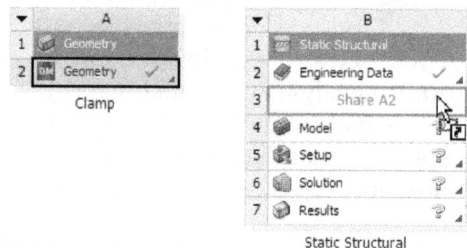

*Figure 6-8 Sharing the geometry with the **Geometry** cell of the **Static Structural** analysis system*

Creating a New Material

The default material that is available in ANSYS Workbench is Structural Steel. In this section, you will create a new material.

1.   Double-click on the **Engineering Data** cell in the **Static Structural** analysis system; the Engineering Data workspace is invoked.

2.   In the **Outline** window, click on the **Click here to add a new material** edit box and enter **Steel** as the name of the new material.

3.   Press ENTER to confirm the name; a new row is created and displays the Steel material along with a question symbol attached to it under the **Contents of Engineering Data** column of the Engineering Data, as shown in Figure 6-9. Also, the **Properties of Outline** window for the Steel material is displayed.

4. Expand the **Physical Properties** toolbox in the **Toolbox** window to display the physical properties of materials.

5. Double-click on **Density** under the **Physical Properties** toolbox, refer to Figure 6-10; **Density** is added to the **Properties of Outline** window displayed for the Steel material.

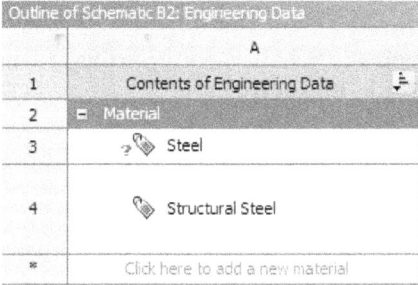

Figure 6-9 *The* *Steel* *material added to the* *Outline* *window*

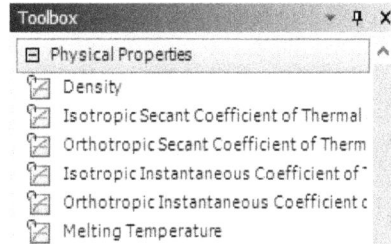

Figure 6-10 *Choosing* *Density* *from the* *Physical Properties* *toolbox*

6. Next, expand the **Linear Elastic** toolbox in the **Toolbox** window; physical properties under this toolbox are displayed.

7. Double-click on the **Isotropic Elasticity** under the **Linear Elastic** toolbox, refer to Figure 6-11; **Isotropic Elasticity** is added to the **Properties of Outline** window.

8. Expand the **Strength** toolbox in the **Toolbox** window; all the properties available under this toolbox are displayed in it.

9. Double-click on the **Tensile Yield Strength** property under the **Strength** toolbox, as shown in Figure 6-12; the **Tensile Yield Strength** node is added to the **Properties of Outline** window.

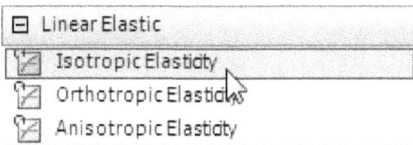

Figure 6-11 *Choosing* *Isotropic Elasticity* *from the* *Linear Elastic* *toolbox*

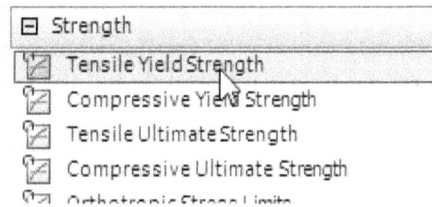

Figure 6-12 *Choosing the* *Tensile Yield Strength* *from the* *Strength* *toolbox*

Note that all the properties that are added to the **Properties of Outline** window have a question symbol attached to them indicating that the values for such properties are yet to be specified.

10. Click on the **Value** field corresponding to the **Density** property in the **Properties of Outline** window and then enter **8100**.

Note

*1. If the **Value** field is blank, it is highlighted in yellow. When a correct value is entered in the **Value** field, the color turns white.*

*2. You can select the desired unit as the value for the properties from the **Unit** drop-down list under the **Unit** field in the **Properties of Outline** window.*

11. Expand the **Isotropic Elasticity** node under the **Property** field in the **Properties of Outline** window; five more rows are added under the **Isotropic Elasticity** node.

12. Click on the down arrow in the **Value** field corresponding to the **Derive from** property; a drop-down list is displayed.

13. Select **Young's Modulus and Poisson's Ratio** from the drop-down list if not already selected, as shown in Figure 6-13.

This option is selected so that the Isotropic Elasticity property of steel can be derived by using the Young's Modulus and the Poisson's ratio. Note that the values for Young's Modulus and the Poisson's ratio are given in the tutorial description at the beginning of this tutorial.

14. Enter **1.9E+11** in the **Value** field corresponding to the **Young's Modulus** property in the **Properties of Outline** window.

15. Enter **0.27** in the **Value** field corresponding to the **Poisson's Ratio** property; the corresponding values of the **Bulk Modulus** and **Shear Modulus** properties are updated in the **Value** field of the **Properties of Outline** window.

16. In the **Properties of Outline** window, enter **2.1E+8** in the **Value** field corresponding to the **Tensile Yield Strength** property; the question symbol placed before **Steel** in the **Outline** window is vanished, indicating that the Steel material can be used.

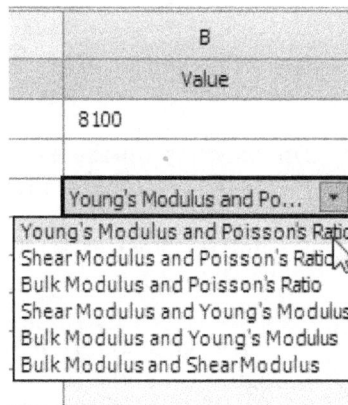

*Figure 6-13 Selecting the **Young's Modulus and Poisson's Ratio** option from the drop-down list*

Applying the Material to the Model

After creating the new material in the Engineering Data workspace, you need to assign this material to the Clamp model.

1. Choose the **Project** button from the **Main** toolbar, as shown in Figure 6-14; the **Project Schematic** window is displayed with the **Clamp** component system and the **Static Structural** analysis system.

Figure 6-14 *Choosing the **Project** button*
*from the **Standard** toolbar*

After creating the new material, you need to apply this material to the model.

2. In the **Workbench** window, choose the **Refresh Project** button from the **Tab** toolbar to refresh the project.

3. Double-click on the **Model** cell of the **Static Structural** analysis system; the **Mechanical** window is displayed.

4. In the **Outline** pane, expand **Geometry** node and select the **Solid** node to display the **Details of "Solid"** window.

5. In the **Details of "Solid"** window, expand the **Material** node if not already expanded to display the **Assignment** dropdown list.

   You will notice in the **Assignment** drop-down list, the structural steel material is applied by default.

6. Next, click on the right arrow on the right of the **Assignment**; **Engineering Data Materials** toolbar is displayed, as shown in Figure 6-15.

7. Select the **Steel** option from the drop-down list, refer to Figure 6-15; the new material is assigned to the model.

8. Close the **Mechanical** window; the **Workbench** window is displayed.

Figure 6-15 *Selecting the **Steel**
option from the drop-down list*

Save the Project and Exit ANSYS Workbench

After assigning material to the model, you now need to exit the ANSYS Workbench session.

1. Choose the **Save Project** button available in the **Main** toolbar in the **ANSYS Workbench**
    window to save the changes made to the project.

2. Choose **File > Exit** to exit the **Workbench** window and close the ANSYS Workbench session.

Tutorial 2

In this tutorial, you will create the beam structure, as shown in Figure 6-16. The dimensions
of the model are given in Figure 6-17. After creating the model you will assign it an
existing Stainless Steel material from the library.   **(Expected time: 30 min)**

Figure 6-16 *Model for Tutorial 2*

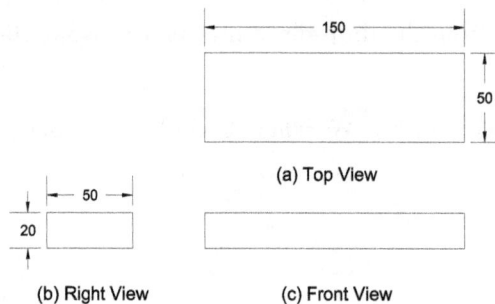

Figure 6-17 *Views and dimensions of the sketch of the model*

The following steps are required to complete this tutorial:

a. Start ANSYS Workbench session and add the **Static Structural** analysis system.
b. Create the model.
c. Add Stainless Steel material to the Engineering Data.
d. Assign the material to the model.

Starting ANSYS Workbench and Adding the Static Structural Analysis System

Before you start the tutorial, you need to start ANSYS Workbench. Next, you need to add **Static Structural** analysis system to the **Project Schematic** window.

1. Start the ANSYS Workbench session.

2. Add the **Static Structural** analysis system to the **Project Schematic** window, refer to Figure 6-18.

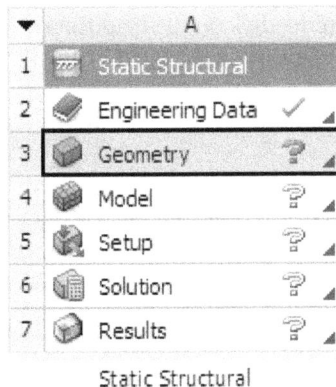

*Figure 6-18 The **Static Structural** analysis system*

3. Choose the **Save Project** tool from the **Main** toolbar; the **Save As** dialog box is displayed.

4. In this dialog box, browse to the location *C:\ANSYS_WB\c06* and then create a folder with the name **Tut02**.

5. In the *Tut02* folder, save the project with the name **c06_ansWB_tut02**.

Creating the Model

After adding the **Static Structural** analysis system to the **Project Schematic** window, you now need to create the model.

1. In the **Static Structural** analysis system, right-click on the **Geometry** cell; Choose **New DesignModeler Geometry**; the **DesignModeler** window gets opened.

2. Choose the **Millimeter** option from the **Units** menu of the **Menu** bar.

3. Next, in the Outline pane of the **DesignModeler** window, select **XYPlane** to make it an active plane.

4. Choose the **Look At** tool in the **Graphics** toolbar; the XY plane is oriented normal to the viewing direction.

5. Draw the sketch shown in Figure 6-17 according to the dimensions given.

6. Extrude the sketch to create the model by using the **Extrude** tool from the **Features** toolbar, refer to Figure 6-16.

7. Exit the **DesignModeler** window; the **Workbench** window is displayed.

Adding Material to the Engineering Data

After a new project is created, you need to add material to it.

1. In the **Workbench** window, double-click on the **Engineering Data** cell in the **Static Structural** analysis system; the Engineering Data workspace is displayed.

2. Choose the **Engineering Data Sources** toggle button; the **Engineering Data Sources** window is added to the workspace.

3. In the **Engineering Data Sources** window, click on the **General Materials** library in the **Data Source** column; the materials in this library are displayed in the **Outline** window, refer to Figure 6-19.

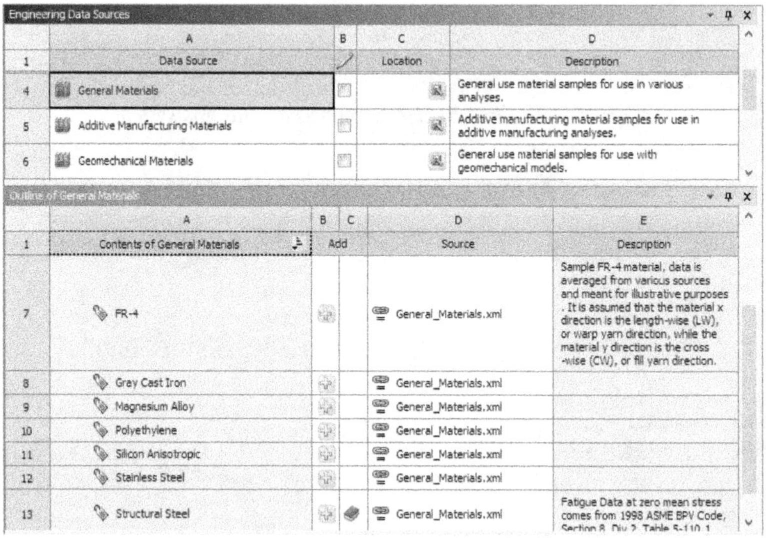

Figure 6-19 *Partial view of the **Engineering Data Sources** displaying various materials in the **General Materials** library*

4. In the **Outline** window, select **Stainless Steel** from the **Contents of Engineering Materials** column; the properties of Stainless Steel material are displayed in the **Properties of Outline** window.

5. Click on the plus (⊞) sign displayed in the **B** column displayed under the **Add** column in the **Outline** window, refer to Figure 6-20; the material from the **Generals** library is added to the Engineering Data. Also, a book icon (📖) is displayed in the **C** column under the **Add** column in the **Outline** window.

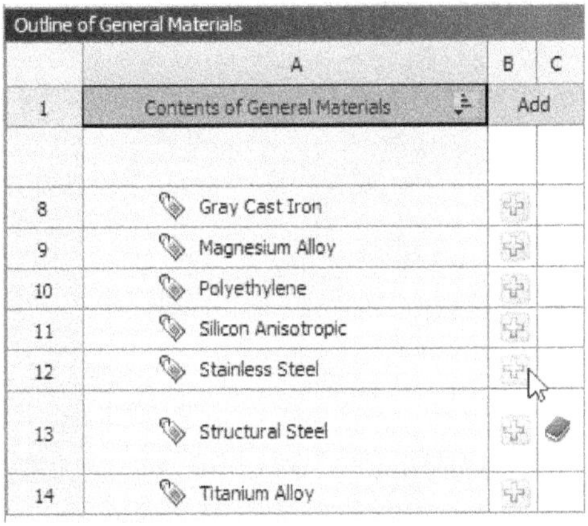

*Figure 6-20 Choosing the plus sign from the **B** column of the **Outline** window*

6. Next, choose the **Engineering Data Sources** toggle button to close the **Engineering Data Sources** window; **Stainless Steel** row is added to the **Outline** window in the **Contents of Engineering Data** section, as shown in Figure 6-21.

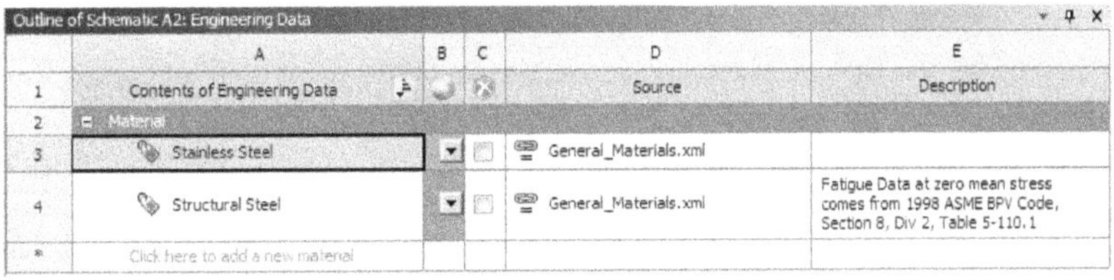

*Figure 6-21 The **Outline** window displaying the newly added Stainless Steel material*

7. Choose the **Project** button in the **Standard** toolbar; the Engineering Data workspace disappears and the **Project Schematic** window with the **Static Structural** analysis system is displayed.

    After the material is added to the Data Source, you now need to apply this new material to the model.

Assigning Material to the Model

You now need to assign the material to the model.

1. Double-click on the **Model** cell in the **Static Structural** analysis system; the **Mechanical** window is displayed.

2. Select **Solid** from the **Outline** pane, as shown in Figure 6-22; the corresponding options are displayed in the **Details of "Solid"** window.

3. In the **Details of "Solid"** window, click on the right arrow displayed on the right of **Assignment** option; **Engineering Data Materials** toolbar is displayed, as shown in Figure 6-23.

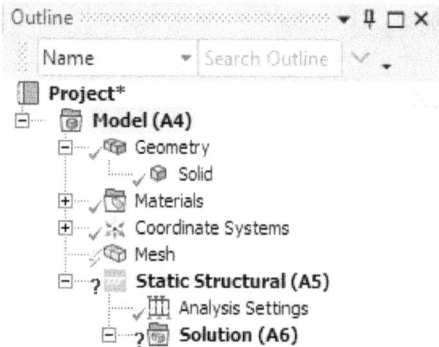

Figure 6-22 *Selecting* ***Solid*** *from the* ***Outline*** *pane*

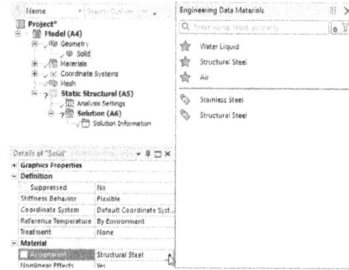

Figure 6-23 *Selecting the* ***Stainless Steel*** *option from the drop-down list*

4.   Select **Stainless Steel** from the drop-down list; the material is applied to the model.

5.   Exit the **Mechanical** window; the **Workbench** window is displayed.

Assigning Material to the Model

You now need to save the project and exit the ANSYS Workbench session.

1.   Choose the **Save Project** button from the **Main** toolbar; the project is saved as *c06_ansWB_tut02*.

2.   Choose **File > Exit** to close the **Workbench** window and exit the ANSYS Workbench session.

Tutorial 3

In this tutorial, you will download the *c06_ansWB_Tut03.zip* file from *www.cadcim.com*. After downloading, you will extract it and then import the *c06_ansWB_tut03.igs* file into ANSYS Workbench. Next, you will apply different materials to different components of the assembly. Figure 6-24 shows the imported assembly. After importing the assembly, rename the components of the assembly in **DesignModeler**, refer to Figure 6-25. Note that the Slider Guide, Base, and Knob are made of Grey Cast Iron, whereas the Slider, Spindle Screw, and Handle are made up of Mild Steel. The properties of Mild Steel and Gray Cast Iron are given next. Also, export Mild Steel material from ANSYS material data library.   **(Expected time: 45 min)**

Properties of Mild Steel:

Density:	7850 Kg/m³
Tensile Yield Strength:	231.94 MPa
Compressive Yield Strength:	407.7 MPa

Properties of Gray Cast Iron:

Density:	7200 Kg/m³
Young's Modulus:	1.1E+11

Poisson's Ratio: 0.28
Bulk Modulus: 8.33E+10

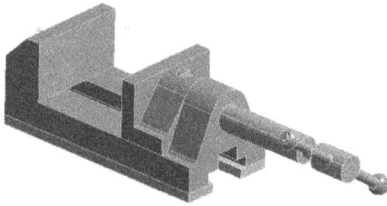

Figure 6-24 Bench Vice Assembly

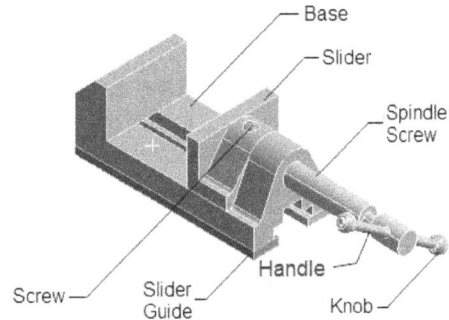

Figure 6-25 Various components of the Bench Vice assembly

The following steps are required to complete this tutorial:

a. Download the part file and import it to ANSYS Workbench.
b. Add materials to the Engineering Data workspace.
c. Assign materials to components.
d. Export a material.
e. Save the project.

Downloading the Part File and Importing it into the Workbench

You need to download the part file from *www.cadcim.com*.

1. Create a folder with the name **Tut03** at the location *C:\ANSYS_WB\C06*.

2. Download the *c06_ansWB_tut03.zip* file from *www.cadcim.com*. The complete path of the file is:

    Textbooks > CAE Simulation > ANSYS > ANSYS Workbench 2023 R2: A Tutorial Approach > Input Files

    After downloading, extract it to save the *c06_ansWB_tut03.igs* to the folder *C:\ANSYS_WB\ c06\Tut03*.

3. Open ANSYS Workbench 2023 R2.

4. Add the **Static Structural** analysis system to the **Project Schematic** window.

5. Right-click on the **Geometry** cell of the **Static Structural** analysis system; a shortcut menu is displayed.

6. From this shortcut menu, choose **Import Geometry > Browse**; the **Open** dialog box is displayed.

7. Browse to *C:\ANSYS_WB\c06\Tut03* and then select **c06_ansWB_tut03.igs.** Next, choose the **Open** button from the **Open** dialog box; the file is imported into ANSYS Workbench. Also, a green tick mark is placed before the **Geometry** cell in the **Static Structural** analysis system.

Adding Materials to the Engineering Data Workspace

Before generating a mesh for the model, you need to define materials for different components of the assembly. In case, you need a material that is not already included in ANSYS Workbench materials libraries, then you need to create that material.

1. Double-click on the **Engineering Data** cell in the **Static Structural** analysis system in the **Project Schematic** window; the **Project Schematic** window is replaced by the Engineering Data workspace.

2. Choose the **Engineering Data Sources** toggle button from the **Tab** toolbar; the **Engineering Data Sources** window is added to the Engineering Data workspace.

3. In the **Engineering Data Sources** window, click on the **General Materials** under the **Data Source** column; the materials included in the **General Materials** library are displayed in the **Outline** window.

4. In the **Outline** window, click on the plus (➕) button available on the right of **Gray Cast Iron** under the **Contents of General Materials** column; the material is added to the Engineering Data and is available to be used.

5. Repeat step 4 to add the Stainless Steel material to the Engineering Data.

   Notice that if Mild Steel material is not available in the library, you need to create a new material in the **General Materials** library, assign it the desired properties and then name it as **Mild Steel**.

6. Again, click on the **Engineering Data Sources** toggle button from the **Tab** toolbar; the **Outline of Schematic A2: Engineering Data** is invoked, as shown in Figure 6-26.

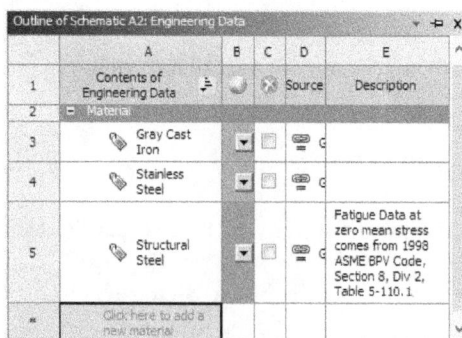

Figure 6-26 *The new field displayed in the*
Outline of General Materials *window*

7. Click on the **Click here to add a new material** field and then enter **Mild Steel** in it; **Mild Steel** material with a question symbol is attached to it added under the **Contents of Engineering Data** column in the **Outline** window. The question symbol indicates that the properties for this material have not been specified.

8. In the **Toolbox** window, expand the **Physical Properties** toolbox if not already expanded.

9. Double-click on **Density** in the **Physical Properties** toolbox; **Density** is ⧉ Density displayed in the **Properties of Outline** window.

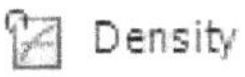

10. Expand the **Strength** toolbox if not expanded.

11. Add Tensile Yield Strength and Compressive Yield Strength properties to the new material.

    Note that the unit of the properties are given in MPa for Tensile Yield Strength and Compressive Yield Strength. First, specify a desired unit for the property from the drop-down list available in the **Unit** column in the **Properties of Outline** window and then enter the value of the property.

12. In the **Properties of Outline** window, specify the values of the parameters as follows:

    Density: 7850 Kg/m^3 **Tensile Yield Strength**: 231.94 MPa

    Compressive Yield Strength: 407.7 MPa

    Notice that **Gray Cast Iron**, **Structural Steel**, **Stainless Steel** and **Mild Steel** materials are added to the **Outline** window, as shown in Figure 6-27. Also, you can verify the properties of **Gray Cast Iron** in the **Properties of Outline** window with the values specified in the problem statement above.

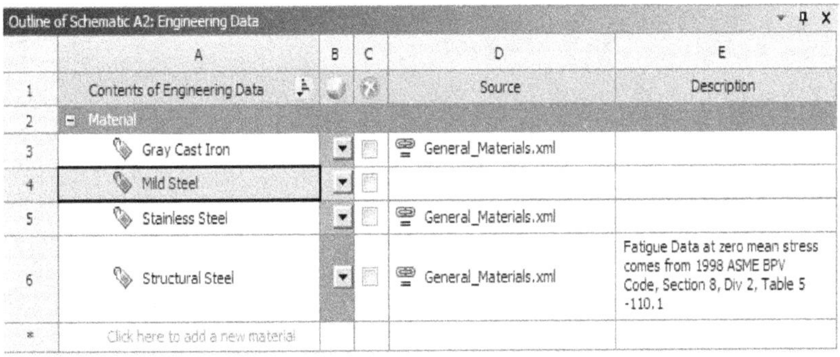

*Figure 6-27   The **Outline** window with the **Mild Steel** material properties displayed in the **Properties** window*

13. Choose the **Project** tab; you are directed to the **Project Schematic** window.

Assigning Material to Components

After all the materials are created and added to the **Outline** window, you now need to assign the materials to different components.

1. Double-click on the **Model** cell in the **Static Structural** analysis system; the **Mechanical** window is displayed.

2. Expand the **Geometry** node in the **Outline** pane; all the components available in the assembly are displayed under the **Geometry** node.

Note

*If name of the parts displayed in the **Geometry** node are different from the names mentioned in Figure 6-25, then rename them accordingly.*

3. Select both **Slider Guides**, **Base**, and **Knob** from the **Outline** window, refer to Figure 6-28; the **Details of "Multiple Selection"** window is displayed with the options related to these geometries

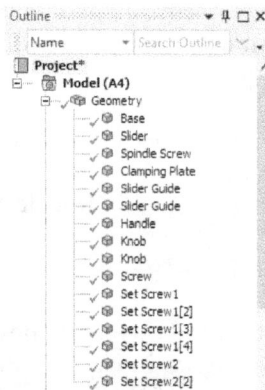

*Figure 6-28 Selecting the components from the **Outline** window*

4. In this window, select **Gray Cast Iron** from the **Assignment** drop-down list to assign the material to the selected components, as shown in Figure 6-29.

    The **Assignment** drop-down list displays all the materials that are added to the Engineering Data.

5. Similarly, select all instances of **Screw** and **Clamping Plate** from the Outline pane and then apply Stainless Steel material to them, refer to Figure 6-29.

6. Next, select **Slider**, **Spindle Screw**, and **Handle** from the Outline pane and apply Mild Steel material to them.

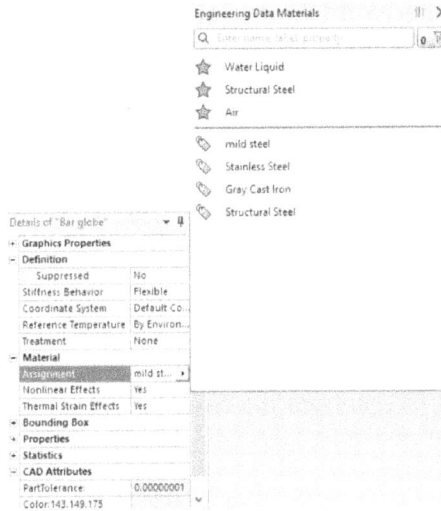

*Figure 6-29 Selecting the **Gray Cast Iron** option from the drop-down list*

7. Exit the **Mechanical** workspace; the **Project Schematic** window is displayed.

8. Choose the **Update Project** button from the **Tab** toolbar to update the project. [Update Project]

Exporting a Material from Material Data Library

After assigning materials to different components of the assembly, you need to export the mild steel material from Engineering Data workspace.

1. Double-click on the **Engineering Data** cell in the **Static Structural** analysis system in the **Project Schematic** window; the **Project Schematic** window is replaced by the **Engineering Data** workspace.

2. In the **Outline** window, select the **Suppression** (⊡) check box available on the right of **Gray Cast Iron** under the **Suppression** column, refer to Figure 6-30, indicating that Gray Cast Iron material is suppressed.

3. Repeat step 2 to suppress materials: **Stainless Steel** and **Structural Steel**, refer to Figure 6-30.

*Figure 6-30 The **Outline** window displaying suppressed materials*

4. In the **Outline** window, select **Mild Steel** under the **Contents of Engineering Data** column and choose **File > Export Engineering Data** to export the material from Engineering Data workspace; the **Save As** dialog box is displayed.

5. Browse to the folder *C:\ANSYS_WB\c06\Tut03* and then save the file with the name *Mild Steel*.

6. Choose the **Project** tab; you are directed to the **Project Schematic** window.

Saving the Model

After exporting the material, you need to save the model. Follow the steps given next to save the model.

1. Choose the **Save Project** button from the **Main** toolbar; the **Save As** dialog box is displayed.

2. Browse to the location *C:\ANSYS_WB\c06* and create a folder with the name *Tut03*.

3. Browse to the folder *C:\ANSYS_WB\c06\Tut03* and then save the project with the name *c06_ansWB_tut03*.

4. Exit the **Workbench** window to close the session.

Self-Evaluation Test

Answer the following questions and then compare them to those given at the end of this chapter:

1. All the common materials are included in the _____ library.

2. You can access the material libraries by choosing the _____ toggle button available in the Standard toolbar.

3. You can insert the Engineering Data standalone system by using the shortcut menu displayed when you right-click in the _____ window.

4. In any system, when the upstream data is updated, you can update the project using the _____ button.

5. The Engineering Data component system is a standalone system. (T/F)

6. You can add a new material even without accessing the libraries available in the Engineering Data Sources window of the Engineering Data workspace. (T/F)

7. The default material available in ANSYS Workbench is Mild Steel. (T/F)

8. The Engineering Data cell can be shared with other standalone systems by selecting it and then dragging it to the respective cell of a system. (T/F)

9. You can change the name of a material at any point of time during the project. (T/F)

Review Questions

Answer the following questions:

1. When you double-click on the **Engineering Data** cell, the _____ workspace is displayed.

2. New material properties can be added to the existing materials using the _____ window.

3. The values of material properties are displayed in the _____ window.

4. To make changes in a library, you first need to select the check box corresponding to the library under the **Edit Library** column. (T/F)

5. The Structural Steel material can be found in the **General Materials** library. (T/F)

6. You can add frequently used materials under the **Favorites** library. (T/F)

7. You can create duplicates of a material available in the library. (T/F)

EXERCISE

Exercise 1

Create the model shown in **Figure** 6-31. Its dimensions are given in Figures 6-32 through 6-34. Next, assign the Aluminium Alloy material from the library to the model.

(Expected time: 45 min)

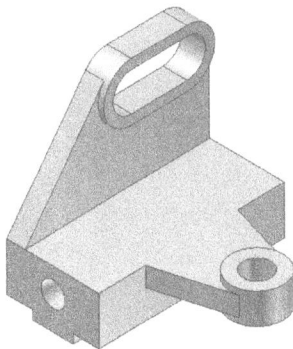

Figure 6-31 Model for Exercise 1

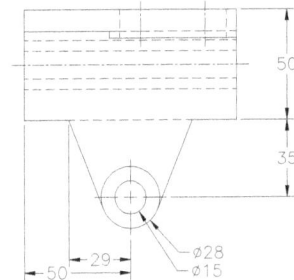

Figure 6-32 Top view of the model

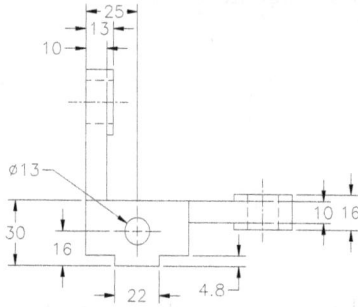

Figure 6-33 Side view of the model

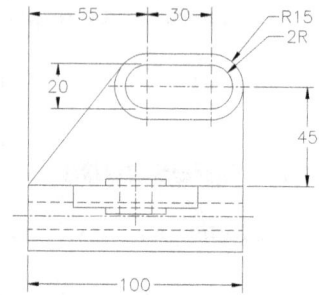

Figure 6-34 Front view of the model

Answers to Self-Evaluation Test
1. General Materials, 2. Engineering Data Sources, 3. Project Schematic, 4. Refresh Project,
5. T, 6. T, 7. F, 8. T, 9. T

Chapter 7

Generating Mesh - I

Learning Objectives

After completing this chapter, you will be able to:

• *Understand the concepts of generating a mesh*
• *Generate meshes for complex models*
• *Generate section views of models*
• *Refine the mesh*
• *Optimize the design of a model*
• *Check the mesh quality*

INTRODUCTION

A mesh is the discretization of a component into a number of small elements of defined size. As discussed in Chapter 1 of this book, finite element analysis would divide the geometry into various small number of elements. These elements are connected to each other at points called nodes. Each node may have two or more than two elements connected to it. A collection of these elements is called a mesh. Figure 7-1 shows a solid model and Figure 7-2 shows mesh created on the solid model.

Figure 7-1 A Solid model

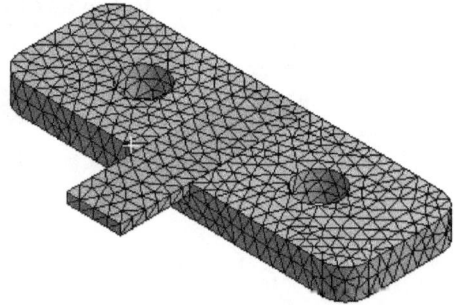

Figure 7-2 Mesh created on the solid model

Meshing is a very important part of pre-processing in any FEA software. In ANSYS Workbench, there are many tools and options available to help you create an effective mesh. An effective mesh is the one that requires less computational time and gives maximum accuracy. In ANSYS Workbench, you can generate mesh with the default settings available when you start the software. Default settings are the ones that are provided by the system based on the geometry to be meshed. You can also set parameters as per your requirement to generate a mesh.

The default mesh generated in the software may produce coarse mesh. Therefore, you need to use the advanced settings to produce a finer mesh. Figures 7-3 and 7-4 show a model with coarse mesh and finer mesh respectively.

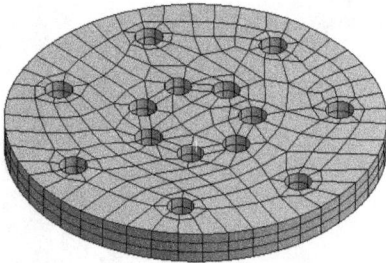

Figure 7-3 A model with coarse mesh

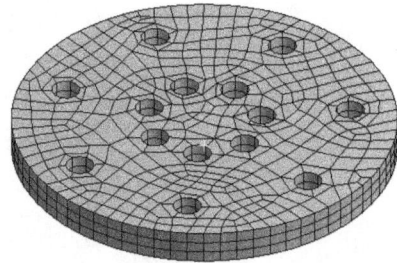

Figure 7-4 Model with fine mesh

In ANSYS Workbench, you can generate a mesh either in the **Mechanical** window (which will be discussed later) or in the **Meshing** window. The **Meshing** window can be invoked by double-clicking on the **Mesh** cell of the **Mesh** component system, refer to Figure 7-5. To invoke the **Mechanical** window, double-click on the **Model** cell of the analysis system, refer to Figure 7-6.

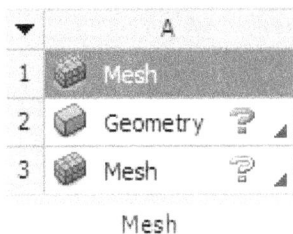

Figure 7-5 *The* **Mesh** *component system*

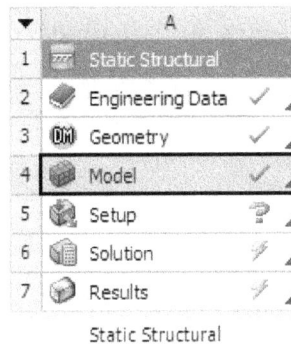

Figure 7-6 *The* **Model** *cell highlighted in the analysis system*

The **Meshing** window and its components are shown in Figure 7-7.

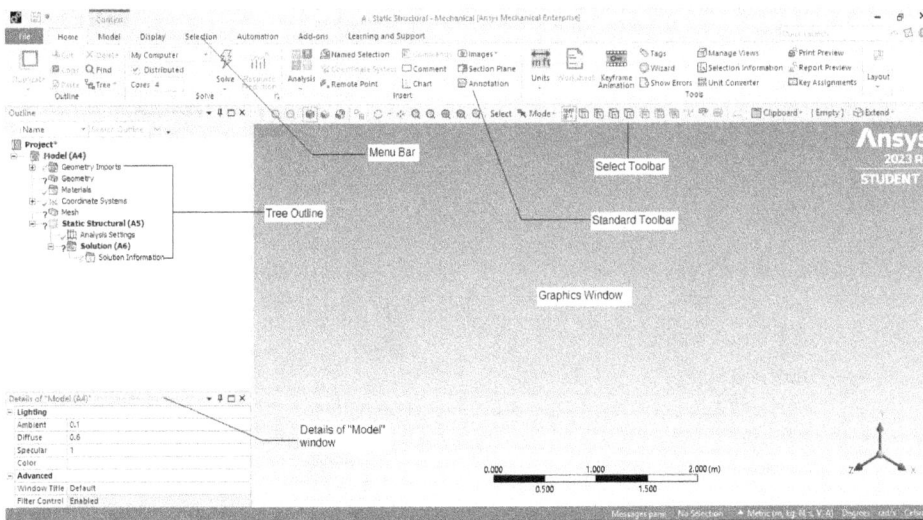

Figure 7-7 *The* **Meshing** *window with various components*

The options available in the **Mechanical** window are similar to the ones available in the **Meshing** window. However, in **Mechanical** window, along with the meshing tools, analysis tools are also available, which can be used for carrying out various analyses in ANSYS Workbench.

The accuracy of the results of an analysis depends a lot on the mesh quality of the model. Ideally, the results obtained from a finite element analysis are more accurate with increased number of elements. However, increased number of elements also increase the processing time required to run an analysis. Therefore, it is always advised to find a balance between the accuracy of results and the processing time required to run the analysis.

Figure 7-8 shows a model constrained around its circumference and loaded centrally. Total Deformation of the model can be determined after the Static Structural analysis is carried out on it. Refer to the model in Figure 7-3 with coarse mesh quality. After performing a Static Structural analysis on this model, the corresponding Legend and the deformed model will be as shown in Figures 7-9 and 7-10, respectively.

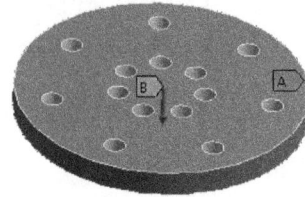

Figure 7-8 *Applied constraints and loads*

Similarly, refer to the model Figure 7-4 with fine mesh. After performing a Static Structural analysis on this model, the corresponding Legend and the deformed model are shown in Figures 7-11 and 7-12, respectively.

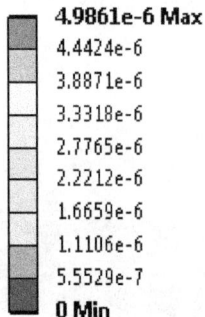

Figure 7-9 *Total deformation contours*

Figure 7-10 *Total deformation on the model*

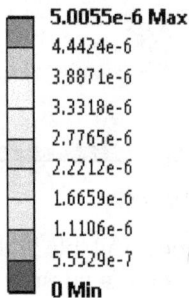

Figure 7-11 *Total deformation contours*

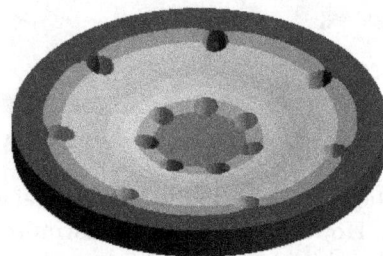

Figure 7-12 *Total deformation on the model*

To understand the importance of generating a mesh which gives better results, compare Figures 7-9 and 7-11. You will observe that the maximum value of Total Deformation in Figure 7-9 is **4.98E-6**, whereas in Figure 7-11, the maximum value of Total Deformation is **5E-6**. Notice that the Total deformation obtained in the model with the finer mesh is more compared to the model with coarse mesh. To understand the importance of generating an effective mesh, you can further reduce the size of the elements, carry out the analysis, obtain the results, and compare the data. Improving the mesh quality is important in the cases where the result obtained are of significance.

Refining the Mesh

After generating mesh for any model, you can increase the number of elements by using various tools and options available in ANSYS Workbench. In the **Details of "Mesh"** window, the **Element Size** option can be used for changing the mesh type from coarse to fine and vice-versa. By default, the element size in ANSYS Workbench is based upon the size of the geometry. But it can be changed by specifying a value in the **Element Size** edit box displayed in the **Details of "Mesh"** window, as shown in Figure 7-13. The number of elements increases as the element size decreases. This means a large element size would produce coarse mesh whereas small element size would produce finer mesh. Finer the mesh, more will be the computational time. To find out about the number of elements created after generating a mesh, you need to expand the **Statistics** node in the **Details of "Mesh"** window, as shown in Figure 7-14. Figure 7-15 shows a model meshed with three different element size values. You will learn about other tools used for generating a mesh later in this chapter.

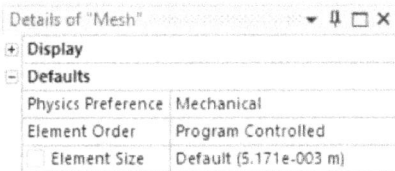

*Figure 7-13 The **Details of "Mesh"** window displaying the **Element Size** edit box*

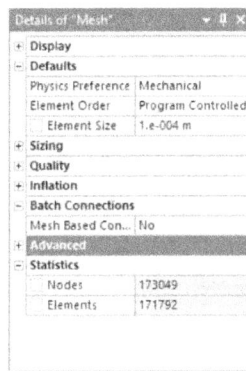

*Figure 7-14 The **Details of "Mesh"** window displaying the number of elements*

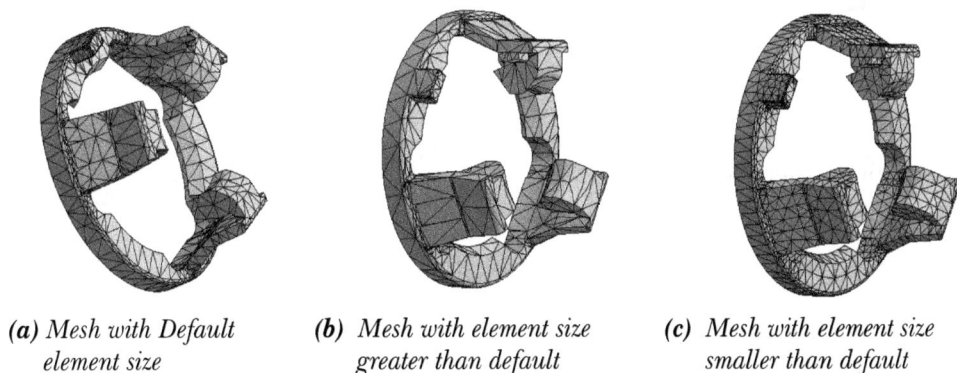

(a) *Mesh with Default element size*

(b) *Mesh with element size greater than default*

(c) *Mesh with element size smaller than default*

Figure 7-15 Meshes created with varying element size

The Decision Making to Find Optimum Results

To achieve optimum results in an analysis, you need to discretize the model into the required number of elements. To do so, decide the number of times upto which a mesh has to be refined to reduce the element size. The decision is very difficult and is based on the experience of the design engineer. For a real complex model, where time is not the primary constraint but accuracy is, an FEA engineer would refine the mesh as long as the analysis shows improved results. However, the time is also an important factor to find a perfect mesh for a model. Figure 7-16 shows a model with its boundary and loading conditions and Figure 7-17 compares the data gathered from various analyses run on this model. In this figure X coordinate represents the element count and the Y coordinate represents the Total Deformation evaluated.

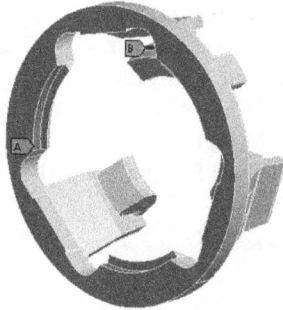

Figure 7-16 *Model with boundary and loading conditions*

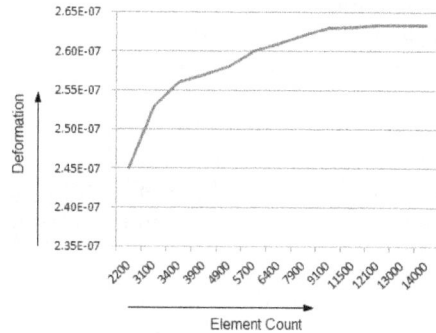

Figure 7-17 *Graph showing the relationship between element count and deformation*

The graph in Figure 7-17 shows the deformation results achieved from an analysis. The X axis represents the number of elements in the model and Y axis represents the Deformation obtained against a particular element count. As you can see that when the element count is 2200, the value of Deformation achieved is 2.45 E-07. As the number of elements increases, the deformation also increases. Note that the maximum value of Deformation is achieved where the element count is 11,500. Hence, it is obvious that meshing the component beyond this point is not required. Meshing the model further contributes to the increase of the runtime of the analysis.

TUTORIALS

Tutorial 1

In this tutorial, you will create the model of a rectangular plate with three holes, refer to Figure 7-18. Next, you will open the **Meshing** window and then generate a mesh with a **Default element size**. The dimensions of the model are shown in Figure 7-19.

(Expected time: 30 min)

The following steps are required to complete this tutorial:

a. Open **ANSYS Workbench** window and add the **Mesh** component system.
b. Create the model.
c. Generate a mesh for the model.
d. Create section views.
e. Save the project and exit ANSYS Workbench.

Figure 7-18 *Rectangular plate with three equispaced holes*

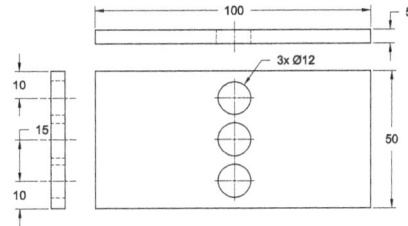

Figure 7-19 *Dimensions of the model for Tutorial 1*

Opening ANSYS Workbench and Adding the Geometry Component System

Before you start the tutorial, it is important that you first open ANSYS Workbench and then add a component system to the **Project Schematic** window.

1. Start an ANSYS Workbench session and then add the **Mesh** component system to the **Project Schematic** window.

2. Create a new folder with the name **c07** at the location *C:\ANSYS_WB*. Now, open the *c07* folder and then create a sub folder in it with the name **Tut01**.

3. Save the project with the name **c07_ansWB_tut01** in this folder.

4. Rename the **Mesh** component system as **Plate with holes**, as shown in Figure 7-20.

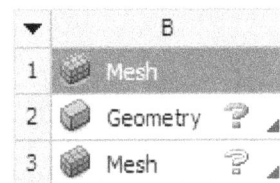

Figure 7-20 *The **Plate with holes** component system*

Creating the Model

After a new project is created, you need to open the **DesignModeler** window to create the model.

1. Right-click on the **Geometry** cell in the **Plate with holes** component system and choose **New DesignModeler Geometry**; the **DesignModeler** window is displayed.

2. Choose **Millimeter** option from the **Units** menu of the **Menu Bar**.

3. In the **DesignModeler** window, select **XYPlane** from the **Tree Outline** pane to specify it as the sketching plane.

4. Orient the plane normal to the viewing direction by using the **Look At** tool.

5. Choose the **Sketching** tab to invoke the **Sketching** mode.

6. Create a rectangle and dimension it, as shown in Figure 7-21.

7. Change *H1* to **50** and *V2* to **5** in the **Details View** window. Next, extrude the sketch up to a depth of 100 by using the **Extrude** tool, then choose the **Generate** tool.

8. Switch to the **Modeling** mode and then change the view to Isometric. Figure 7-22 shows the Isometric view of the Base Plate.

9. Create a new plane on the top face of the extruded feature and orient the view normal to the viewing direction.

Figure 7-21 Rectangle created on the XY plane

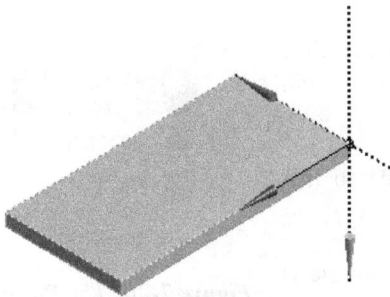

10. Switch to the **Sketching** mode and draw three circles on the top face of the plate. For dimensions of the holes refer to Figure 7-19.

11. Extrude the sketch in such a manner that material is removed from the base plate, as shown in Figure 7-23.

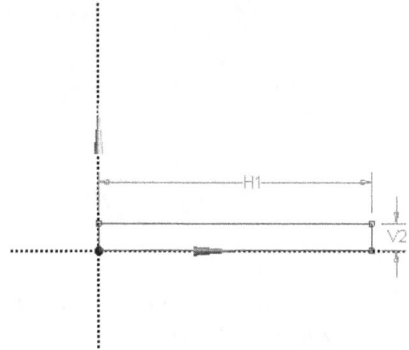

Figure 7-22 Creating a new plane on the top face of the Base Plate

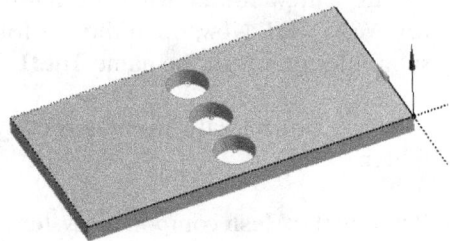

*Figure 7-23 Creating three holes using the **Extrude** tool*

12. Exit the **DesignModeler** window to return to the **Workbench** window.

Generating Mesh for the Plate

Now, you need to go to the **Meshing** window to generate the mesh for the Base Plate.

1. Double-click on the **Mesh** cell in the **Plate with holes** component system and wait for sometime; the **Meshing** window is displayed. Also, you will notice that **Mesh** is displayed in the Outline pane with a yellow thunderbolt attached to it indicating that this field needs to be satisfied.

Note
In this tutorial, you will use the default settings provided in ANSYS Workbench for generating a mesh.

2. Click on **Mesh** in the Outline pane and then right-click to display the shortcut menu.

3. Choose the **Generate Mesh** option from the shortcut menu displayed; the **ANSYS Workbench Mesh Status** window is displayed. After sometime, the **ANSYS Workbench Mesh Status** window disappears and the mesh is generated, refer to Figure 7-24.

Note
*Note that a green tick mark is displayed next to **Mesh** in the Outline pane indicating that this field is satisfied.*

Creating Section Views

After the mesh is generated in the graphics screen, you need to create section view to visualize the element types created.

1. Choose the **Section Plane** tool from the **Insert** panel of the **Home** tab; the **Section Planes** window is displayed, as shown in Figure 7-25.

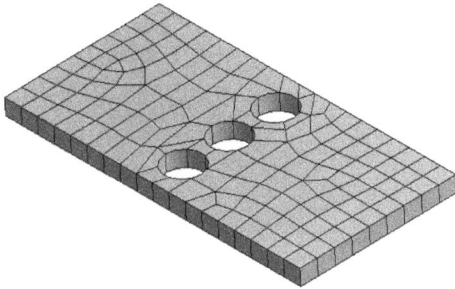

Figure 7-24 Mesh generated for the model

*Figure 7-25 The **Section Planes** window displayed*

2. Right-click in the graphics screen and then choose **View > Top** from the shortcut menu displayed; the view of the model is set to Top.

Now, you need to create a section such that the section plane cuts the three holes and the plate.

3. Move the cursor to a position similar to the one shown in Figure 7-26.

4. Click, hold, and drag the cursor to a position similar to the one shown in Figure 7-27 and then release; the model is sectioned, as shown in Figure 7-28. Also, the **Section Plane 1** check box is selected and is displayed in the **Section Planes** window.

Note that section planes are used to section a model.

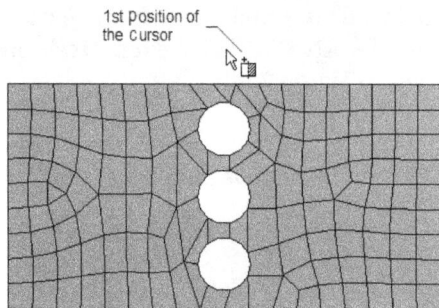

Figure 7-26 *First position of the cursor for creating section planes*

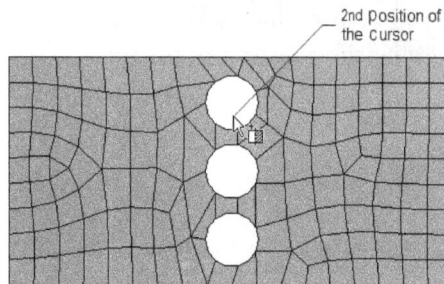

Figure 7-27 *Second position of the cursor for creating section planes*

5. To have a better view of the model and its elements, change the view to Isometric as shown in Figure 7-29.

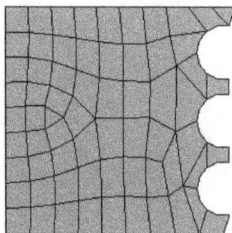

Figure 7-28 *Sectioned model*

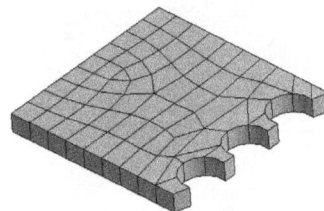

Figure 7-29 *Isometric view of the model*

Note

*1. To add more slice planes, choose the **New Section Plane** button again and then create a slicing plane in the graphics screen by following the steps explained earlier in this tutorial. When you do so, more nodes will be added to the **Section Planes** window.*

*2. To view the complete model again and temporarily hide the section plane, clear the check box corresponding to the **Section Plane 1** in the **Section Planes** window.*

*3. You can also permanently delete the section plane created. To do so, select the **Section Plane1** in the **Section Planes** window and then choose the cross (X) button displayed in the toolbar of the window; the **Section Plane1** is deleted and the model retrieves its complete view.*

6. Close the **Mesh** window; the **Workbench** window is displayed.

Saving the Project and Exiting the Workbench Window

After meshing is done, you need to save the project again and exit the **Workbench** window.

1. Choose the **Save Project** button from the **Standard** toolbar to save the project.

2. Close the **Workbench** window.

Tutorial 2

In this tutorial, you will download the *c07_ansWB_tut02.zip* from *www.cadcim.com* and extract it to save the *c07_ansWB_tut02.stp* file in the project folder. After extracting, import the **STEP** file into ANSYS Workbench. Next, you will generate a mesh and then optimize the model for further use. The model after importing into ANSYS Workbench is shown in Figure 7-30.

(Expected time: 45 min)

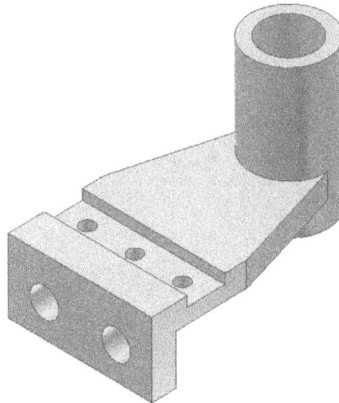

Figure 7-30 Model for Tutorial 2

The following steps are required to complete this tutorial:

a. Download the part file and import it into ANSYS Workbench.
b. Generate a mesh for the model.
c. Create the section of the model.
d. Optimize the model.
e. Create a symmetrical model.
f. Change global mesh control settings.
g. Save the project.

Downloading the Part File and Importing it into the Workbench

Before you start the tutorial, you need to download the part file and import into ANSYS Workbench.

1. Create a folder with the name **Tut02** at the location *C:\ANSYS_WB\c07*.

2. Download the file *c07_ansWB_tut02.zip* from *www.cadcim.com*. The complete path for downloading the file is:

    Textbooks > CAE Simulation > ANSYS > ANSYS Workbench 2023 R2: A Tutorial Approach > Input Files

    After the file is downloaded, extract it to save the *c07_ansWB_tut02.stp* file at the location *C:\ANSYS_WB\c07\Tut02*.

After extracting the file, you need to open ANSYS Workbench and import the file into it.

3.   Open ANSYS Workbench 2023 R2.

4.   Add the **Static Structural** analysis system to the **Project Schematic** window.

5.   Right-click on the **Geometry** cell of the **Static Structural** analysis system; a shortcut menu is displayed.

6.   Choose the **Import Geometry** option; a flyout is displayed, refer to Figure 7-31.

7.   From this flyout, choose the **Browse** option, refer to Figure 7-31; the **Open** dialog box is displayed.

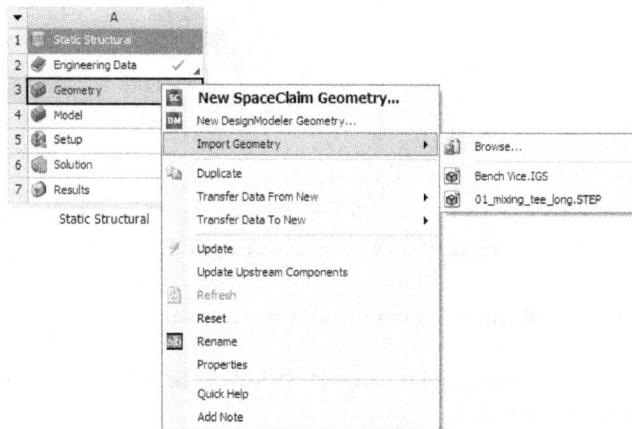

Figure 7-31 *Choosing the **Browse** option from the flyout*

8.   Browse to the location *C:\ANSYS_WB\c07\Tut02* and then select **c07_ansWB_tut02.stp**. Next, choose the **Open** button from the **Open** dialog box; the file is imported into the **Workbench** window.

     Notice that a green tick mark is placed corresponding to the **Geometry** cell in the **Static Structural** analysis system indicating that the geometry for the model is imported into ANSYS Workbench.

9.   Choose the **Save Project** tool from the **Standard** toolbar; the **Save As** dialog box is displayed.

10. Browse to the location *C:\ANSYS_WB\c07\Tut02* and save the project with the name **c07_ansWB_tut02**.

Generating a Mesh for the Model

After the file is imported, you need to generate mesh of the imported geometry.

1.   Double-click on the **Model** cell in the **Static Structural** analysis system; the **Mechanical** window is displayed.

Note

*A yellow thunderbolt symbol is attached to **Mesh** in the Tree Outline indicating that there is no mesh attached to the geometry.*

Now, you will first generate a mesh using the default settings and later use the advanced meshing tools to generate a better mesh.

2. Orient the model by using the tools available in the **Graphics** toolbar.

3. Select **Mesh** from the Outline pane; corresponding options are displayed in the **Details of "Mesh"** window, as shown in Figure 7-32.

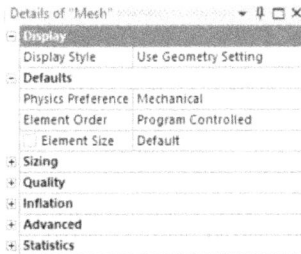

*Figure 7-32 The **Details of "Mesh"** window*

Based on the requirement of analysis, you need to select an option from the **Physics Preference** drop-down list in the **Details of "Mesh"** window. By default, **Mechanical** is selected in this drop-down list, refer to Figure 7-32. Other option in this drop-down list are **Electromagnetics**, **CFD**, **Nonlinear**, **Explicit**, and **Hydrodynamics**.

Leave rest of the value set as default.

4. Right-click on **Mesh** in the Tree Outline; a shortcut menu is displayed, refer to Figure 7-33.

5. Choose the **Generate Mesh** option from the shortcut menu, as shown in Figure 7-33; the **ANSYS Workbench Mesh Status** window is displayed. After sometime, the **ANSYS Workbench Mesh Status** window is closed and the mesh view of the model is displayed in the graphics screen, as shown in Figure 7-34.

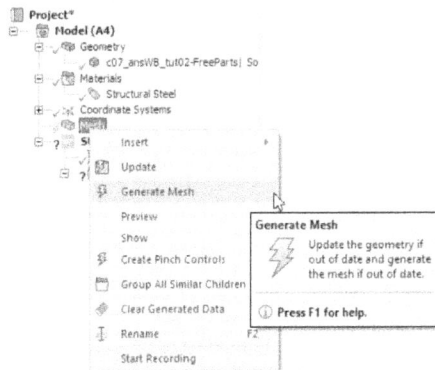

*Figure 7-33 Choosing the **Generate Mesh** option*

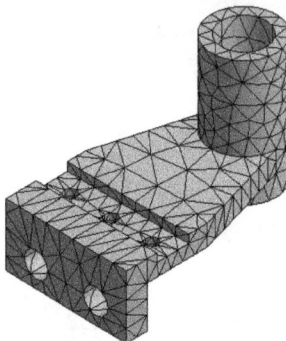

Figure 7-34 *The mesh view of the model*

Note
*The mesh preview is displayed in the graphics screen as long as **Mesh** is selected in the **Outline** pane.
On selecting any other entity in the **Outline** pane, the mesh preview will not be displayed.*

A green tick mark is placed corresponding to **Mesh** in the Outline pane which indicates
that the mesh for the model is created successfully. Note that while generating a mesh for
a model, it is important to understand the effect it causes on the number of elements and
nodes that are created as a result of the meshing process. It is recommended that you keep a
note of the number of elements that are created after generating each mesh. You can repeat
meshing operation to achieve better results.

To get the details about the number of elements and nodes
created after the model is discretized, expand the **Statistics**
node in the **Details of "Mesh"** window, refer to Figure 7-35.

After the mesh is generated, the number of elements will be
1359 and the number of nodes will be 2945. You can use this
data later to compare the effectiveness of the mesh.

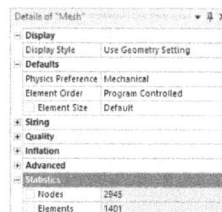

Figure 7-35 *The **Statistics**
node in the **Details of "Mesh"**
window*

Note
*The values displayed in the **Statistics** node of the **Details of
"Mesh"** window may vary in your system.*

Creating the Section View

The quality of mesh is not viewed while working on a flat surface. It is generally viewed at
curved and cylindrical faces. In this tutorial, you will view the quality of mesh by creating a
section of the model.

1. Orient the model by using the tools in the **Graphics** toolbar, as shown in Figure 7-36.

2. Choose the **Section Plane** tool from the **Insert** panel of the **Home** tab; the **Section
    Planes** window is displayed.

3. Place the cursor at the position shown in Figure 7-36.

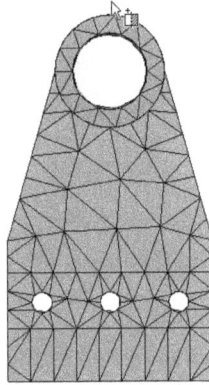

Figure 7-36 *The oriented view of the model*

4. Next, click and drag the cursor downward; the meshed model is sectioned, as shown in Figure 7-37.

5. Use the tools available in the **Graphics** toolbar to orient the view of the model to Isometric, as shown in Figure 7-38.

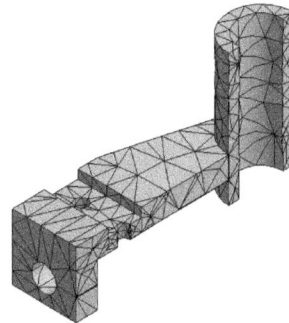

Figure 7-37 *View after the model is sectioned*

Figure 7-38 *Isometric view of the sectioned model*

6. Now, clear the **Section Plane 1** check box in the **Section Planes** window to view the complete meshed view of the model, refer to Figure 7-34.

When the automatic method is used for meshing, different algorithms are applied at different locations of the model, depending upon its complexities. The meshing generates hexahedrals, sweep, or tetrahedrons depending upon the model.

Notice three small holes on the meshed model. Now, if you try to create a finer mesh on this model, the number of elements will increase, which in turn, will increase the runtime of analysis. To decrease the runtime of the analysis, you can fill the holes in the model by using the tools available in the **DesignModeler** window.

7. Exit the **Mechanical** window; the **Workbench** window is displayed.

Optimizing the Model

After the mesh is generated, you need to remove the irregularities in the model by using the **DesignModeler** window.

1. Right-click on the **Geometry** cell and choose the **Edit Geometry in DesignModeler**; the DesignModeler window is displayed. Make sure you set the units to millimeter in the **ANSYS Workbench** dialog box.

2. Choose the **Generate** tool from the **Features** toolbar to generate the model in the graphics screen.   ⸱⫶ Generate

3. Choose the **Fill** tool from the **Tools** menu; **Fill1** is attached to the Outline pane. Also, the corresponding options are displayed in the **Details View** window and you are prompted to select surfaces of the cavity to fill.   🔲 Fill

4. In the **Details View** window, select the **Faces** selection box to display the **Apply** and **Cancel** buttons.

5. Press CTRL and select the three faces of the lower groove and three small holes, as shown in Figure 7-39.

6. Choose the **Apply** button in the **Faces** selection box; **6** is displayed in the **Faces** selection box.

7. Choose the **Generate** tool from the **Features** toolbar to create the fill feature; the preview of the fill is displayed in the graphics screen, as shown in Figure 7-40.

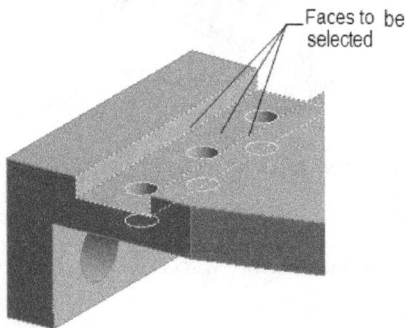

Figure 7-39 Partial view of the model with the faces selected for fill operation

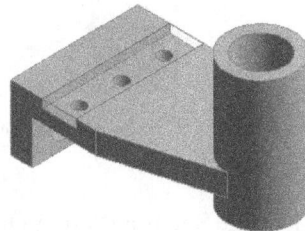

Figure 7-40 Preview of the fill feature displayed in the graphics screen

8. Next, choose the **Boolean** tool from the **Create** menu; **Boolean1** with a yellow thunderbolt is attached to the Outline pane. Also, the corresponding options are displayed in the **Details View** window. And, you are prompted to select tool bodies for the operation.

    Notice that in the **Details View** window, the **Apply** and **Cancel** buttons are displayed in the **Tool Bodies** selection box.

9. Select the fill feature created, refer to Figure 7-41.

10. Now, hold the CTRL key and then select the main body, as shown in Figure 7-42.

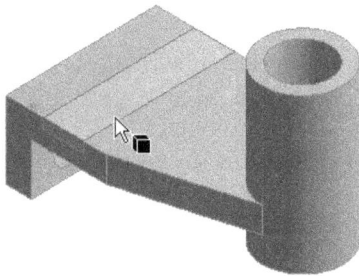

Figure 7-41 *Selecting the fill feature*

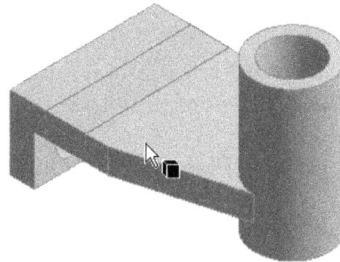

Figure 7-42 *Selecting the main body*

11. Select **Unite** from the **Operation** drop-down list in the **Details View** window if not already selected.

12. Choose the **Apply** button in the **Tool Bodies** selection box; **2 Bodies** is displayed in the **Tool Bodies** selection box indicating that two bodies are selected for the operation.

13. Next, choose the **Generate** tool from the **Features** toolbar; the fill body and the main body merge into a single body, as shown in Figure 7-43.   `Generate`

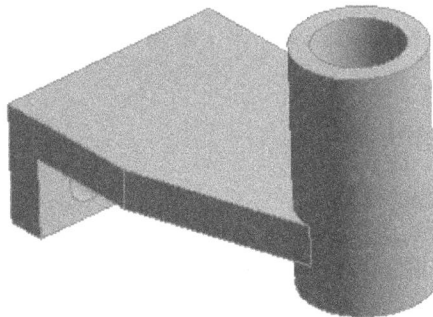

Figure 7-43 *The single body created after uniting two solid bodies*

Figure 7-43 shows the model with the three holes and the groove cut removed. Now, you can generate an effective mesh on it.

Creating a Symmetrical Model

The model is symmetrical in nature, therefore to save processing time, you will cut it into two portions.

1. Right-click in the graphics screen and then choose **View > Back View** from the shortcut menu displayed to orient the model, refer to Figure 7-44.

2. Create a new plane at an offset value **30** mm from **XYPlane** and choose the **Symmetry** tool from the **Tools** menu; **Symmetry1** is attached to the Outline pane and corresponding options are displayed in the **Details View** window. Also, you are prompted to select the symmetry plane.

3. In the **Details View** window, select the **Symmetry Plane1** selection box to display the **Apply** and **Cancel** buttons, if not already displayed.

4. Next, click on **Plane1** in the Outline pane to select the **Plane1** as the section plane.

5. Choose the **Apply** button available in the **Symmetry Plane1** selection box; the Plane1 is selected as the plane for symmetry creation.

6. Choose the **Generate** tool; the symmetrical half of the model is generated, as shown in Figure 7-45.

Figure 7-44 *Choosing the* *Back View* *option from the shortcut menu*

Figure 7-45 *Back view of the symmetrical model*

7. Exit the **DesignModeler** window.

Setting Global Mesh Controls

In this section, you will set global mesh controls for meshing in the **Details of "Mesh"** window. Next, you will change the global mesh control settings to create an effective mesh.

1. Open the **Mechanical** window by double-clicking on the **Model** cell of the **Static Structural** analysis system; the **ANSYS Workbench** message box is displayed as shown in Figure 7-46. Choose **Yes** from this message box; the **Mechanical** window is displayed.

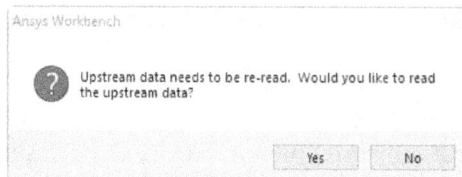

Figure 7-46 *The ANSYS Workbench message box*

2. Orient the model using the tools available in the **Graphics** toolbar, refer to Figure 7-34.

Note
*As the symmetrical model was created in the **DesignModeler** window, **Symmetry** will be added to the Outline pane of the **Mechanical** window. Also, **Named Selection** will be displayed in the Outline pane.*

3. Select **Mesh** in the Outline pane; the **Details of "Mesh"** window is displayed.

4. In the Outline pane, right-click on **Mesh**; a shortcut menu is displayed. $\mathcal{G}$ Generate Mesh
Choose the **Generate Mesh** option.

5. In the **Details of "Mesh"** window, expand the **Sizing** node.

6. In the **Sizing** node, select the **No** option from the **Use Adaptive Sizing** drop-down list. Then select **Yes** from the **Capture Proximity** drop-down list as shown in Figure 7-47.

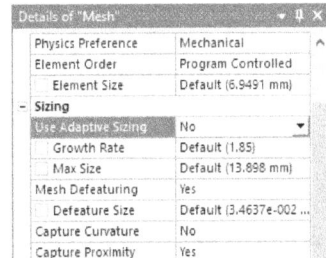

The default selection in this drop-down list is **Yes**. This option is used to generate fine mesh for the model that has small details like fillets and curves.

*Figure 7-47 Selecting the **No** option from the **Use Adaptive Sizing** drop-down list*

7. Enter **2** in the **Proximity Gap Factor** edit box in the **Details of "Mesh"** window.

Note
*You can also use the spinner available in the **Num Cells Across Gap** edit box to set the value.*

The **Proximity Gap Factor** edit box is available only when the **Capture Proximity** parameter set to **Yes** and the **Capture Proximity and Capture Curvature** option are available only when the **Use Adaptive Sizing** parameter set to **No**, refer to Figure 7-48. The value in this edit box signifies the minimum number of cells that will be available in small gaps while meshing. The default value in the **Proximity Gap Factor** edit box is 3. You can enter any number between 1 to 100 in this edit box. Retain the default value and then right-click on **Mesh** in the Outline pane; a shortcut menu is displayed. Choose the **Generate Mesh** option.

Details of "Mesh"	
± Display	
− Defaults	
Physics Preference	Mechanical
Element Order	Program Controlled
Element Size	Default (6.4452 mm)
− Sizing	
Use Adaptive Sizing	No
Growth Rate	1.2
Max Size	Default (12.89 mm)
Mesh Defeaturing	No
Capture Curvature	No
Capture Proximity	Yes
Proximity Min Size	Default (6.4452e-002 mm)
Proximity Gap Factor	2.0
Proximity Size Sources	Faces and Edges
Bounding Box Diagonal	128.9 mm

*Figure 7-48 The **Details of "Mesh"** window with the*
***Proximity Gap Factor** edit box highlighted*

Figure 7-49 shows a model meshed with the **Use Adaptive Sizing** parameter set to **Yes** and all the other parameters left as default. The number of nodes created when this drop-down list is set to **No** is 2890 and the number of elements created is 1359. Figure 7-50 shows the same model with the **Use Adaptive Sizing** parameter set to **No** and the **Capture Proximity** parameter set to **Yes** and all the other parameter left as to default. The number of nodes and elements when this option is selected are 23835 and 14911, respectively.

Also, observe the size of elements at the cylindrical curvature in Figures 7-49 and 7-50. You will notice that component shown in Figure 7-50 has more number of elements as compared to the model shown in Figure 7-49. Also, the model in Figure 7-50 has finer elements at curve and fillets.

*Figure 7-49 Section of the meshed model created with the **Use Adaptive Sizing** parameter set to **Yes***

*Figure 7-50 Section of the meshed model with the **Use Adaptive Sizing** parameter set to **No** and **Capture Proximity** parameter set to **Yes***

8. In the **Outline** pane, right-click on **Mesh**; a shortcut menu is displayed.   ⅀ Generate Mesh
    Choose the **Generate Mesh** option; the mesh is generated with all other options left as default, as shown in Figure 7-51.

9. Now, invoke the **Section Plane** tool from the **Standard** toolbar.

10. Orient the model using the tools available in the **Graphics** toolbar, as shown in Figure 7-52.

Figure 7-51 The mesh generated

Figure 7-52 Oriented view of the model

11. Move the cursor to the location shown in Figure 7-53.

12. Now, click and drag the cursor upward; the sectioned view of the meshed model is displayed. Next, change the view to Isometric, as shown in Figure 7-54.

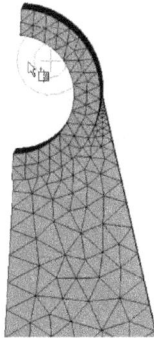

Figure 7-53 Cursor position

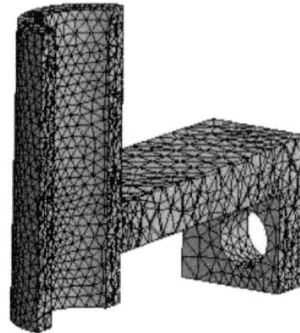

Figure 7-54 Isometric view of the sectioned model

13. From the **Section Planes** window, choose the **Show Whole Elements** button; whole elements are displayed in the model, as shown in Figure 7-55.

14. Now, clear the **Section Plane 1** check box in the **Section Planes** window to view the complete meshed view of the model.

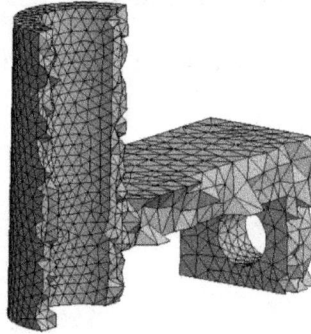

Figure 7-55 Partial view of the model displaying whole elements along the section plane

Checking a Mesh Metric

The **Mesh Metric** option is used to view mesh metric information and thereby evaluate the mesh quality. Once you have generated a mesh, you can choose to view information about any of the following mesh metrics: **Element Quality**, **Aspect Ratio** for triangles or quadrilaterals, **Jacobian Ratio** (MAPDL, corner nodes, or Gauss points), **Warping Factor**, **Parallel Deviation**, **Maximum Corner Angle**, **Skewness**, **Orthogonal Quality**, and **Characteristic Length**. Selecting **None** turns off mesh metric viewing. To access the mesh metric information, you need to perform the following steps.

1.   Select **Mesh** in the Outline pane; the **Details of "Mesh"** window is displayed.

2.   In the **Details of "Mesh"** window, expand the **Quality** node.

3.   In the **Quality** node, select **Element Quality** from the **Mesh Metric** drop-down list.

     By default, the **Min**, **Max**, **Average**, and **Standard Deviation** values for the selected metric are reported under the **Mesh Metric** in the **Details of "Mesh"** window, as shown in Figure 7-56.

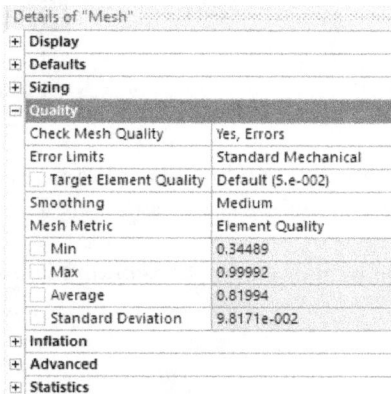

*Figure 7-56 The **Details of "Mesh"** window with details of mesh metric value*

In addition, the **Mesh Metric** window showing bar graphs is displayed under the Geometry window, as shown in Figure 7-57. For this illustration, the **Element Quality** mesh metric was selected in the **Details of "Mesh"** window, so the bar graph displays the minimum to maximum **Element Quality** values over the entire mesh. The graph is labeled with color-coded bars for each element shape represented in the model's mesh, and can be manipulated to view specific mesh statistics of interest. The X-axis represents the value of the selected mesh metric, in this case- Element Quality, and the Y-axis represents the number of elements, refer to Figure 7-57.

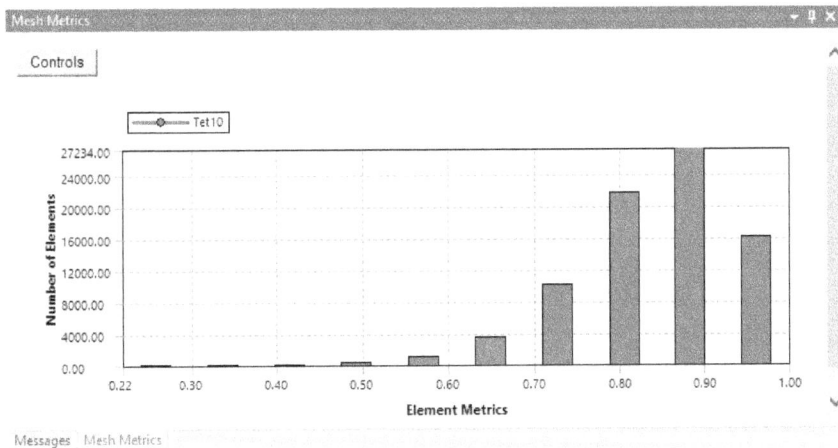

Figure 7-57 *The* **Mesh Metric** *bar graph window*

When you click on an individual bar on the graph (or in the white space above the bar), the view in the Geometry window changes. The geometry becomes transparent and only those elements meeting the criteria values corresponding to the selected bar are displayed, as shown in Figure 7-58.

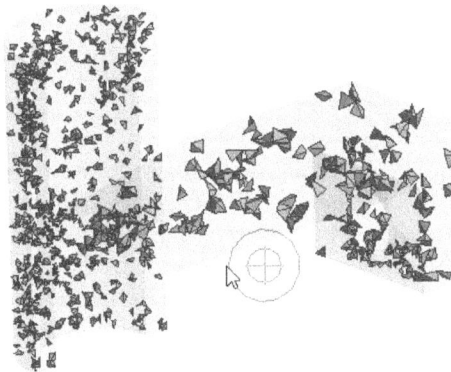

Figure 7-58 *Transparent view of the Geometry after selecting individual bar from the* **Mesh Metric** *window*

If you click and hold the cursor on an individual bar or column, you see a tooltip showing the metric value associated with the bar, along with either a number of elements or the percent of total volume represented by the elements (depending on the settings done for the **Y-Axis** option). For example, refer to Figure 7-59, 0.57 is the mid-point of the range of metric values covered by the selected bar, and there are 972 elements with values that fall within that range. The 972 elements are displayed in the Geometry window.

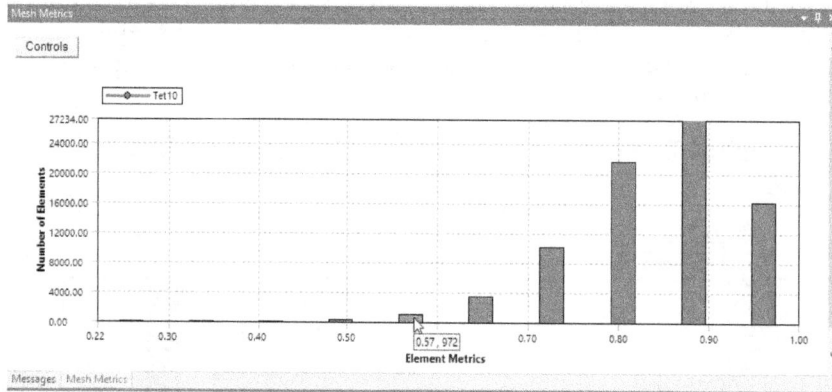

Figure 7-59 Individual bars showing the Metric value associated with the bar

Note

*1. The **Mesh Metric** option is used to display the selected mesh metric without considering the qualifying criteria of the elements acceptability.*

2. To select multiple bars, hold the CTRL key and click all desired bars. All elements corresponding to all selected bars are displayed in the Geometry window.

*3. To change the bar graph settings, choose the **Control** button from the **Mesh Metric** bar graph window.*

*4. Similarly, you can check the mesh quality for different criteria such as **Aspect Ratio**, **Wraping Factor**, **Skewness**, and so on by selecting the option from **Mesh Metric** drop-down list .*

4. Select the **None** option from the **Mesh Metric** drop-down list to turn off the Mesh Metric bar graph window.

Depending upon the accuracy and time allotted for analyzing this particular component, the mesh quality is decided. For example, if you set the value in the **Proximity Gap Factor** edit box to **4**, more cells will be created around curved surfaces, refer to Figure 7-60. The table given next shows the number of elements and the Equivalent Stress generated when different values are specified in the **Proximity Gap Factor** edit box under the boundary and loading conditions shown in Figure 7-61. Depending upon the severity of the analysis and time permitted to run it, you can choose a procedure to mesh the model. When you need to achieve extremely accurate results, you need to generate very small elements around curved faces and fillets.

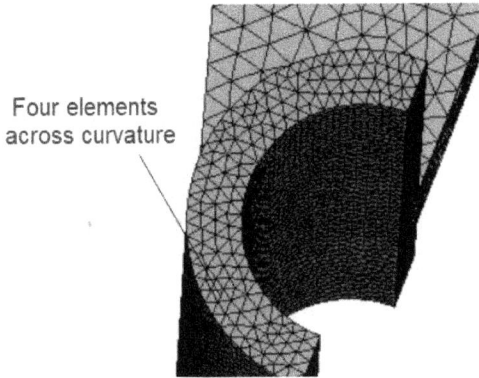

Figure 7-60 Elements around curved surfaces

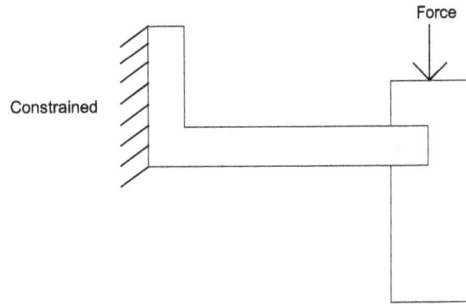

Figure 7-61 Boundary and loading conditions

Values in the Proximity Gap Factor	Number of elements	Equivalent Stress
Default (3)	29388	2.550 E8
1	1865	1.843 E8
2	9105	2.130 E8
6	204769	3.214 E8
10	911274	4.735 E8

Generally depending upon the requirement of the results, a mesh can be generated by changing the default mesh control settings. However, when better results and more user defined mesh is needed, you can introduce various mesh controls that are available within ANSYS Workbench. By using mesh controls, you can control types and shapes of elements, set mesh types, and so on.

14. Exit the **Mechanical** window.

Save the Project

After the **Mechanical** window is closed, you now need to save the project and exit the **Workbench** window.

1. Save the project and exit the **Workbench** window.

Tutorial 3

In this tutorial, you will first download the *c07_ansWB_tut03.zip* file from *www.cadcim.com*. After downloading the file, you will extract it and save the *c07_ansWB_tut03.stp* file into the specified folder. Next, you will create an effective mesh for the model. To simplify the model, you will create a half section of the model. Figure 7-62 shows the model and Figure 7-63 shows the symmetrical half of the model. Use an appropriate method to refine the mesh.

(Expected time: 2 hr)

Figure 7-62 *Model for Tutorial 3*

Figure 7-63 *Section view of the model displaying its intricate features*

The following steps are required to complete this tutorial:

a. Download and import the geometry into ANSYS Workbench.
b. Generate mesh.
c. Optimize the model.
d. Create the symmetrical Model.
e. Generate mesh of the sectioned model.
f. Create a section to visualize mesh quality.
g. Inserting Local Mesh Controls.
h. Save the Project and Exit ANSYS Workbench.

Downloading and Importing the Geometry into ANSYS Workbench

Before starting the tutorial, you need to download the file *c07_ansWB_tut03.zip* from *www.cadcim.com* and then extract it to save the *c07_ansWB_tut03.stp* file.

1. Create a folder named **Tut03** at the location *C:\ANSYS_WB\c07*.

2. Download the stp file *c07_ansWB_tut03.zip* from *www.cadcim.com*. The complete path for the file to be downloaded is:

    Textbooks > CAE Simulation > ANSYS > ANSYS Workbench 2023 R2: A Tutorial Approach > Input Files

    After downloading the zip file, extract it and save the *c07_ansWB_tut03.stp* file at location *C:\ANSYS_WB\c07\Tut03*.

3. Start ANSYS Workbench 2023 R2 from the Start menu; the **Workbench** window is displayed.

4. Add the **Static Structural** analysis system to the **Project Schematic** window.

5. Right-click on the **Geometry** cell of the **Static Structural** analysis system and then choose **Import Geometry > Browse** from the shortcut menu displayed, as shown in Figure 7-64; the **Open** dialog box is displayed.

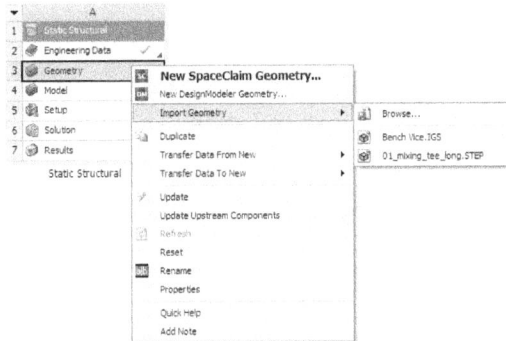

Figure 7-64 *Choosing the **Browse** option from the shortcut menu*

6. Browse to the location *C:\ANSYS_WB\c07\Tut03* and then select **c07_ansWB_tut02.stp**. Next, choose the **Open** button from the **Open** dialog box; the file is imported into the **Workbench** window. Also, a green tick mark is placed corresponding to the **Geometry** cell in the **Static Structural** analysis system indicating that a geometry is specified for the analysis.

7. In the **Workbench** window, choose the **Save project** button; the **Save As** dialog box is displayed.

8. In this dialog box browse to the location *C:\ANSYS_WB\c07\Tut03* and then save the project with the name **c07_ansWB_tut03**.

Generating a Mesh for the Model

Once the file is imported, you now need to assign a material to it and then generate a mesh.

1. Right-click on the **Model** cell in the **Static Structural** analysis system and then choose **Edit** from the shortcut menu displayed, as shown in Figure 7-65; the **Mechanical** window is displayed.

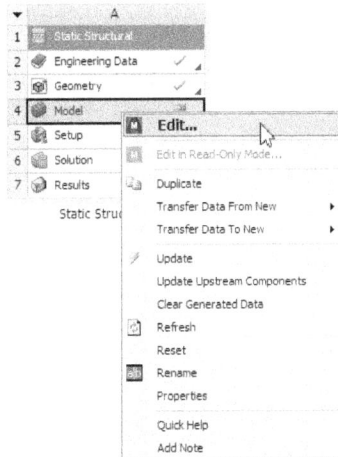

Figure 7-65 *Choosing the **Edit** option from the shortcut menu*

2. In the **Mechanical** window, right-click on **Mesh** in the Outline pane; a shortcut menu is displayed. Also, the **Details of "Mesh"** window is displayed.

3. In the Outline pane, right-click on **Mesh**; a shortcut menu is displayed.   🗲   Generate Mesh Choose the **Generate Mesh** option from it to generate the mesh with default settings provided by ANSYS Workbench, refer to Figure 7-66.

4. In the **Details of "Mesh"** window, expand the **Statistics** node to view the number of nodes and elements in the model.

   After generating the mesh by using the default mesh control settings, the approximate number of elements and nodes generated in the model will be 98175 and 161039, respectively, refer to Figure 7-67.

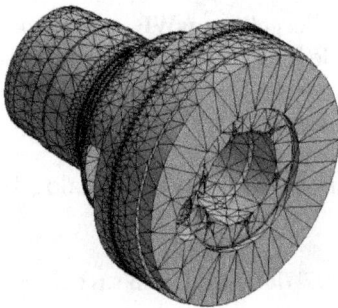

Figure 7-66 *Meshed model*

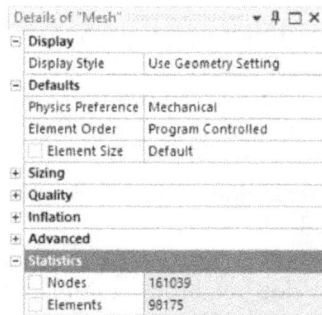

Figure 7-67 *The **Details of "Mesh"** window with the **Statistics** node expanded*

Note

The number of elements and nodes in your system may vary from this tutorial.

You will also notice small features in the model such as holes, chamfers, fillets, and so on, as shown in Figure 7-68. The mesh generated using default settings generates elements for the small features in the model. This increases the element count that in turn increases the computing time. For an effective analysis, you need to reduce the complexities in the model. Filling the holes and deleting the chamfers and fillets will make the model more simpler.

5. Exit the **Mechanical** window; the **Workbench** window is displayed.

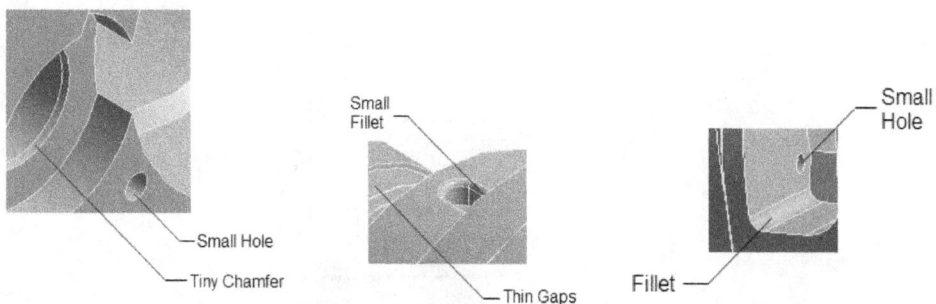

Figure 7-68 *Small features available in the model*

Optimizing the Model

After generating the mesh using the default settings, you need to optimize the model.

1. In the **Project Schematic** window, right-click on the **Geometry** cell and choose **Edit Geometry in DesignModeler**; the **DesignModeler** window gets opened.

2. Choose the **Millimeter** option from the **Units** menu of the **Menu Bar**.

3. In the Outline pane, the **Import1** node is attached with a yellow thunderbolt indicating that you need to generate the geometry to proceed further.

4. Now, choose the **Generate** button in the **Features** toolbar; the model is displayed in the graphics screen. Also, a green tick mark is displayed before **Import1** in the Outline pane.

5. Change the view to Isometric. You will notice that there are six holes on the front of the model, as shown in Figure 7-69 and they need to be eliminated.

Figure 7-69 *Holes on the front of the model*

6. Choose the **Fill** tool from the **Tools** menu; **Fill1** with a yellow thunderbolt is added to the Outline pane. Also, corresponding options of the **Fill** tool are displayed in the **Details View** window and you are prompted to select faces that form holes or cavities.

    The **Fill** tool is used to fill depression lines, uneven surfaces, dents, and holes in the model. To create a fill feature, you need to select surrounding surfaces in such a manner that a frozen or dead material is filled in the selected area.

7. Make sure that **By Cavity** option is selected in the **Extraction Type** drop-down list in the **Details View** window.

    The **By Cavity** option is selected by default in the **Details View** window. This option is used to create fills around selected surfaces. The **By Cavity** option is used with solid bodies only. ANSYS will display a warning if this option is used for surfaces.

You can select the **By Caps** option from the **Extraction Type** drop-down list for the Computational Fluid Dynamics analysis. On selecting this option, a replica of the fluid is created in an enclosure. You can use this option for both solids and surfaces.

8. In the **Details View** window, expand the **Details of Fill1** node if not already expanded.

9. Click on the **Faces** selection box to display the **Apply** and **Cancel** buttons, if they are not already displayed.

10. Now, choose the **Faces** tool from the **Select** toolbar.

11. Now, select all the holes on the front face of the model by using the CTRL key, as shown in Figure 7-70.

12. In the **Details View** window, choose the **Apply** button to confirm the selection; **6** is displayed in the **Faces** selection box.

13. After the holes are defined, choose the **Generate** tool; the fills are created in holes, as shown in Figure 7-71.

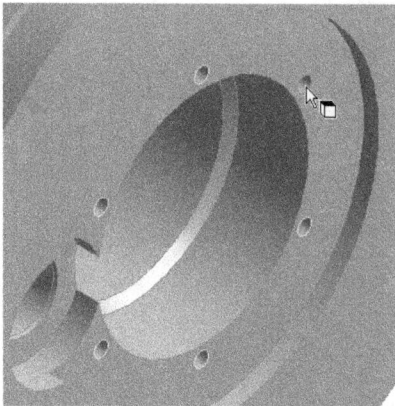

Figure 7-70 Holes to be selected   *Figure 7-71* Fills created

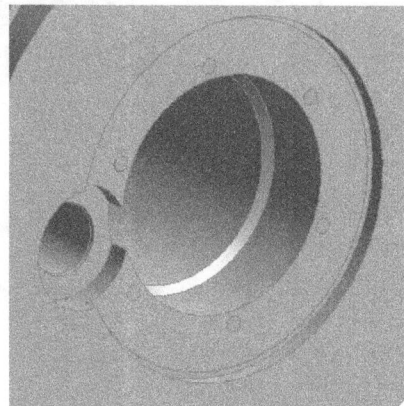

After creating the fills, you need to merge these fills with the main body.

14. Choose the **Boolean** tool from the **Create** menu; **Boolean1** is added to the Outline pane. Also, the corresponding options of the **Boolean** tool are displayed in the **Details View** window.

    The **Boolean** tool is used to unite, subtract, and intersect existing bodies.

15. Select the **Unite** option from the **Operation** drop-down list of the **Details View** window if not already selected, as shown in Figure 7-72.

    The **Unite** option is used for merging solid bodies.

16. Click on the **Tool Bodies** selection box; the **Apply** and **Cancel** buttons are displayed.

17. Select the six fills created, as shown in Figure 7-73.

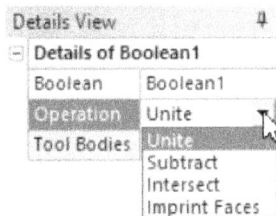

*Figure 7-72 Selecting the **Unite** option from the **Operation** drop-down list*

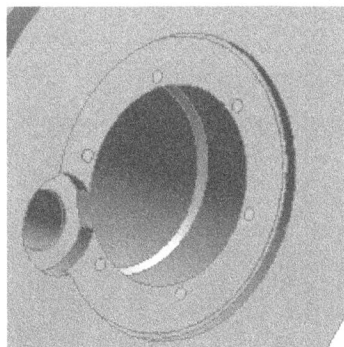

Figure 7-73 Selecting the holes with the fills

18. Next, select the main body, as shown in Figure 7-74, and then click on the **Apply** button to confirm selection. The **Tool Bodies** selection box now displays **7 Bodies**. Also, the color of the model changes to cyan.

19. Choose the **Generate** tool from the **Features** toolbar; the fills in the holes and the main body merge into a single body and the holes with the fills are no more visible, as shown in Figure 7-75.

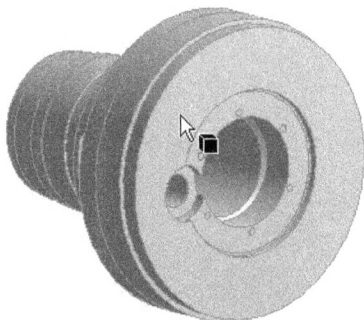

Figure 7-74 Selecting the main body

Figure 7-75 Model after merging the holes with the main body

Now, you need to create fills for the holes of smaller diameter.

20. Invoke the **Fill** tool again from the **Tools** menu; **Fill2** is attached to the Outline pane.

21. Select all the holes of smaller diameter, as shown in Figures 7-76.

22. Now, choose the **Apply** button from the **Faces** selection box; **8** is displayed in the **Geometry** selection box.

23. Make sure that the **By Cavity** option is selected in the **Extraction Type** drop-down list in the **Details View** window. Next, choose the **Generate** tool from the **Features** toolbar; the fills are created in the selected cavities.

Figure 7-76 Holes selected for fill operation

After the fills are created, you need to merge them with the main body to create a single body.

24. Choose the **Boolean** tool from the **Create** menu; **Boolean2** is attached to the Outline pane.

25. Select the holes on which fills were created and then select the main body. Next, choose the **Apply** button from the **Tool Bodies** selection box; **9 Bodies** is now displayed in the **Tool Bodies** selection box.

26. Choose the **Generate** tool from the **Features** toolbar to generate the features.

The model after filling the holes and merging them with the main body is shown in Figures 7-77 and 7-78.

Figure 7-77 Back View of the model after holes are filled and then merged with the main body

Figure 7-78 Front View of the model after holes are filled and then merged with the main body

After optimizing the holes, you now need to look into some other details in the models.

There are slots on the top of the model and these slots contain small fillets of 1 mm radius. Removing these small details will help improve the topology for meshing and will reduce the element count. There are four slots on the top of the model, as shown in Figure 7-79.

27. Invoke the **Face Delete** tool from the **Create** menu; **FDelete1** with a yellow   Face Delete
thunderbolt is attached to the Outline pane.

28. Select the fillets present on the slot, as shown in Figure 7-80, and then choose the **Apply** button in the **Faces** selection box; **4** is displayed in the **Faces** selection box indicating that four faces have been selected for the **Face Delete** operation.

Figure 7-79 *Slot with fillets displayed*

Figure 7-80 *Selecting fillets*

29. Choose the **Generate** tool available in the **Features** toolbar; the fillets on the selected slot are deleted. Also, a green tick mark is placed before **FDelete1** in the Outline pane.

30. Similarly, by using the **Face Delete** tool, delete the fillets available on the other three slots in the model.

After the fillets are removed from the slots present in the model, you now need to fill the slots to simplify the geometry.

31. Choose the **Fill** tool from the **Tools** menu; **Fill3** is attached to the Outline pane.

32. Select the **Faces** selection box to display **Apply** and **Cancel** buttons, if not already displayed.

33. Select the five faces of the slot, as shown in Figure 7-81.

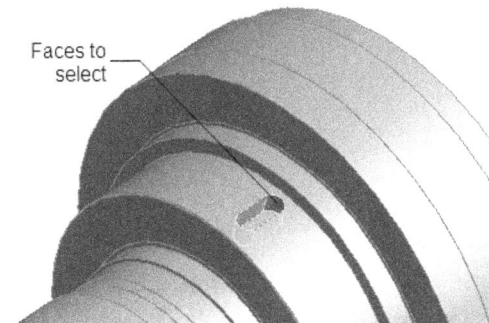

Figure 7-81 *Faces selected for the fill feature*

34. Similarly, select all other faces of the remaining slots.

35. Now, choose the **Apply** button from the **Faces** selection box; **20** is displayed in the **Faces** selection box.

36. Right-click on the graphics screen; a shortcut menu is displayed. Next, choose ⌐⌐ Generate **Generate** from it to create the fill feature.

    Now, you need to unite the fills with the main body.

37. Choose the **Boolean** tool from the **Create** menu; **Boolean3** is attached to the Outline pane. Also, a yellow thunderbolt is displayed before **Boolean3** in the Outline pane.

38. Select the slots on which fills are created and then select the main body.

39. In the **Details View** window, choose **Apply** from the **Tool Bodies** selection box; **5 Bodies** is displayed in the **Tool Bodies** selection box.

40. Choose the **Generate** tool to merge the fills in the slots and the main body as one component.

    Next, you will fill the groove in the model, as shown in Figure 7-82.

41. Use the tools available in the **Graphics** toolbar to zoom closer to the groove cut on the cylindrical part of the model.

42. Choose the **Fill** tool from the **Tools** Menu; **Fill4** is attached to the Outline pane.

43. In the **Details View** window of the **Fill** tool, select the **Faces** selection box to display the **Apply** and **Cancel** buttons.

44. Now, select the faces shown in Figure 7-83.

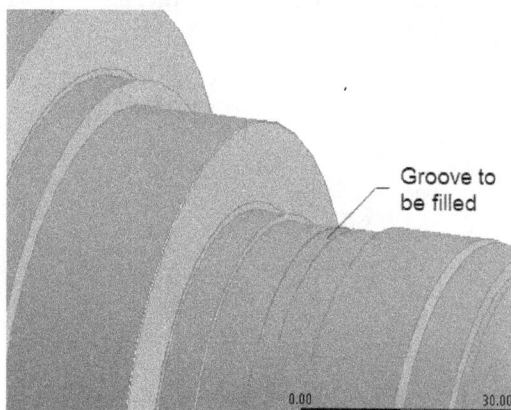

Figure 7-82 Groove to be filled

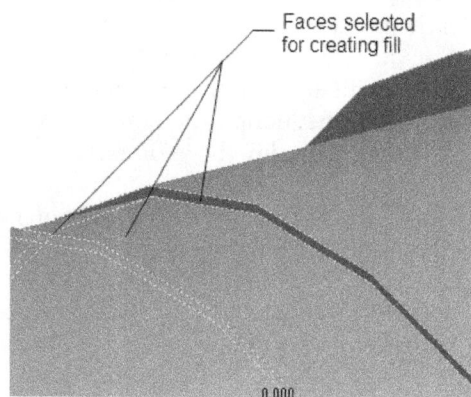

Figure 7-83 Selecting faces

45. Now, choose the **Apply** button from the **Details View** window; **3** is displayed in the **Faces** selection box.

46. Choose the **Generate** tool from the **Features** toolbar; the fill feature is created and a green tick mark is placed before **Fill4** in the Outline pane.

47. Similarly, fill all the grooves in the cylindrical feature of the model, as shown in Figure 7-84.

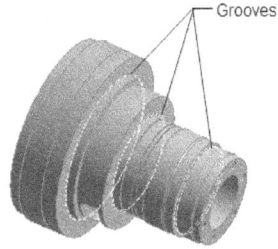

Figure 7-84 Grooves to be filled highlighted in green

Note

*1. To have access to all the faces of the groove, use the tools available in the **Graphics** toolbar.*

2. To create a fill between two walls of varying height, select the wall upto which you want to create the fill feature.

*3. In the Outline pane, **Fill5** represents all the fills created in Step 47.*

48. Now, after the grooves on the cylindrical features are filled, merge them with the main body by using the **Boolean** tool from the **Create** menu.

Next, you have to delete small fillets and chamfers to make the model simpler, as shown in Figure 7-85.

49. Use the **Delete Faces** tool from the **Create** menu to delete small fillets and chamfers in the model, refer to Figure 7-85.

The model after optimization should look like the one shown in Figure 7-86.

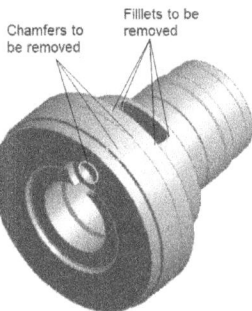

Figure 7-85 Fillets and chamfers to be removed

Figure 7-86 Model after optimization

Creating the Symmetrical Model

Note that the model is symmetrical in nature. Therefore a symmetrical half can be created to increase the efficiency of meshing. It will decrease the element count and also improves visibility of the model.

1. Invoke the **Symmetry** tool from the **Tools** menu; **Symmetry1** gets attached   ⚠ Symmetry
   to the Outline pane and you are prompted to select a symmetry plane.

2. Click in the **Symmetry Plane1** selection box to display the **Apply** and **Cancel** buttons if
   not displayed.

3. In the Outline pane, select **YZPlane**; a preview of the plane is displayed in the graphics
   screen, as shown in Figure 7-87.

4. Choose the **Apply** button from the **Symmetry Plane1** selection box.

   Now, you will generate the symmetrical half of the model.

5. Choose the **Generate** tool from the **Features** toolbar; the half section of the   ⧸ Generate
   model is created, as shown in Figure 7-88.

6. Close the **DesignModeler** window; the **Workbench** window is displayed.

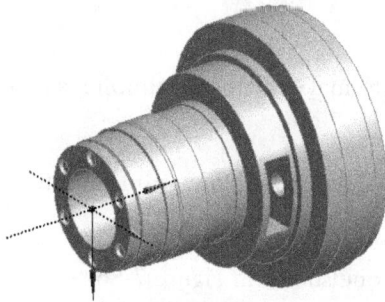

*Figure 7-87 Preview of the plane for creating
the section of the model*

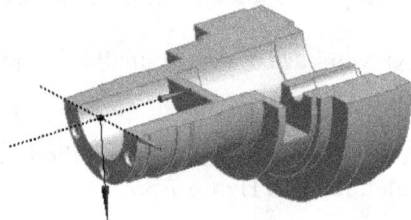

Figure 7-88 Half-section of the model

Generating Mesh of the Sectioned Model

After the model is optimized, you need to generate mesh for the optimized model.

1. In the **Project Schematic** window of the **Workbench** window, right-click on the **Model** cell
   and then choose the **Refresh** tool from the shortcut menu; the project is updated with the
   optimized geometry data.

2. Double-click on the **Model** cell of the **Static Structural** analysis system to open the
   Mechanical window.

3. In the Outline pane, right-click on **Mesh**; a shortcut menu is displayed.
   Choose the **Generate Mesh** option from it to generate the mesh with default   ▦ Generate Mesh
   settings; the mesh is generated on the sectioned model, as shown in Figure 7-89.

The element count after generating the mesh with default settings is 9362. Note that the mesh generated in this model will not be uniform.

Note

If all the imperfections or minute details are removed, the element count in your system will display 9362. However, if your model is not optimized accurately like the model shown in Figure 7-89, the element count may vary in your case.

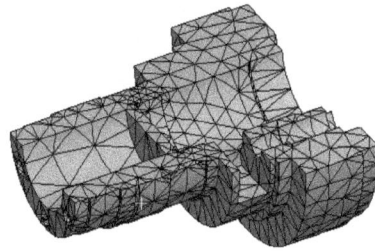

Figure 7-89 Mesh generated on the sectioned model

Creating a Section to Visualize Mesh Quality

To have a better visibility of the volume cells, you need to visualize the model through a section plane.

1. Right-click in the graphics screen and then choose **View > Left** from the shortcut menu displayed; the model is oriented as shown in Figure 7-90.

2. Choose the **Section Plane** tool from the **Standard** toolbar; the **Section Planes** window is displayed on the left in the graphics screen.

3. In the graphics screen, place the cursor as shown in Figure 7-90 and drag it in such a manner that section of the model is created, as shown in Figure 7-91.

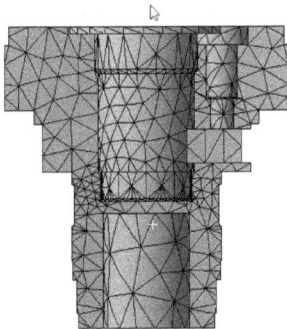

Figure 7-90 The model after orientation

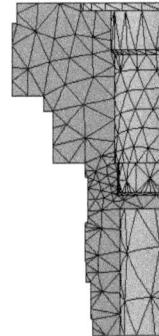

Figure 7-91 Sectioned model

4. Choose the **Show Whole Elements** button from the **Section Planes** window to visualize the volume cells, as shown in Figure 7-92.

5. Now, choose the **Show Whole Elements** button from the **Section Planes** window again to turn the visibility of the volume cells off.

The visualization of volume cells will help you understand the effect of generating a better mesh later in this tutorial.

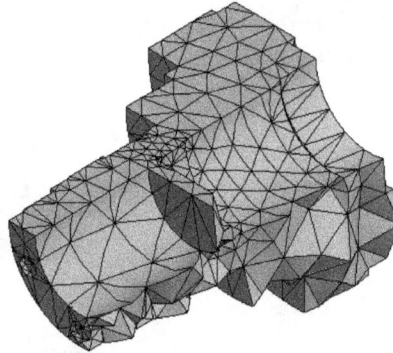

Figure 7-92 Volume cells created on the mesh with default settings

Inserting Local Mesh Controls

Now, you will generate a mesh by using the local mesh control settings.

An effective mesh can be generated by changing the settings for a particular zone or region of the model. These settings are also known as local mesh control settings. These options can be accessed from the shortcut menu displayed when you right-click on the **Mesh** node in the Outline pane, as shown in Figure 7-93.

1. Right-click on the **Mesh** node in the Outline pane; a shortcut menu is displayed, refer to Figure 7-93.

2. Next, choose **Insert > Method** from the shortcut menu, refer to Figure 7-93; **Automatic Method** with a question symbol is added under the **Mesh** node in the Outline pane. Also, the **Details of "Automatic Method"** window is displayed.

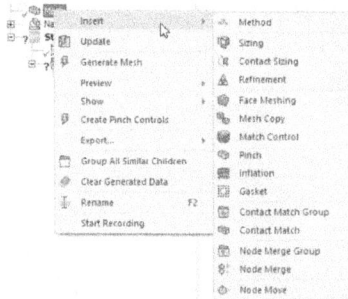

Figure 7-93 Various local mesh control options in the shortcut menu

There are various methods that can be used to create a mesh. First, by specifying the shape of elements. Second, by using an algorithm. The algorithms are listed in the **Method** option drop-down list available in the **Details of "Automatic Method"** window. When a method control is inserted into a mesh, it appears as a new instance under the **Mesh** node in the Outline pane.

3. In the **Details of "Automatic Method"** window, click on the **Geometry** selection box to display the **Apply** and **Cancel** buttons.

4. Choose the **Body** tool from the **Select** toolbar if not already.

5. Select the model and then choose the **Apply** button in the **Geometry** selection box; the **Geometry** selection box displays **1 Body** in the selection box.

6. In the **Method** drop-down list, select the **Tetrahedrons** option, as shown in Figure 7-94; the **Algorithm** drop-down list is displayed below the **Method** drop-down list. Also, the **Details of "Automatic Method"** window is replaced by the **Details of "Patch Conforming Method"** window.

The **Tetrahedrons** method is used when all the elements required are tetrahedral. This method is applied only in case of solid bodies.

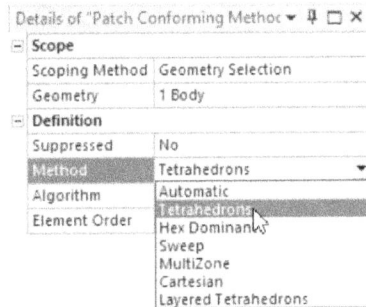

*Figure 7-94 Selecting the **Tetrahedrons** option from the **Method** drop-down list*

7. In the **Details of "Patch Conforming Method"** window, select the **Patch Independent** option from the **Algorithm** drop-down list, as shown in Figure 7-95; the **Patch Conforming** node is replaced by **Patch Independent**. Also, the **Details of "Patch Conforming Method"** window is replaced by **Details of "Patch Independent"** window.

8. In the **Details of "Patch Independent"** window, expand the **Advanced** node and specify **.004** in the **Max Element Size** edit box.

When the Patch Independent algorithm is used, smaller elements are created at regions of higher importance. Similarly, it creates larger elements at regions of less importance. Using this algorithm the model is discretized in an effective way.

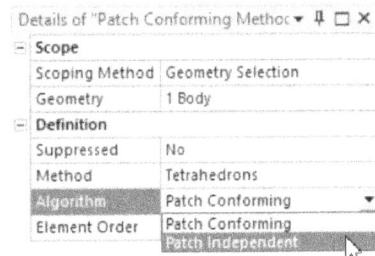

*Figure 7-95 Selecting the **Patch Independent** option from the **Algorithm** drop-down list*

9. In the **Min Size Limit** edit box, enter **.003**.

The **Min Size Limit** edit box is displayed only when **Proximity and Curvature** is selected in the **Refinement** drop-down list. Specify a value in this edit box to prevent the software to create elements of smaller size than the specified value.

10. Now, enter **2** in the **Num Cells Across Gap** edit box.

   Note
   Depending on the requirement, different parts and bodies in a system can have different algorithms.

11. Accept all other default options in the **Details of "Mesh"** window and then right click on the **Mesh** node in the **Outline** window and choose the **Generate Mesh** button; the mesh for the model is created, as shown in Figure 7-96.

   Figure 7-97 shows the sectioned view of the mesh displaying the volume cells.

Figure 7-96 Mesh generated with the specified settings

Figure 7-97 Sectioned view of the model displaying volume elements

12. Exit the **Mechanical** window; the **Workbench** window is displayed.

Save the Project and Exit ANSYS Workbench

After generating the mesh, you need to save the project and exit the ANSYS Workbench session.

1. Save the model by choosing the **Save Project** button from the **Standard** toolbar; the model is saved with the name *c07_ansWB_tut03*.

2. Exit the **Workbench** window to end the ANSYS Workbench session.

Self-Evaluation Test

Answer the following questions and then compare them to those given at the end of this chapter:

1. The _____ tool is used to create a symmetry of the model.

2. You can create the section views of a mesh by using the _____ tool.

3. When you do not specify any method for generating a mesh in ANSYS Workbench, by default, the _____ method is selected.

4. The _____ tool is used to generate a mesh for a model.

5. When a new method is applied on the model, **Automatic Method** is displayed under the **Mesh** node in the Outline pane. (T/F)

6. There are two algorithms available for a Tetrahedron method. (T/F)

7. The **Patch Independent** option is selected by default in the **Algorithm** drop-down list. (T/F)

8. In ANSYS Workbench, you cannot view the wireframe mode of the model. (T/F)

9. All the options available in the **Mesh** node of outline pane are global mesh control options. (T/F)

10. By default, the **By Cavity** option is selected in the **Extraction Type** drop-down list in the **Details View** window while creating fill features of surfaces. (T/F)

Review Questions

Answer the following questions:

1. Which of the following tools is used to change the size of elements in a particular region of a model?

    (a) **Refinement**   (b) **Mapped Face Meshing**
    (c) **Pinch**   (d) **Generate Mesh**

2. Which of the following methods is used to introduce Patch Independent algorithm into meshing.

    (a) **Hex Dominant**   (b) **Automatic**
    (c) **Sweep**   (d) **Tetrahadrons**

3. By default, ANSYS Workbench generates the mesh with _____ and _____ models wherever possible, but generates _____ in more complicated areas.

4. You can control the sizing of the mesh by using the options available in the **Details of "Mesh"** window. (T/F)

5. For complex models, tetrahedrals elements are generated. (T/F)

6. You can view the total number of elements in the **Statistics** node in the **Details of "Mesh"** window. (T/F)

7. You can rename **Mesh** available in the Outline pane. (T/F)

EXERCISES
Exercise 1

Open the project created in Exercise 1 of Chapter 5. You will first open the file and save it with the name *c07_ansWB_Exr01*. The model and its dimensional values are given in Figures 7-98 through 7-101. Share the model with a Static Structural analysis system and then generate a mesh for the model with the default settings. Next, change the global and local mesh settings as given next: **(Expected time: 1 hr)**

        Global mesh control settings:
        Use Adaptive Sizing Function set to **Capture Proximity**
        The **Number of cells Across Gap** should be 4

        Local mesh control settings
        Introduce **Refinement** at the small duct at the front of the model.

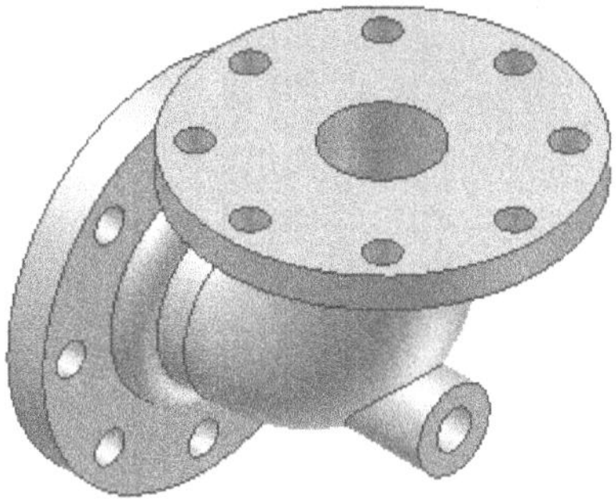

Figure 7-98 *Model for Exercise*

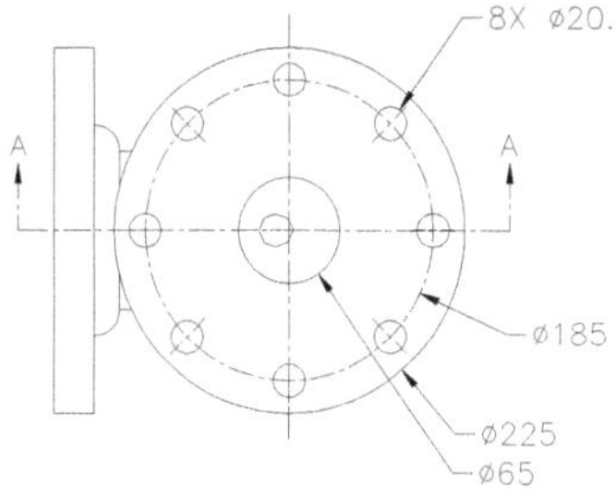

Figure 7-99 *Top view of the model with the hidden lines suppressed for clarity*

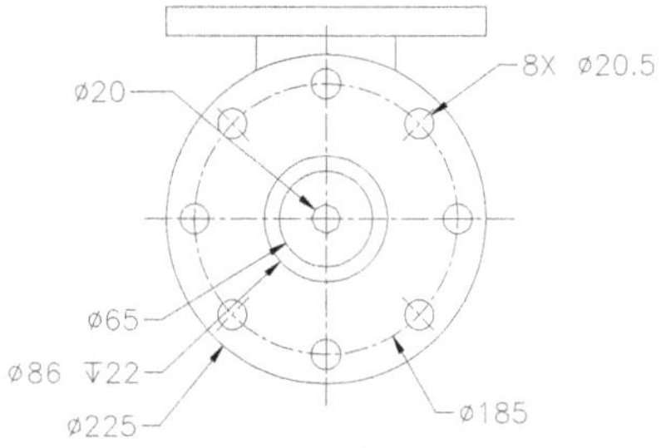

Figure 7-100 *Left side view of the model with the hidden lines suppressed for clarity*

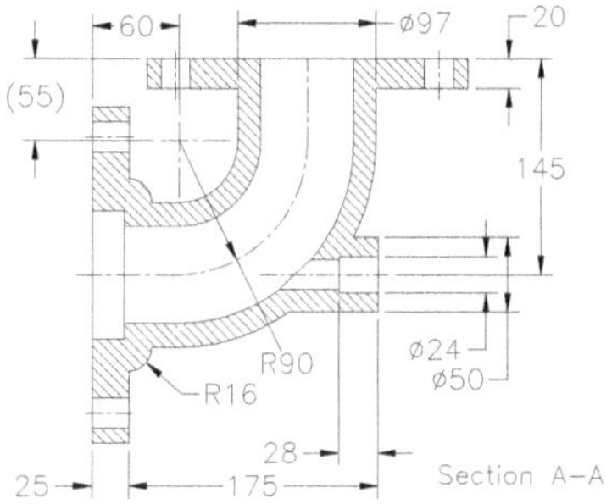

Figure 7-101 *Sectioned view of the model*

Answers to Self-Evaluation Test

1. Symmetry, 2. New Section Plane, 3. Automatic, 4. Generate Mesh, 5. T, 6. T, 7. F, 8. F, 9. F, 10. F

Chapter 8

Generating Mesh - II

In the previous chapter, you used some of the tools and global and local mesh control settings to generate a mesh. Also, you used various tools available in the **DesignerModeler** to optimize a model. In this chapter you will use some more tools and more global and local mesh control settings. Also, you will use tools and options to generate meshes for assemblies and surface models.

TUTORIALS

Tutorial 1

In this tutorial, you will generate a mesh for the model shown in Figure 8-1 and then apply face sizing controls on all the curved faces of the model. This can be accessed by downloading the zip file *c08_ansWB_tut01.zip* from *www.cadcim.com* and then extracting it to the desired folder.

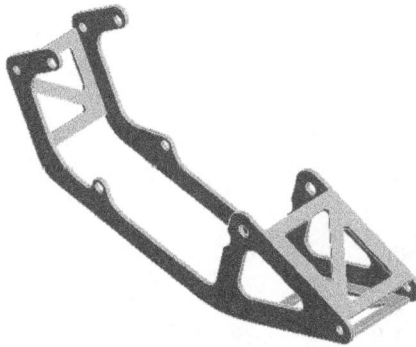

Figure 8-1 *Isometric view of the model for Tutorial 1*

The following steps are required to complete the tutorial:

a. Download the part file.
b. Start a new project and import the file in ANSYS Workbench.
c. Set global mesh controls and generate the mesh.
d. Set local mesh controls and generate the mesh.
e. Refine the mesh.
f. Save the project.

Downloading the Part File

Before starting the tutorial, you need to create two folders and download the file from *www.cadcim.com*.

1. Create a new folder with the name **c08** at the location *C:\ANSYS_WB*.

2. Next, create another folder named as **Tut01** at the location *C:\ANSYS_WB\c08*.

3. Download the file *c08_ansWB_tut01.zip* from *www.cadcim.com*. The complete path to download the file is as follows:

    Textbooks > CAE Simulation > ANSYS > ANSYS Workbench 2023 R2: A Tutorial Approach > Input Files

After downloading, extract the zip file to save the stp part file *c08_ansWB_tut01.stp* at the location *C:\ANSYS_WB\c08\Tut01*.

Importing the File in ANSYS Workbench

After downloading the file from *www.cadcim.com*, you need to import the file into ANSYS Workbench.

1. Start ANSYS Workbench to display the **Workbench** window.

2. In the **Workbench** window, add the **Modal** analysis system into the **Project Schematic** window.

3. Right-click on the **Geometry** cell of the **Modal** analysis system to display a shortcut menu.

4. Choose the **Import Geometry** option from the shortcut menu; a flyout is displayed.

5. From this flyout, choose the **Browse** option; the **Open** dialog box is displayed.

6. Browse to the location *C:\ANSYS_WB\c08\Tut01* and then open *c08_ansWB_tut01.stp*; a green tick mark is placed in the **Geometry** cell of the **Modal** analysis system indicating that the geometry is specified for the analysis.

7. In the **Workbench** window, choose the **Save Project** button from the **Standard** toolbar; the **Save As** dialog box is displayed.

8. In this dialog box, browse to the location *C:\ANSYS_WB\c08\Tut01* and then save the project with the name **c08_ansWB_tut01**.

Setting Global Mesh Controls and Generating Mesh

After the file is imported, you need to set the global mesh control settings to generate a mesh.

1. In the **Project Schematic** window, double-click on the **Model** cell; the **Mechanical** window along with the model is displayed, refer to Figure 8-2.

Note
The orientation of the model in Figure 8-2 has been changed for better visibility of the model.

2. Select **Mesh** in the **Outline** pane; the **Details of "Mesh"** window is displayed, refer to Figure 8-3.

3. In the **Details of "Mesh"** window, expand the **Defaults** node if not already expanded.

4. Enter **0.006** in the **Element Size** edit box.

5. Next, expand the **Sizing** node if not already expanded.

Notice that very small fillets and holes are displayed in the model. To generate a mesh to suit the model with rounds, fillets, and curves, you need to change the global mesh control settings.

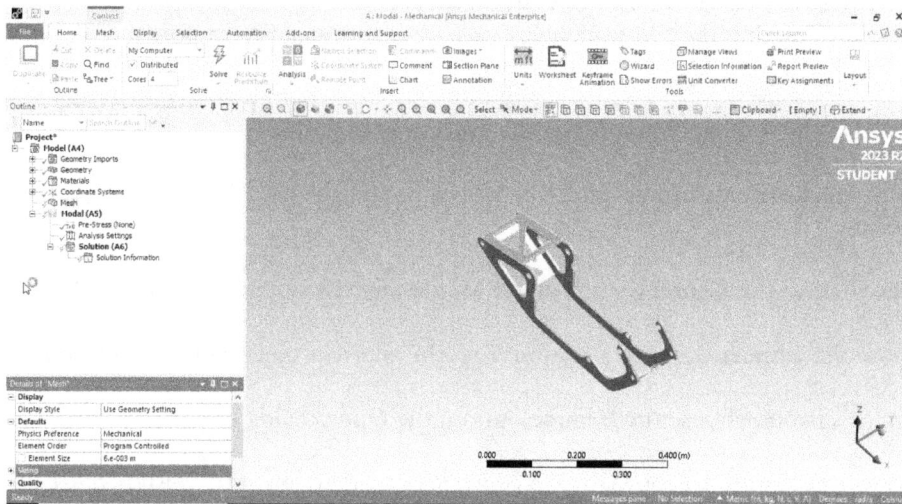

*Figure 8-2 The **Mechanical** window with the engine mount model displayed*

6. Select the **No** option from the **Use Adaptive Sizing** drop-down list.

7. Specify **0.009** in the **Max Size** edit box.

8. Choose the **Yes** option from the **Capture Proximity** drop-down list.

9. Specify **0.004** in the **Proximity Min Size** edit box.

10. Specify **2** in the **Proximitiy Gap Factor** edit box.

11. Set remaining values to default, refer to Figure 8-3.

12 Right-click on **Mesh** in the **Outline** pane to display a shortcut menu.

13. Next, choose the **Generate Mesh** option from the shortcut menu; the mesh with the specified global control settings is generated, as shown in Figure 8-4.

Note
1. The number of elements and nodes after generating the mesh are 59,549 and 105,953.

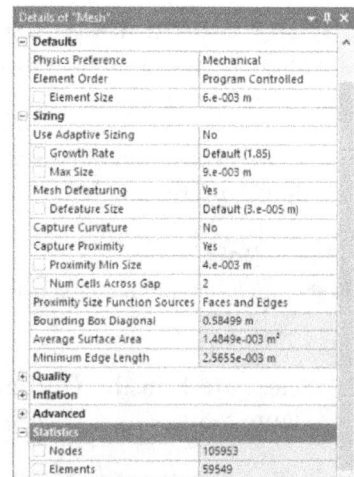

Figure 8-3 Partial view of the Details of "Mesh" window

2. It is always important to keep a track of the number of elements so that you can later compare the quality of the mesh with the element count and decide whether you need to further refine the mesh.

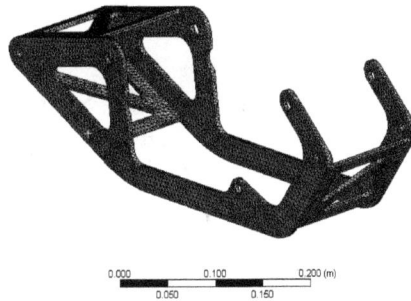

Figure 8-4 *Mesh generated with global mesh control settings*

Setting Local Mesh Controls and Generating a Mesh

Global mesh controls are the settings that are applied to the whole component considered for meshing. However, if required, you can generate mesh of varying refinement on different regions by setting the local mesh controls.

1. In the **Outline** pane, right-click on **Mesh**; a shortcut menu is displayed. Choose the **Insert** option and click on the **Sizing** option, as shown in Figure 8-5; **Sizing** is added under the **Mesh** node in the **Outline** pane. Also, a question symbol is attached to **Sizing** indicating that values for this field are yet to be satisfied.

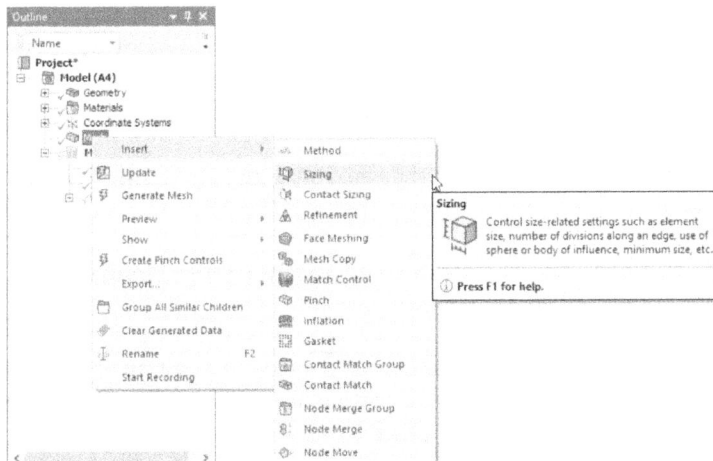

Figure 8-5 *Choosing the Sizing option from the flyout*

> **Note**
> *You can also choose the Sizing option from the Mesh contextual tab in the Menu bar.*

2. Select **Sizing** from the **Outline** pane, if it is not already selected; the **Details of "Sizing"** window is displayed. Click on the **Geometry** selection box, the **Apply** and **Cancel** buttons are displayed in the **Geometry** selection box indicating that you need to select the faces, edges, or bodies to be sized.

3. Choose the **Face** tool from the **Select** toolbar.

4. Select all the cylindrical faces, as shown in Figure 8-6. Next, choose the **Apply** button from the **Geometry** selection box in the **Details of "Sizing"** window; **63 Faces** is displayed in the **Geometry** selection box indicating that 63 faces have been selected for sizing.

    Notice that in the **Outline** pane, **Sizing** is replaced by **Face Sizing**. Also, the **Details of "Face Sizing"** window is displayed below the **Outline** pane.

    Note
    The number of faces selected for face sizing may vary as there are many small cylindrical faces in the model. Therefore, you need to be very careful while selecting the faces.

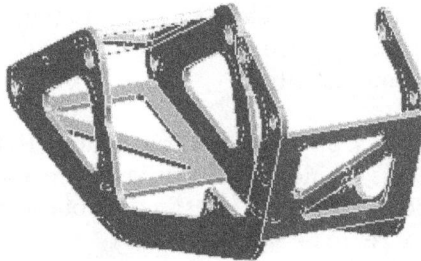

Figure 8-6 Selecting cylindrical faces in the component

5. In the **Details of "Face Sizing"** window, expand the **Definition** node if not already expanded.

6. In the **Type** drop-down list, make sure that the **Element Size** option is selected. Next, specify **0.002** as the element size in the **Element Size** edit box.

7. Choose the **Update** tool from the **Mesh** contextual tab in the **Menu** bar; the mesh with local mesh control is generated, as shown in Figure 8-7.

    Note that the total number of elements in this case is 81,360 which is more than 59,549. Also, if you compare Figures 8-4 and 8-7, you will notice that even though the element count does not seem to grow very fast numerically, the actual number of elements in the model has increased, refer to Figure 8-7.

8. Create a section view of the model and change the view to Isometric, as shown in Figure 8-8.

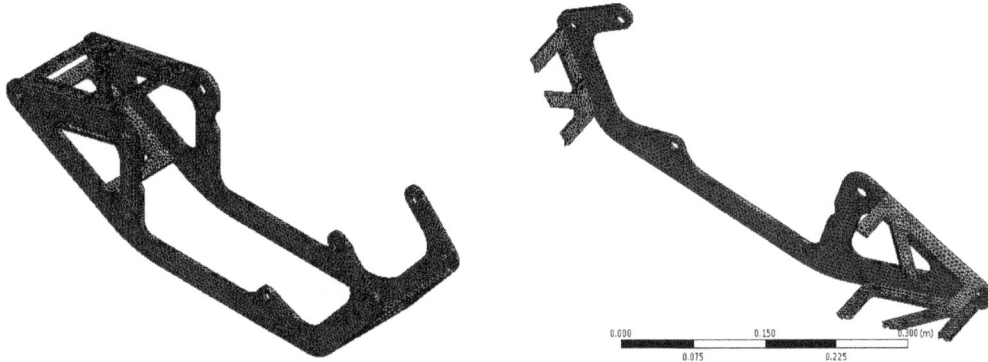

Figure 8-7 *Mesh generated with local settings*

Figure 8-8 *Sectioned isometric view of the model with the elements displayed*

You can have a close look at the shape and sizes of the elements by using the tools available in the **Graphics** toolbar. You will notice that the sizes of elements have changed. You can also compare the mesh result of Figure 8-9 with that of Figure 8-4 to notice the difference.

9. Exit the section view.

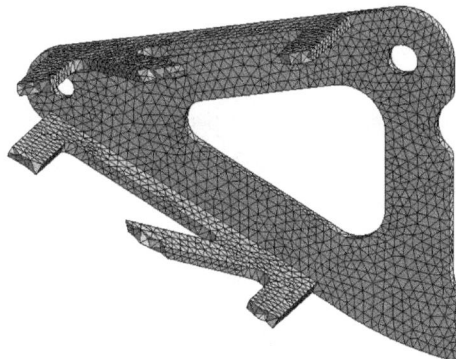

Figure 8-9 *Partial view of the sectioned model with a closer look at the elements formed*

Refining the Mesh

After changing the mesh locally by introducing sizing into it, you can further refine the mesh wherever required by using various mesh controls available in the **Mesh** contextual tab of the **Menu** bar.

1. Right-click on the **Mesh** node to display a shortcut menu.

2. Choose **Insert > Refinement** from the shortcut menu; **Refinement** is added under the **Mesh** node in the **Outline** pane with a question symbol attached to it.

3. Select the **Refinement** node to display the **Details of "Refinement"** window.

Refinement is applied to edges, faces, and vertices. It cannot be applied to the whole body. It is a local mesh control and by applying it to a face, edge, or vertex, you can further reduce the size of the elements. After applying refinement control to a face, edge, or vertex if you delete **Refinement** from the **Outline** pane, the mesh of that particular region will be restored to its previous state.

4. In the **Details of "Refinement"** window, expand the **Definition** node if not already expanded.

5. In the **Definition** node, specify **1** in the **Refinement** edit box if not already specified.

6. Select the **Geometry** selection box in the **Scope** node of the **Details of "Refinement"** window; the **Apply** and **Cancel** buttons are displayed.

7. From the **Select** toolbar, choose the **Face** tool to select faces to refine.

8. Select the face shown in Figure 8-10. Next, choose the **Apply** button from the **Geometry** selection box in the **Details of "Refinement"** window; **1 Face** is displayed in the **Geometry** selection box indicating that 1 face has been selected to apply refinement.

Figure 8-10 *The face to be selected*

9. Right-click on the **Mesh** node to display a shortcut menu. Choose the **Update** tool to generate the mesh with more local mesh controls applied. Figure 8-11 shows partial view of the model with the mesh generated and refine control applied on it.

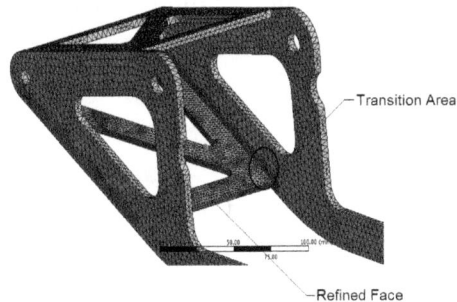

Figure 8-11 Mesh generated on the face with refinement
control applied on it

10. Close the **Mechanical** window; the **Workbench** window is displayed.

   You will notice that the face where refinement was applied has relatively smaller element
   size as compared to the other areas of the model.

Saving the Model

   After meshing the model, you need to exit the **Mechanical** window and save the project.

1. In the **Workbench** window, choose the **Save Project** button to save the model with the name
    c08_ansWB_tut01.

2. Exit the **Workbench** window to end the session.

Tutorial 2

In this tutorial, you will generate mesh for different components of the Bench Vice
assembly created in Tutorial 3 of Chapter 6. Figure 8-12 shows the Bench Vice assembly.
The model is not considered for any analysis. However, you need to add **Static Structural**
analysis system to the **Project Schematic** window just for a better understanding
of the process of generating mesh for assemblies.   **(Expected time: 2 hr)**

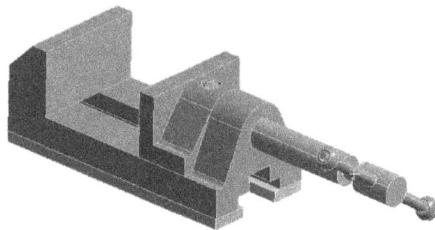

Figure 8-12 Bench Vice Assembly

The following steps are required to complete this tutorial:

a. Open the existing project.
b. Generate mesh for the assembly component.
c. Introduce local mesh controls.
d. Save the project.

Opening the Existing Project

Before starting the tutorial, you need to open the *c06_ansWB_tut03* file from the *C:\ANSYS_WB\c06\Tut03* folder and then save it with a new name.

1. Start ANSYS Workbench to display the **Workbench** window.

2. Choose the **Open** button from the **Standard** toolbar; the **Open** dialog box is displayed.

3. Browse to the location: *C:\ANSYS_WB\c06\Tut03* and open the *c06_ansWB_tut03* file; the **Static Structural** analysis system is displayed in the **Project Schematic** window.

4. Double-click on the name field of the **Static Structural** analysis system and rename it to **Bench Vice Assembly**, as shown in Figure 8-13.

5. In the **Workbench** window, choose the **Save Project As** button from the **Standard** toolbar to invoke the **Save As** dialog box.

6. In this dialog box, browse to the folder *C:\ANSYS_WB\c08* and then create the folder named as **Tut02** in it.

*Figure 8-13 Renaming the analysis system to **Bench Vice Assembly***

7. Next, browse to the *Tut02* folder and then save the project with the name **c08_ansWB_tut02**.

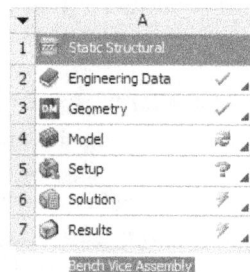

Setting Contacts and Generating Mesh

Now, you need to set the contacts among the components and also generate mesh for all the components of the assembly.

1. Double-click on the **Model** cell of the **Bench Vice Assembly** analysis system; the **Mechanical** window is displayed.

2. Expand the **Connections** node to display the **Contacts** node, as shown in Figure 8-14.

3. Expand the **Contacts** node under the **Connections** node to display the list of contacts available for the assembly, as shown in Figure 8-15.

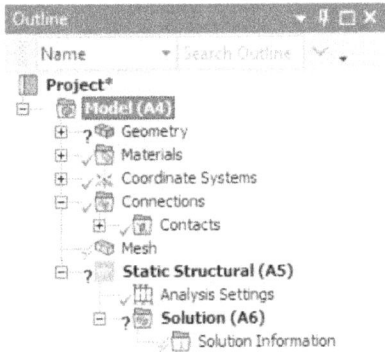

Figure 8-14 The **Connections** node in the **Outline** pane

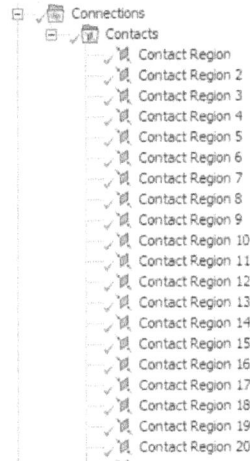

Figure 8-15 Partial view of the **Contacts** node under the **Connections** node

4.   Right-click on the **Contacts** node and then choose the **Rename Based on Definition** option from the shortcut menu displayed; the contacts are renamed, refer to Figure 8-16.

Note

1. When you expand the **Contacts** *node in the* **Outline** *pane, the default names of the contacts are displayed as* **Contact Region**, **Contact Region 2**, *and so on depending upon the number of contacts. When you right-click on any name and then choose the* **Rename Based on Definition** *option from the shortcut menu displayed, the names of the contacts will change according to the type of contact specified in the* **Details** *window. Also, the names of the components will be displayed.*

2. In the **Outline** *pane, expand the* **Geometry** *node; a list of components in the assembly is displayed. To rename a component, right-click on it in the* **Outline** *pane; a shortcut menu is displayed. From this shortcut menu, choose* **Rename**; *the name of the component gets highlighted. Enter a new name to rename it.*

Whenever you import an assembly into ANSYS Workbench, the contact situation arises. This is because when an assembly is imported, all the relations among various components of the assembly are converted into contacts. These contacts are then displayed in the **Contacts** node under the **Connections** node in the **Outline** pane. If you have provided names to all the components of the assembly, corresponding names and the contact type will be displayed under the **Contacts** node.

After you renamed all the contact regions based on the names of the components, you will observe that the type of contact applied among the components is bonded.

5.   Select **Bonded - Slider Guide To Slider Base** from the **Contacts** node, as shown in Figure 8-16; the **Details of "Bonded - Slider Guide To Slider Base"** window is displayed, as shown in Figure 8-17.

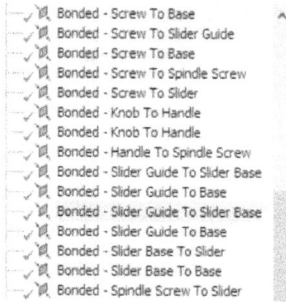

Figure 8-16 *Selecting* **Bonded - Slider Guide To Slider Base** *from the* **Contacts** *node*

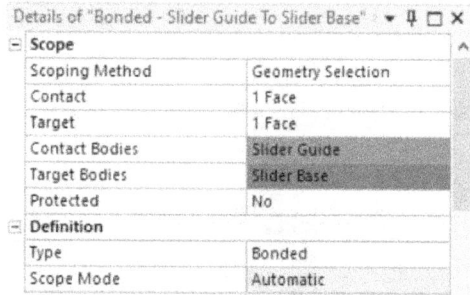

Figure 8-17 *Partial view of the* **Details of "Bonded - Slider Guide To Slider Base"** *window*

When a contact is selected from the **Outline** pane, the assembly becomes transparent, except the components that form the contact, refer to Figure 8-18. The surfaces in contact are assigned a particular color in the Graphics screen to distinguish the components and surfaces under consideration. In Figure 8-18, notice that the surface of the Slider Guide that is in contact with the surface of the Slider Base is highlighted in red in the Graphics screen. Similarly, the corresponding surface of the Slider Base is marked in blue, as shown in Figure 8-19.

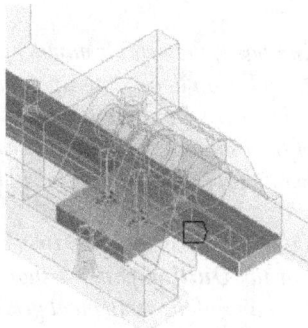

Figure 8-18 *Partial view of the model displaying the surface of the Slider Guide*

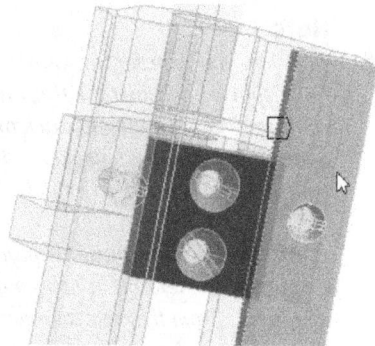

Figure 8-19 *Partial view of the assembly displaying the color assigned to the Slider Base*

6. In the **Details of "Bonded - Slider Guide To Slider Base"** window, expand the **Definition** node if not already expanded.

7. In this window, select the **Frictional** option from the **Type** drop-down list; some more options are added to the **Advanced** node. Also, name of the contact in the **Outline** pane is updated to **Frictional- Slider Guide To Slider Base**.

8. Specify **0.6** as the coefficient of friction in the **Friction Coefficient** edit box in the **Definition** node.

Note
The coefficient of friction is considered to be 0.6 between the mild steel and gray cast iron surfaces.

9. Similarly, edit all the contacts in the **Outline** pane. For reference, use the data in Table 8-1.

Table 8-1 *Various contact definitions*

Contact Between	Contact to be Placed	Coefficient of Friction
Slider Guides To Base	Bonded	
Screws To Slider Guide	Bonded	
Knobs To Handle	Bonded	
Handle to Spindle Screw	Bonded	
Slider Base to Slider	Bonded	
Slider Base To Base	Frictional	0.6
Slider Guides To Slider Base	Frictional	0.6
Screws To Slider Base	Bonded	
Spindle Screw To Slider	Bonded	
Spindle Screw To Base	Frictional	0.6
Screw To Spindle Screw	Bonded	
Slider to Base	Frictional	0.6
Screws To Slider	Bonded	
Screws To Base	Bonded	
Screws To Slider Guide	Bonded	

Notice that, a yellow thunderbolt symbol is attached to **Mesh** in the **Outline** pane, refer to Figure 8-14, which indicates that a mesh of the assembly has to be generated. To do so, follow the procedure explained next.

10. Right-click on **Mesh** in the **Outline** pane; a shortcut menu is displayed.

11. Choose the **Generate Mesh** option from the shortcut menu; mesh is generated   ⚡ Generate Mesh
    according to the default settings. The mesh generated is shown in Figure 8-20.

Note that there are approximately 5989 elements generated by using the default settings.

Note
*1. To check the quality of the mesh, you can use the tools available in the **Graphics** toolbar.*

2. In ANSYS Workbench, there are many ways by which you can improve the mesh quality. Setting the global mesh controls help you generate a mesh that would be same for all the instances of mesh generation. However, local mesh controls help you achieve better quality of mesh wherever needed. There are many tools available to refine the mesh in a particular region.

Figure 8-21 shows partial view of the assembly with the mesh generated.

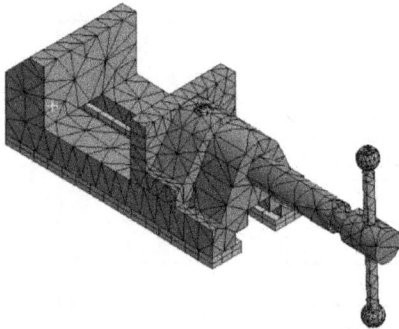

Figure 8-20 Mesh generated for the assembly

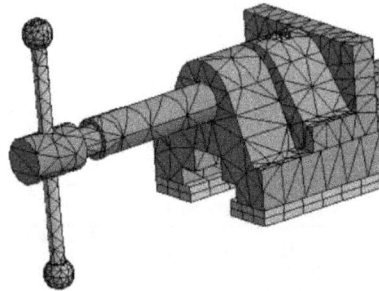

Figure 8-21 Partial view of the model showing the mesh generated

12. Select the **Mesh** node in the **Outline** pane; the **Details of "Mesh"** window is displayed.

13. In the **Details of "Mesh"** window, expand the **Defaults** node in the **Details of "Mesh"** node if not already expanded, and then specify **20 mm** in the **Element Size** edit box.

14. In the **Details of "Mesh"** window, expand the **Sizing** node if not already expanded.

15. Choose the **No** option from the **Use Adaptive Sizing** drop-down list and choose the **Yes** option from the **Capture Proximity** drop down list, as shown in Figure 8-22; a list of options are added to the **Details of "Mesh"** window.

16. Specify **2** in the **Proximity Gap Factor** edit box under the **Sizing** node and set other options to default.

17. Right-click on **Mesh** in the **Outline** pane; a shortcut menu is displayed. Choose the **Update** option; the mesh is generated with changed global mesh control settings, as shown in Figure 8-23.

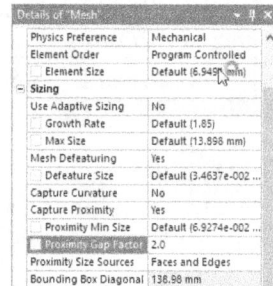

Figure 8-22 Choosing the **Yes** option from the **Capture Proximity** drop-down list

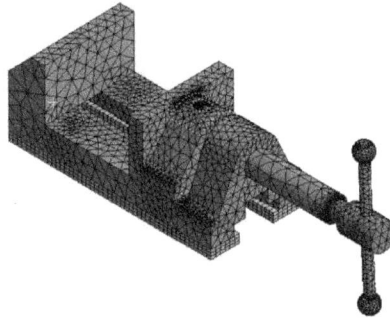

Figure 8-23 *Mesh generated with changed global mesh control settings*

Note

*The default value in the **Proximity Gap Factor** edit box available in the **Sizing** node is **3**. In this tutorial, you have specified the value 2 which means that ANSYS will maintain 2 cells across thin regions of the model, refer to Figure 8-23. The approximate number of elements generated will be 68808.*

You will notice that the mesh generated with changed global mesh control settings is finer than the previous mesh, refer to Figure 8-20. As the Bench Vice assembly is used to hold objects, therefore the main concentration of stresses should be on the Base and the Slider. As a result, these two components should have a better mesh as compared to other components such as the Handle, Knobs, and so on where the effect of the pressure applied is comparatively less.

Based on the results obtained from the current mesh, you can apply a mesh method or any other local mesh controls to get better results. As the Base and the Slider are the main components in this assembly, you will apply the **Patch Independent** algorithm to these components only.

18. In the **Outline** pane, right-click on **Mesh**; a shortcut menu is displayed. Choose the **Insert** option and click on the **Method** option, as shown in Figure 8-24; **Automatic Method** is attached with a question symbol, under **Mesh** in the **Outline** pane, as shown in Figure 8-25. Also, the **Details of Automatic "Mesh"** window is displayed.

19. In the **Details of "Automatic Method"** window, click on the **Geometry** selection box to display the **Apply** and **Cancel** buttons, if they are not already displayed.

20. Press and hold the Ctrl key and select the Base and the Slider in the Graphics screen; they turn green.

21. In the **Details of "Automatic Method"** window, choose the **Apply** button in the **Geometry** selection box; the Base and the Slider components are specified as the geometries on which Patch Independent algorithm is applied. Also, **2 Bodies** is displayed in the **Geometry** selection box indicating that 2 bodies have been selected for the operation.

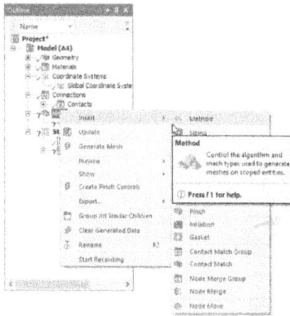

*Figure 8-24 Choosing the **Method** option from the **Mesh Control** drop-down*

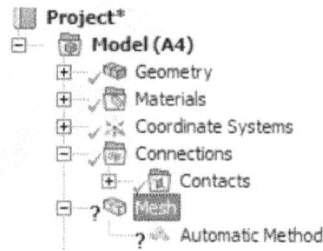

*Figure 8-25 The **Automatic Method** node under the **Mesh** node*

22. Expand the **Definition** node in the **Details of "Automatic Method"** window and then select the **Tetrahedrons** option from the **Method** drop-down list, as shown in Figure 8-26; the **Algorithm** drop-down list is displayed with the **Patch Conforming** option selected by default.

23. Select the **Patch Independent** option from the **Algorithm** drop-down list, as shown in Figure 8-27; **Automatic Method** is replaced by **Patch Independent** in the **Outline** pane. Also, the **Details of "Patch Independence"** window is displayed.

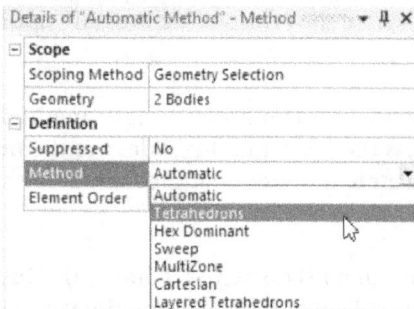

*Figure 8-26 Selecting the **Tetrahedrons** option from the **Method** drop-down list*

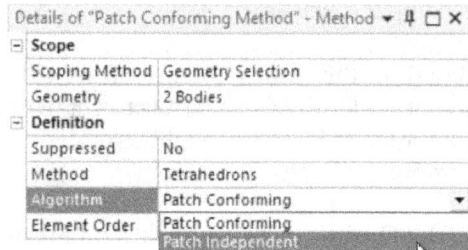

*Figure 8-27 Selecting the **Patch Independent** option from the **Algorithm** drop-down list*

To have a better idea of the effects of meshing using the **Patch Independent** algorithm, you need to generate a mesh with the default settings in the **Details of "Patch Independent"** window.

24. Right-click on **Mesh** in the **Outline** pane; a shortcut menu is displayed. Choose the **Update** button to update the mesh with the **Patch Independent** algorithm; the mesh with **Patch Independent** algorithm is generated, as shown in Figure 8-28.

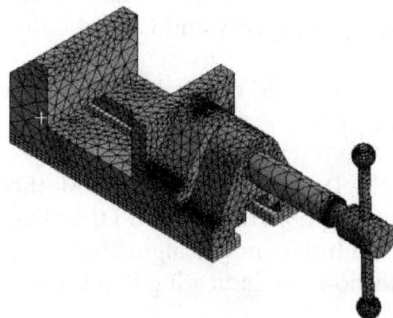

Figure 8-28 Mesh generated with default algorithm settings

Note
*1. You cannot preview a mesh by using the **Preview Mesh** tool when the **Patch Independent** algorithm is selected in the **Algorithm** drop-down list.*

2. Approximate number of elements generated is 84146.

25. Choose the **Body** tool from the **Select** toolbar and then select the Base and the Slider components from the Graphics screen.

26. Right-click on the Base component in the Graphics screen and then choose **Hide All Other Bodies** from the shortcut menu displayed; all the components of the assembly except the Base and the Slider components become invisible.

27. Right-click on the Graphics screen and then choose **View > Top** from the shortcut menu displayed; the model is oriented as shown in Figure 8-29.

28. Choose the **Section Plane** tool from the **Insert** panel of the **Home** tab; the **Section Planes** window is displayed on the left in the Graphics screen.

29. Choose the **New Section Plane** tool from the **Section planes** window, then in the Graphics screen, place and drag the cursor in such a manner that a section is created, as shown in Figure 8-30.

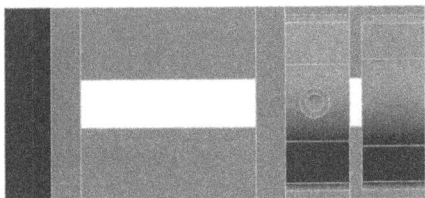

Figure 8-29 Top view of the model   *Figure 8-30 Section created*

30. Orient the model using the tools available in the **Graphics** toolbar, refer to Figure 8-31.

When you create a slice plane, the cells are visible through the sectioned plane. You can turn on or off the display of the cells using the **Show Whole Elements** toggle button in the **Section Planes** window after selecting the **Mesh** node from the **Outline** pane. The model after turning off the display is shown in Figure 8-32.

31. Select **Patch Independent** under the **Mesh** node in the **Outline** pane; the **Details of "Patch Independent"** window is displayed.

32. In this window, expand the **Advanced** node if not already expanded.

33. In the **Max Element Size** edit box, replace the default value by 6 mm.

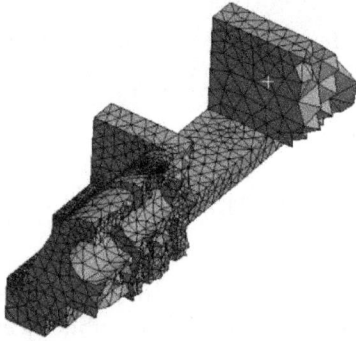

Figure 8-31 Sectioned model with whole elements displayed

Figure 8-32 Elements displayed through the section plane and with the **Show Whole Elements** toggle button turned off

The **Max Element Size** edit box is used to specify a value for the maximum size of the element in a model. The default value in this edit box depends on whether an option other than **NO** is selected in the **Use Adaptive Sizing Function** drop-down list in the **Sizing** node of **Details of "Mesh"** window.

34. Enter **30** in the **Feature Angle** edit box.

35. Specify **2** as the minimum size of the elements in the **Min Size Limit** edit box.

36. Specify **3** in the edit box **Num cell Across Gap** if not already specified.

37. Select the **On** option from the **Mesh Based Defeaturing** drop-down list under the **Advanced** node, refer to Figure 8-33; the **Defeature Size** edit box is displayed.

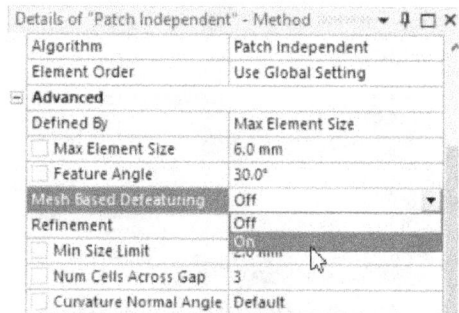

Figure 8-33 Selecting the **On** option from the **Mesh Based Defeaturing** drop-down list

38. Enter **0.0** in the **Defeature Size** edit box to set it to default.

Defeaturing is used to automatically defeature small features and undesirable geometry according to the **Defeature Size**.

39. Select the **On** option from the **Smooth Transition** drop-down list.

> **Note**
> *Make sure that the **Proximity and Curvature** option is selected in the **Refinement** drop-down list of the **Advanced** node.*

40. Select and then right-click on the Base in the Graphics screen; a shortcut menu is displayed.

41. Choose the **Show All Bodies** option from this shortcut menu; all the bodies are displayed in the Graphics screen.

42. In the **Section Planes** window, clear the **Section Plane1** check box; complete view of the model is restored.

43. Right-click on **Mesh** in the **Outline** pane; a shortcut menu is displayed. Choose the **Update** option; the mesh is generated with local mesh control settings, as shown in Figure 8-34.

Figure 8-35 shows the **Details of "Patch Independent"** window, after specifying all the parameters. The element count after generating the mesh is 1,18,908 which is more than the previous element count of 84,146.

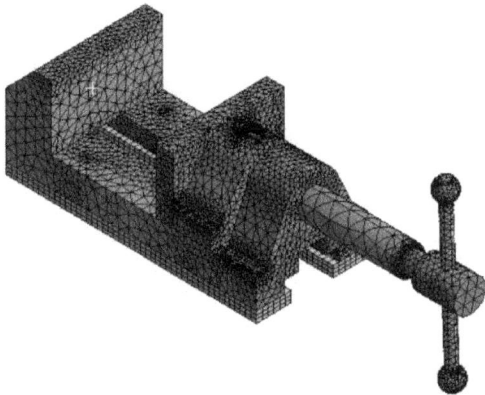

Figure 8-34 *Mesh generated with the **Patch Independent** method applied to the Base and the Slider*

Figure 8-35 *The **Details of "Patch Independent"** window*

The local mesh controls are introduced when further refinement of meshing is needed. The decision of applying local sizing controls depends upon the accuracy of the results required and time allotted for the pre-processing of components.

Introducing More Local Mesh Controls

After the Patch Independent algorithm is applied to the Base and the Slider, you may need to further refine the mesh in the areas where the quality of the mesh is not achieved as per the requirement.

1. Select the **Mesh** node in the **Outline** pane to display the **Details of "Mesh"** window.

2. Right-click on the **Mesh** node to display a shortcut menu. Choose **Insert > Sizing**, refer to Figure 8-36; **Sizing** is added under the **Mesh** node. Also, the **Details of "Sizing"** window is displayed and you are prompted to select faces, edges, or bodies to apply sizing controls.

When you select a face or many faces to apply sizing control, the number of faces and the area of the selected faces are displayed in the status bar. Also, when sizing is applied to a face, the transition of faces to adjoining faces will be smooth. You can select vertices, edges, faces, and bodies to apply sizing.

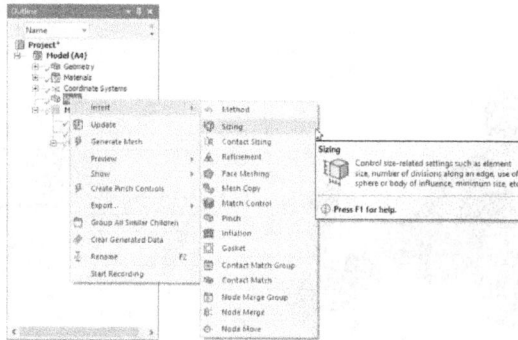

*Figure 8-36 Choosing the **Sizing** tool from the drop-down*

3. From the **Select** toolbar, choose the **Face** tool to select faces for sizing.

> **Note**
> *Sizing controls can also be applied to edges, vertices, and bodies.*

4. In the **Details of "Sizing"** window, select the **Geometry** selection box to display the **Apply** and **Cancel** buttons if they are not already displayed.

5. Next, press and hold the Ctrl key and then select the faces of the Base and the Slider components in the Graphics screen, refer to Figure 8-37.

6. Choose the **Apply** button from the **Geometry** selection box in the **Details of "Sizing"** window; **2 Faces** is displayed in the **Geometry** selection box indicating that

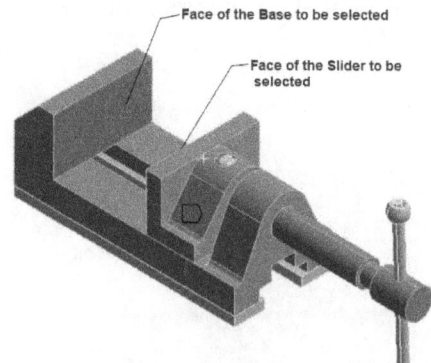

Figure 8-37 Selecting faces to apply sizing

the two faces are selected to apply face sizing. Also, in the **Outline** pane, **Sizing** is replaced by **Face Sizing** with a green tick mark on it.

7. Select **Face Sizing** in the **Outline** pane if not already selected; the **Details of "Face Sizing"** window is displayed.

8. Expand the **Definition** node in the **Details of "Face Sizing"** window if not already expanded.

9. Select the **Element Size** option from the **Type** drop-down list in the **Definition** node of the **Details of "Size"** window if not selected by default.

10. In the **Element Size** edit box, enter **2 mm** as the element size to be applied on the faces of the Base and the Slider. Also, make sure that the **Soft** option is selected from the **Behavior** drop-down list in the **Definition** node.

Note
*If you specify an element size of **2 mm** in the **Element Size** edit box of the **Details of "Face Sizing"** window, the element size will remain strictly to 2 mm wherever possible.*

11. Right-click on **Mesh** in the **Outline** pane; a shortcut menu is displayed. Choose the **Update** button; the mesh is generated, as shown in Figure 8-38.

    Note that the element count after the mesh is generated is 1,26,433 which is more than the previous element count of 84,146.

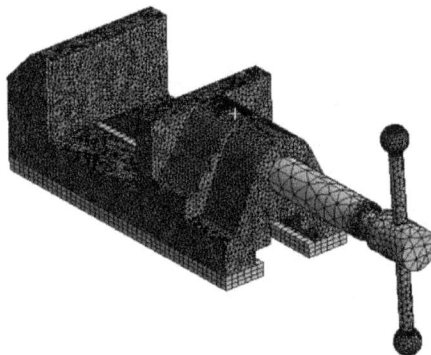

Figure 8-38 Mesh generated after sizing

12. Close the **Mechanical** window.

Saving the Model

After meshing the model and closing the **Mechanical** window, save the model in the specified folder.

1. Choose the **Save Project** button from the **Standard** toolbar to save the project.

2. Exit the **Workbench** window.

Tutorial 3

In this tutorial, you will create the surface model shown in Figure 8-39. The dimensions and views of the model are shown in Figures 8-40 and 8-41. After creating the model, you will generate a mesh for the surface model. The boundary and loading conditions are shown in Figure 8-42. Also, to understand the importance of geometry optimization, you will run a static structural analysis, optimize the model, and then evaluate the results.

(Expected time: 3 hr)

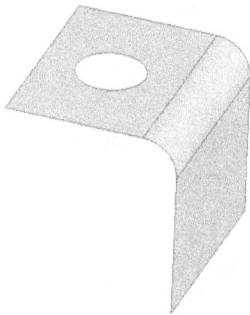

Figure 8-39 *Model for Tutorial 3*

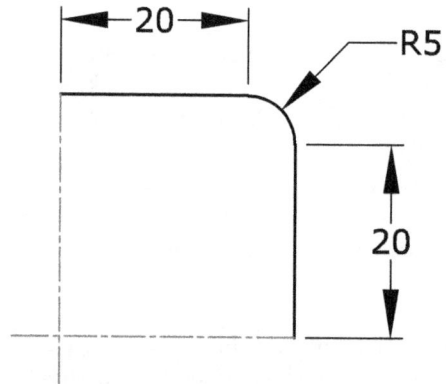

Figure 8-40 *Front view of the model*

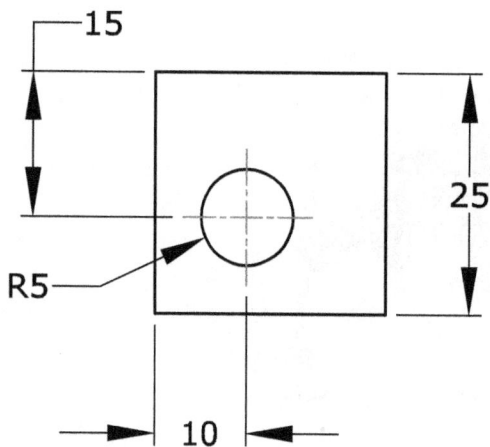

Figure 8-41 *Top view of the model*

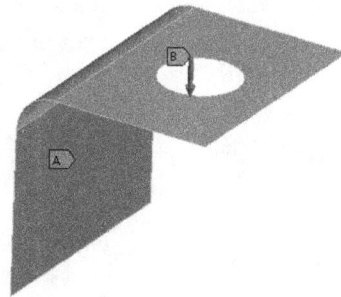

Figure 8-42 *Boundary and loading conditions for the model*

The following steps are required to complete the tutorial:

a. Start ANSYS Workbench and add an analysis system.
b. Create the base feature.
c. Create the hole feature.
d. Generate the mesh.
e. Set global and local mesh control settings.
f. Modify the geometry.
g. Set the boundary and loading conditions.
h. Set the results.
i. Optimize the results.
j. Save the project.

Starting ANSYS Workbench and Adding the Static Structural Analysis System

First, you need to start ANSYS Workbench and then add an analysis system to the **Project Schematic** window.

1. Start ANSYS Workbench to display the **Workbench** window.

   After invoking the **Workbench** window, you need to add an appropriate analysis system or a component system to the **Project Schematic** window.

2. Expand the **Analysis Systems** node and drag the **Static Structural** analysis system to the **Project Schematic** window; the **Static Structural** analysis system is added to the **Project Schematic** window.

   Note
   The selection of an analysis system depends on the type of analysis you want to perform.

3. Rename the **Static Structural** analysis system to **Surface_Mesh**.

4. In the **Workbench** window, choose the **Save Project** button from the **Standard** toolbar; the **Save As** dialog box is displayed.

5. In the **Save As** dialog box, browse to the location C:\ANSYS_WB\c08.

6. In this location, create a folder with the name **Tut03** and then choose the **Save** button from the **Save As** dialog box.

7. Save the project with the name **C08_ansWB_Tut03** at the location *C:\ANSYS_WB\c08\Tut03*.

Creating the Base Feature

After you have finished adding the analysis system, you now need to create the model using the **DesignModeler** window.

1. In the **Project Schematic** window, right-click on the **Geometry** cell in the **Surface_Mesh** component system and then select **New DesignModeler Geometry**; the **DesignModeler** window is displayed.

2. Choose the **Millimeter** option from the **Units** menu of the **Menu** bar.

    Note that the default plane selected in the **DesignModeler** window is the XY plane. Since the sketch is to be created on the XY plane, therefore you need not specify the plane before creating the sketch.

3. Choose the **Look At** tool from the **Graphics** toolbar; the XY plane is oriented normal to the viewing direction.

4. Create the sketch on the XY plane, as shown in Figure 8-43. For dimensions, refer to Figure 8-40.

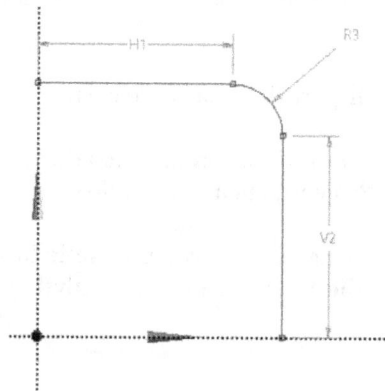

Figure 8-43 Fully constrained sketch on the XY Plane

5. Change the view to Isometric by using the **ISO** tool from the **Graphics** toolbar.

6. Choose the **Extrude** tool from the **Features** toolbar; the **Sketching** mode is closed and the **Modeling** mode is invoked. Also, **Extrude1** is attached to the **Outline** pane and the corresponding **Details View** window is displayed.

7. Click on the **Geometry** selection box in the **Details View** window; the **Apply** and **Cancel** buttons are displayed. As there is only one sketch in the Graphics screen, it gets automatically selected for the extrude operation.

8. Select **Sketch1** under the **XYPlane** node in the **Outline** pane if not already selected. Next, choose the **Apply** button from the **Geometry** selection box to specify the sketch for extrusion.

9. In the **Details View** window, specify **25** as the length of extrusion in the **FD1, Depth (>0)**, edit box, refer to Figure 8-44.

10. Choose the **Generate** tool from the **Features** toolbar; the sketch is extruded, as shown in Figure 8-45.

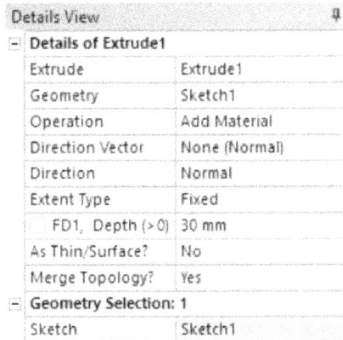

Details View	
Details of Extrude1	
Extrude	Extrude1
Geometry	Sketch1
Operation	Add Material
Direction Vector	None (Normal)
Direction	Normal
Extent Type	Fixed
FD1, Depth (>0)	30 mm
As Thin/Surface?	No
Merge Topology?	Yes
Geometry Selection: 1	
Sketch	Sketch1

Figure 8-44 The *FD1, Depth (>0)* edit box in the *Details View* window

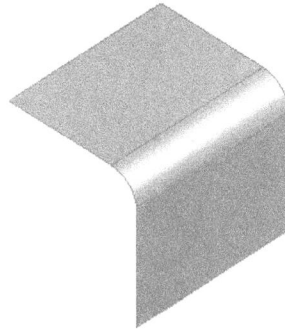

Figure 8-45 Extruded surface model

A surface model created in 3D modeling software does not exist practically in the real world. To use a surface model, you need to convert it into a 3D model by providing it a thickness in the **DesignModeler** window. In case the surface model is modeled in any other CAD package, you can give it a thickness in the same package where it was modeled. Otherwise, you can import it to ANSYS and then provide it a thickness while generating a mesh.

In ANSYS, the surfaces are modeled with plate/shell elements. You can consider generating a mesh with plate/shell elements if the thickness of the plate is in a ratio of 10:1 with the smallest dimension of the model.

Creating the Hole Feature

After the model is created, you need to create a hole feature on the top surface of the model.

1. Choose the **Face** tool from the **Select** toolbar and then select the top face of the model.

2. Next, choose the **New Plane** tool from the **Active Plane/Sketch** toolbar to create a new plane on the selected face; preview of the new plane is displayed on the model, as shown in Figure 8-46.

3. Choose the **Generate** tool from the **Features** toolbar; the plane is generated.

4. Choose the **Look At** tool from the **Graphics** toolbar to orient the model normal to your viewing direction.

5. Create a circle on the top face of the plane. For dimensions, refer to **Figure 8-41**.

6. Next, choose the **Extrude** tool and then create a circular cut, as shown in Figure 8-47.

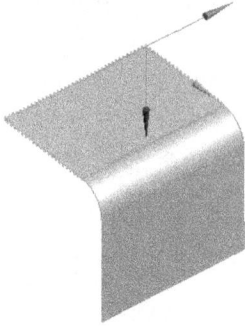

Figure 8-46 Preview of the work plane

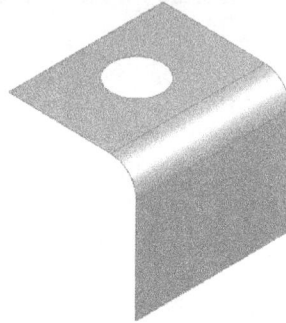

Figure 8-47 Circular cut created on the top face of the model

7.   Exit the **DesignModeler** workspace; the **Workbench** window is displayed.

Generating the Mesh

After creating the hole feature, you need to generate an effective mesh for it.

1.   In the **Project Schematic** window, double-click on the **Model** cell of the **Surface_Mesh** analysis system to display the **Mechanical** window.

2.   Expand the **Geometry** node in the **Outline** pane; **Surface Body** is displayed with a question symbol attached to it indicating that immediate action is required.

3.   Select **Surface Body** in the **Outline** pane; the **Details of "Surface Body"** window is displayed. Also, the surface body is displayed in green in the Graphics screen.

4.   In the **Details of "Surface Body"** window, expand the **Definition** node if not already expanded, as shown in Figure 8-48.

     The **Thickness** edit box in the **Details of "Surface Body"** window is displayed in yellow, indicating that there is no value attached to it.

5.   Enter **0.2** in the **Thickness** edit box.

6.   Right-click on **Mesh** in the **Outline** pane and then choose the **Generate**   ⋛ Generate Mesh
     Mesh option from the shortcut menu displayed; the mesh with default
     global mesh control settings is generated, as shown in Figure 8-49.

Details of "Surface Body"	▼ ⊣ ☐ ✕
+ Graphics Properties	
− Definition	
Suppressed	No
Dimension	3D
Stiffness Behavior	Flexible
Coordinate System	Default Coordinate...
Reference Temperature	By Environment
Thickness	0.2 mm
Thickness Mode	Manual
Offset Type	Middle
Treatment	None
− Material	
Assignment	Aluminum Alloy

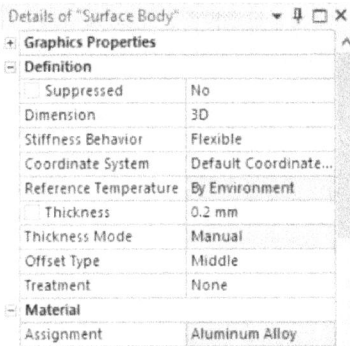

Figure 8-48 The *Details of "Surface Body"* window

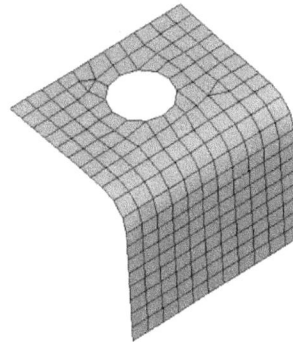

Figure 8-49 *Mesh generated with default global mesh control settings*

It is always recommended to generate mesh with ANSYS default settings. This helps in understanding the behavior of the elements with respect to the model.

7. Select **Mesh** in the **Outline** pane; the **Details of "Mesh"** window is displayed, as shown in Figure 8-50.

    Notice that in the **Sizing** node of the **Details of "Mesh"** window, **NO** is selected in the **Use Adaptive Sizing** drop-down list. Also, the **Element Size** edit box displays **Default** in it. For better results, you need to change few settings in the **Details of "Mesh"** window.

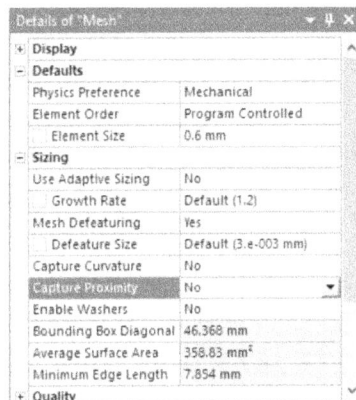

Details of "Mesh"	▼ ⊣ ✕
+ Display	
− Defaults	
Physics Preference	Mechanical
Element Order	Program Controlled
Element Size	0.6 mm
− Sizing	
Use Adaptive Sizing	No
Growth Rate	Default (1.2)
Mesh Defeaturing	Yes
Defeature Size	Default (3.e-003 mm)
Capture Curvature	No
Capture Proximity	No
Enable Washers	No
Bounding Box Diagonal	46.368 mm
Average Surface Area	358.83 mm²
Minimum Edge Length	7.854 mm
+ Quality	

Figure 8-50 The *Details of "Mesh"* window

Note

*After the default material is selected for the analysis, you now need to change the material of the surface model to aluminium alloy. You can access this material from the materials library in the **Engineering Data Sources** window.*

8. Assign the **Aluminum Alloy** material to the model in the **Mechanical** window.

Setting the Global and Local Mesh Controls

In the previous section, you had created the finite element model with default global mesh control settings. Now, you will change these settings for the model.

1. As there is a circular cutout and a bend in this model, you need to have finer mesh around the curves. To do so, select the **No** option from the **Use Adaptive Sizing** drop-down list and select the **Yes** option from the **Capture Proximity** drop-down list in the **Details of "Mesh"** window.

2. Specify **4** in the **Proximitiy Gap Factor** edit box.

3. Enter **0.4** in the **Proximity Min Size** edit box.

4. Enter **0.6** as the size of the element in the **Element Size** edit box.

    Leave all other options as set to default.

5. Right-click on **Mesh** in the **Outline** pane; a shortcut menu is displayed. Choose the **Update** button; the mesh is updated, as shown in Figure 8-51.

Figure 8-51 Mesh generated for the model

Depending upon the requirement of the analysis such as the time required to run the analysis and accuracy expected out of the analysis, you may need to introduce local mesh control for a particular region or for the complete body.

After specifying the global mesh control settings, the next step is to apply local mesh control settings to the model.

6. Right-click on **Mesh** in the **Outline** pane and then choose **Insert > Method** from the shortcut menu displayed; **Automatic Method** is added under the **Mesh** node with a question symbol attached to it. Also, the **Details of "Automatic Method"** window is displayed, as shown in Figure 8-52.

7. Click on the **Geometry** selection box to display the **Apply** and **Cancel** buttons.

    Note
    *The **Geometry** selection box in this window is highlighted in yellow, indicating that the geometry to be considered for a method control is yet to be selected.*

8. Choose the **Body** tool from the **Select** toolbar and then select the body, as shown in Figure 8-53.

9. Choose the **Apply** button from the **Geometry** selection box; the body is selected for applying the method. Also, a green tick mark is placed before **Automatic Method** in the **Outline** pane.

Notice that the **Quadrilateral Dominant** option is displayed in place of **Automatic Method** in the **Method** drop-down list in the **Details of "Automatic Method"** window.

The **Quadrilateral Dominant** option is selected by default in the **Method** drop-down list. This option is used for surface bodies only. When the **Quadrilateral Dominant** option is selected, the body is meshed with free quadrilaterals. Also, the **Free Face Mesh Type** drop-down lists is displayed, refer to Figure 8-52. You can use the options available in these drop-down lists to control element shapes and sizes in a geometry.

10. In the **Method** drop-down list, select the **Triangles** option, refer to Figure 8-54; the **Automatic Method** in the **Outline** pane is replaced by **All Triangles Method**, as shown in Figure 8-55.

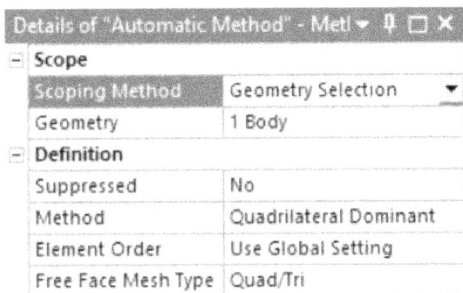

*Figure 8-52 The **Details of "Automatic Method"** window*

Figure 8-53 Selected surface body in the Graphics screen

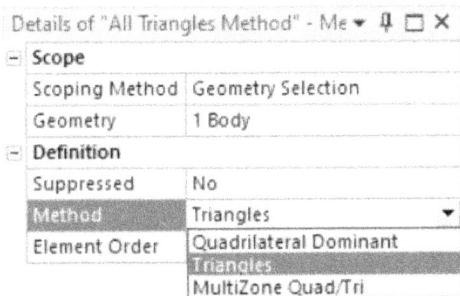

*Figure 8-54 Selecting the **Triangles** option from the **Method** drop-down list*

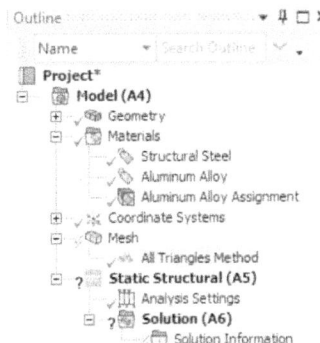

*Figure 8-55 The **All Triangles Method** displayed in the **Outline** pane*

11. Right-click on **Mesh** in the **Outline** pane; a shortcut menu is displayed. Choose the **Update** button, the mesh is updated, as shown in Figure 8-56.

    Notice that the total number of elements created in this model are 8188.

Node Move

The **Node Move** is a mesh edit tool and used to select and then manually move a specific node on the mesh to improve the local mesh quality.

Figure 8-56 *Mesh generated with all traingular elements*

1. Right-click on **Mesh** in the **Outline** pane and then choose **Insert > Node Move** from the shortcut menu displayed; **Mesh Edit** is added under the **Mesh** node as shown in Figure 8-57 . Also, the **Details of "Mesh Edit"** window is displayed.

2. Expand the **Mesh Edit** node to view the **Node Move** object in the **Outline.** Also, click on **Node Move** object to display **Details of "Node Move" window,** as shown in Figure 8-58.

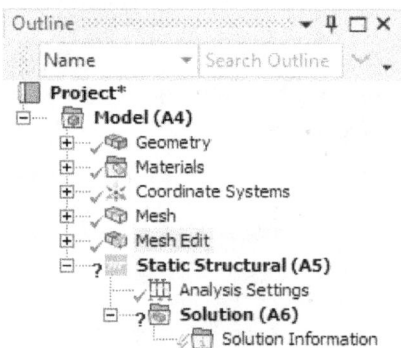

Details of "Node Move"	
Scope	
Node Move Method	Manual
Scoping Method	Geometry Selection
Geometry	All Bodies
Displacement File	
Definition	
Number of Moves	2
Number of Nodes	2
Information	Hold F4 Key to Move

Figure 8-57 *The **Mesh Edit** node in outline*

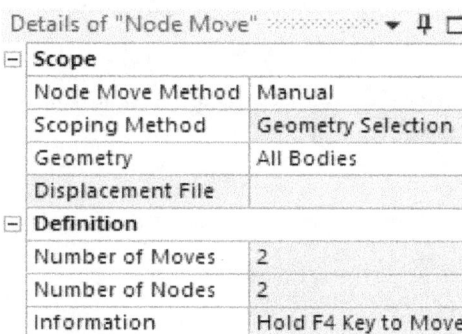

Figure 8-58 *The **Details of "Node Move"** window*

3. In the **Outline** pane, click on **Mesh** node. Next, select **Element Quality** from **Display Style** drop-down list under **Details of "Mesh"** window; the graphics screen is displayed with mesh style according to the **Element Quality** criterion along with corresponding contour, as shown in Figure 8-59.

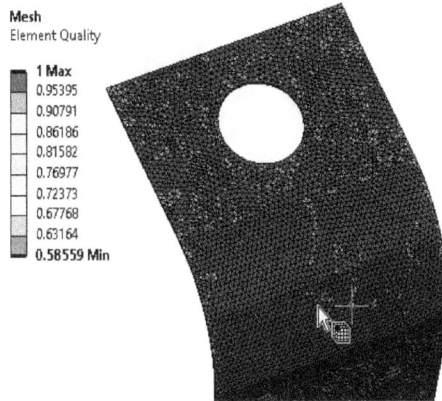

Figure 8-59 *The **Element Quality** contour*

Note that from Figure 8-59, contour indicates maximum and minimum value of element quality. The maximum value of element quality is 1 and minimum value is 0.65454. Next, you can move the node of the elements of lower element quality value to reach as much as closer to the elements of the maximum element quality value

4.   Next, click on **Node Move** in the Outline(**Mesh** contextual tab in the **Menu** bar has changed to **Node Move**). Choose the **Min** tool from the **Display** panel of the **Node Move** tab; the graphics window is displayed the minimum value element over the mesh.

5.   Now, choose the **Node** tool from the **Select** toolbar to enable selection of nodes.

6.   Select corresponding node of the minimum value element as shown in Figure 8-60, and start moving in a such way that red color of element is changed to any other color, refer to Figure 8-61.

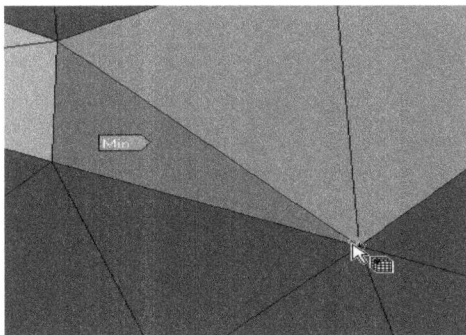

Figure 8-60 *Partial view node selection of corresponding minimum vlue element*

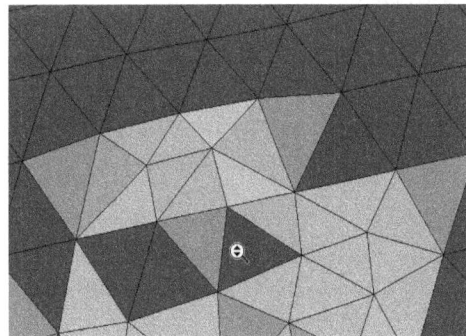

Figure 8-61 *Displaying color changes of elements after node move.*

7. Similarly, select all other corresponding nodes of minimum value elements and move the node to improve the minimum value of the **Element Quality** contour, refer Figure 8-62 .

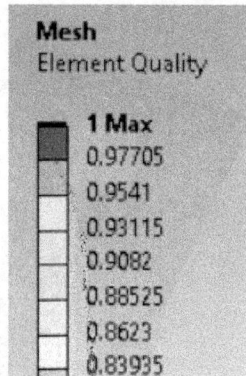

Figure 8-62   Element
***Quality** after Node move.*

To cancels the last node movement performed on the mesh, select **Last** tool from the
Undo panel of the **Node** Move tab.

Similarly, you can cancel all node movements that you have performed on the mesh.
To do this, choose the **All** tool from the **Undo** panel of the **Node Move** tab.

Note that the minimum value of the selected mesh criteria after nodes move is 0.79823
which is more than 0.58559, refer to Figure 8-59 and Figure 8-62 . Similarly, you can select
another mesh criteria and improve the local mesh quality.

8.   In the **Outline** pane, click on **Mesh** node. Next, select **Use Geometry Setting** from **Display
Style** drop-down list under **Details of "Mesh"** window to display default geometry setting

9.   Exit the **Mechanical** window; the **Workbench** window is displayed.

Optimizing the Geometry

In Figure 8-42, you can notice that to apply load on the model, you first need to create
the patch on the circular hole.

1.   In the **Workbench** window, double-click on the **Geometry** cell of the **Surface_Mesh** analysis
system; the **DesignModeler** window is displayed.

2.   Choose the **Surface Patch** tool from the **Tools** menu; **SurfPatch1** is added to the **Outline**
pane. Also, the corresponding **Details View** window is displayed and you are prompted to
select a loop of edges to create the patch.

3.   Next, select the **Patch Edges** selection box in the **Details View** window to display the **Apply**
and **Cancel** buttons, if they are not already displayed.

4.   Select the circular edge, as shown in Figure 8-63.

5. Choose the **Apply** button to confirm the selection; the edge turns blue and **1** is displayed in the **Patch Edges** selection box, indicating that one loop has been selected for applying the patch.

6. Choose the **Generate** tool from the **Features** toolbar to generate the surface patch, as shown in Figure 8-64.

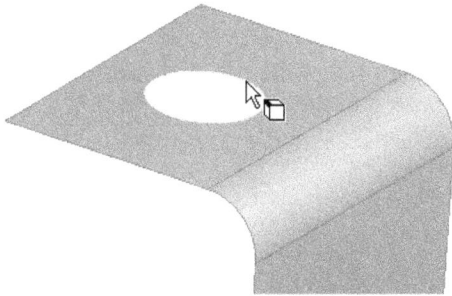

Figure 8-63 Partial view of the model with the circular edge selected

Figure 8-64 Patch created around the selected edge

7. Exit the **DesignModeler** window to display the **Workbench** window.

Setting the Boundary and Loading Conditions

After the patch is created, you now need to set the boundary and loading conditions for the model.

1. In the **Workbench** window, choose the **Update Project** tool from the **Standard** toolbar; the model is updated to the current state.

2. In the **Surface_Mesh** analysis system, double-click on the **Model** cell; the **Mechanical** window is displayed.

3. In the **Outline** pane, right-click on the **Static Structural** node. Next, choose **Insert > Fixed Support** from the shortcut menu displayed; **Fixed Support** is added under the **Outline** pane. Also, the **Details of "Fixed Support"** window is displayed with the **Geometry** selection box displaying the **Apply** and **Cancel** buttons.

4. Choose the **Face** tool from the **Select** toolbar if not already chosen and then select the surface, as shown in Figure 8-65.

Note
You can restrict the movement of a model in all directions by applying Fixed support to it. You can apply Fixed support to edges, faces, and vertices.

5. Next, choose the **Apply** button from the **Geometry** selection box in the **Details of "Fixed Support"** window; the face is selected to apply fixed constraint.

6. Right-click on the **Static Structural** node again and then choose **Insert > Force** from the shortcut menu displayed; **Force** is added under the **Static Structural** node in the **Outline** pane. Also, the corresponding **Details of "Force"** window is displayed.

Force is known as the rate of change of momentum. In ANSYS Workbench, you can apply Force load by using the **Environment** contextual toolbar.

7. In the **Details of "Force"** window, select the **Geometry** selection box to display the **Apply** and the **Cancel** buttons if they are not already displayed.

8. Choose the **Face** tool from the **Select** toolbar and then select the patched surface, as shown in Figure 8-66. Next, choose the **Apply** button in the **Geometry** selection box; **1Face** is displayed in the **Geometry** selection box.

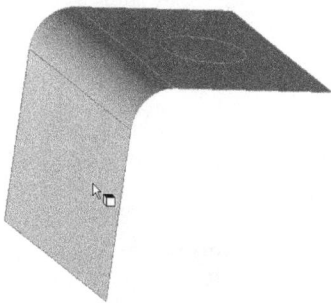

Figure 8-65 Selecting the face to apply Fixed support

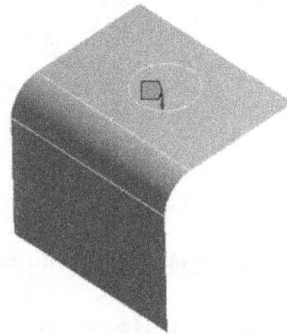

Figure 8-66 Selecting the face to apply Force load

9. In the **Details of "Force"** window, enter **2.5** in the **Magnitude** edit box to specify the magnitude of force.

10. Select the **Direction** selection box to display the **Apply** and **Cancel** buttons.

11. Select the edge of the model, as shown in Figure 8-67 and then choose the **Apply** button; the Force load of **2.5** N is applied on the selected edge.

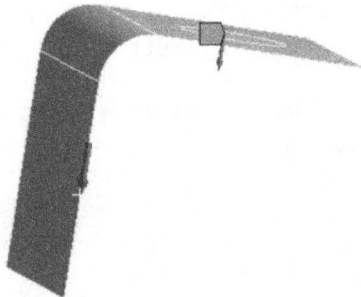

Figure 8-67 Selecting the edge for specifying the direction of force load

Setting the Results

After the boundary and loading conditions are specified in the **Mechanical** window, you need to define the results.

1. Right-click on the **Solution** node in the **Outline** pane to display a shortcut menu.

2. In this shortcut menu, choose **Insert > Deformation > Total**; **Total Deformation** is added under the **Solution** node.

3. Right-click on the **Solution** node again and then choose **Insert > Stress > Equivalent (von-Mises)** from the shortcut menu displayed; **Equivalent Stress** is added under the **Solution** node in the **Outline** pane.

   Notice that **Total Deformation** and **Equivalent Stress** added under the **Solution** node are displayed with yellow thunderbolts which indicate that results are not evaluated yet.

4. Choose the **Solve** tool from the **Standard** toolbar; the **ANSYS Workbench Solution Status** dialog box is displayed and after sometime, a green tick mark is placed before **Total Deformation** and **Equivalent Stress** in the **Outline** pane, indicating that they are evaluated.

5. Select **Total Deformation** in the **Outline** pane; the corresponding contours are displayed in the Graphics screen, as shown in Figure 8-68.

6. Select **Equivalent Stress** in the **Outline** pane; the corresponding contours are displayed in the Graphics screen, as shown in Figure 8-69.

Note
*You can change the display of the result by selecting desired **Contours** and **Edges** from the **Result** contextual tab in the **Menu** bar.*

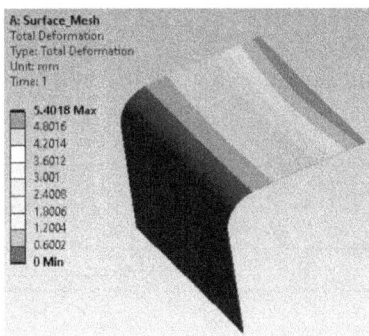

Figure 8-68 Total Deformation contours in the model

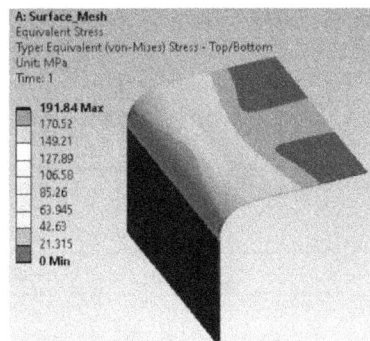

Figure 8-69 Equivalent Stress contours in the model

The following table describes the maximum and minimum value for the **Total Deformation** and **Equivalent Stress** induced in the model.

S.No.	Parameter	Min. Value	Max. Value
1	Total Deformation	0	5.40 mm
2	Equivalent Stress	0	191.84 MPa

After the values are found, the next step is to optimize the results by modifying the meshing parameters. In the next section, you will optimize the model for better results.

Optimizing the Results

To optimize the model for better results, you need to change the global and local mesh settings.

1. In the **Mechanical** window, select the **Mesh** node in the **Outline** pane; the **Details of "Mesh"** window is displayed.

2. In the **Details of "Mesh"** window, expand the **Sizing** node, if it is not already expanded.

3. Enter **2** in the **Proximitiy Gap Factor** edit box.

4. Enter **0.3** in the **Proximity Min Size** edit box.

5. Enter **0.9** in the **Element Size** edit box.

6. Choose the **Solve** tool from the **Standard** toolbar to generate the mesh and update the results in the **Solution** node.

7. Next, select **Total Deformation** from the **Outline** pane to display the contours in the Graphics screen.

8. Similarly, select **Equivalent Stress** from the **Outline** pane; the stress contours are displayed in the Graphics screen.

    The following table describes the maximum and minimum values for the **Total Deformation** and **Equivalent Stress** induced in the model.

S.No.	Parameter	Min. Value	Max. Value
1	Total Deformation	0	5.37 mm
2	Equivalent Stress	0	173.16 MPa

Notice that when the optimized model is meshed the element count has decreased. On comparing the data available in the tables given previously in this section, you will find that there is not much difference in the Total Deformation achieved when the optimized model is used. However, there is a fall in Equivalent Stress value when the element count decreases. Therefore, to save the processing time and keep the model simple, you need to use the optimized model.

9. Exit the **Mechanical** window; the **Workbench** window is displayed.

Saving the Model

1. Choose the **Save Project** button in the **Standard** toolbar to save the project with the name *c08_ansWB_tut03*.

2. Choose **File > Exit** to close the **Workbench** window.

Tutorial 4

In this tutorial, you will download the *c08_ansWB_tut04.zip* file from www.cadcim.com, extract, and save the *c08_ansWB_tut04.stp* file in the project folder. Next, you will generate a mesh for the model shown in Figure 8-70 and then modify the model using the Mid-Surface extraction tool to convert a 3D model into a 2D surface body.

(Expected time: 3 hr)

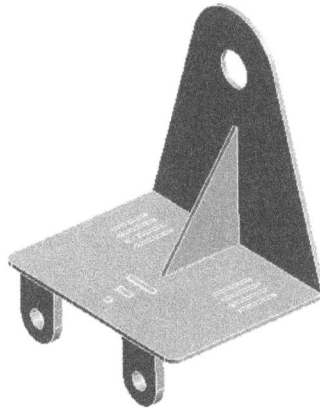

Figure 8-70 *Isometric view of the model for Tutorial 4*

The following steps are required to complete the tutorial:

a. Download the part file.
b. Start a new project and import the file in ANSYS Workbench.
c. Generate the mesh.
d. Modify the model.
e. Generate the mesh again.
f. Save the project.

Downloading the Part File

1. Create a new folder with the name **Tut04** at the location *C:\ANSYS_WB\c08*.

2. Download the file *c08_ansWB_tut04.zip* from *www.cadcim.com*. The complete path to download the file is as follows:

    Textbooks > CAE Simulation > ANSYS > ANSYS Workbench 2023 R2: A Tutorial Approach > Input Files

    After downloading, extract the zip file to save the igs part file *c08_ansWB_tut04.igs* at the location *C:\ANSYS_WB\c08\Tut04*.

Importing the File in ANSYS Workbench

After downloading the file from *www.cadcim.com*, you need to import the file into ANSYS Workbench.

1. Start ANSYS Workbench to display the **Workbench** window.

2. In the **Workbench** window, add the **Static Structural** analysis system into the **Project Schematic** window.

3. Double-click on the name field of the **Static Structural** analysis system and rename it to **Mid-Surface_Mesh.**

4. Right-click on the **Geometry** cell of the **Mid-Surface_Mesh** analysis system to display a shortcut menu.

5. Choose the **Import Geometry** option from the shortcut menu; a flyout is displayed.

6. From this flyout, choose the **Browse** option; the **Open** dialog box is displayed.

7. Browse to the location *C:\ANSYS_WB\c08\Tut04* and then open *c08_ansWB_tut04.igs*; a green tick mark is placed in the **Geometry** cell of the **Mid-Surface_Mesh** analysis system indicating that the geometry is specified for the analysis.

8. In the **Workbench** window, choose the **Save Project** button from the **Standard** toolbar; the **Save As** dialog box is displayed.

9. In this dialog box, browse to the location *C:\ANSYS_WB\c08\Tut04* and then save the project with the name **c08_ansWB_tut04**.

Generating the Mesh

After importing the model, you need to generate an effective mesh for it.

1. In the **Project Schematic** window, double-click on the **Model** cell of the **Mid-Surface_Mesh** analysis system to display the **Mechanical** window.

2. Click on **Mesh** in the **Outline** pane and then right-click to display the shortcut menu.

3. Choose the **Generate Mesh** option from the shortcut menu displayed; the **ANSYS Workbench Mesh Status** window is displayed. After sometime, the **ANSYS Workbench Mesh Status** window disappears and the mesh is generated, as shown in Figure 8-71.

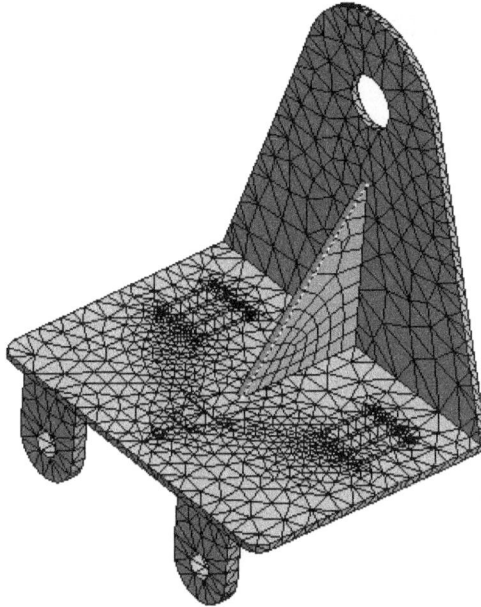

Figure 8-71 Mesh generated for the Model

4. Select **Mesh** in the **Outline** pane; the **Details of "Mesh"** window is displayed. Also, a green tick mark is placed corresponding to **Mesh** in the **Outline** pane indicating that the mesh for the model is created successfully. Note the number of elements that are created after enerating the mesh.

To get the details about the number of elements and nodes created after the model is discretized, expand the **Statistics** node in the **Details of "Mesh"** window, refer to Figure 8-72.

After the mesh is generated, the number of elements will be 5401 and the number of nodes will be 11865. You can usethis data later to compare the effectiveness of the mesh aftergeometry modification.

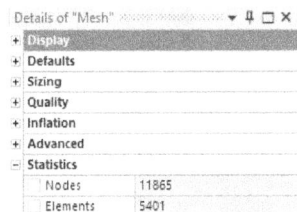

Figure 8-72 The **Statistics** node in the **Details of "Mesh"** window

Note

*The values displayed in the **Statistics** node of the **Details of "Mesh"** window may vary in your system.*

5. Exit the **Mechanical** window; the **Workbench** window is displayed.

Optimizing the Geometry

After the mesh is generated, you need to remove the irregularities in the model by using the **DesignModeler** window.

1. In the **Workbench** window, right-click on the **Geometry** cell and choose the **Edit Geometry in DesignModeler**; the **DesignModeler** window is displayed.

2. Choose the **Generate** tool from the **Features** toolbar to generate the model in the graphics screen.

> **Note**
> *Choose the **Frozen Body Transparency** option from the **View** menu to turn off transparent mode.*

There are three extruded parts on the top surface of the model, refer to Figure 8-73. Removing these small details will help improve the topology for meshing and will reduce the element count.

3. Invoke the **Face Delete** tool from the **Create** menu; **FDelete1** with a yellow thunderbolt is attached to the **Outline** pane.

4. Select the all visible faces present on the extruded parts , as shown in Figure 8-74, and then choose the **Apply** button in the **Faces** selection box; **15** is displayed in the **Faces** selection box indicating that fifteen faces have been selected for the **Face Delete** operation.

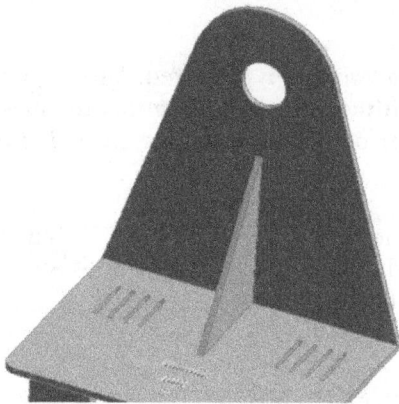

Figure 8-73 Extruded parts on the Model

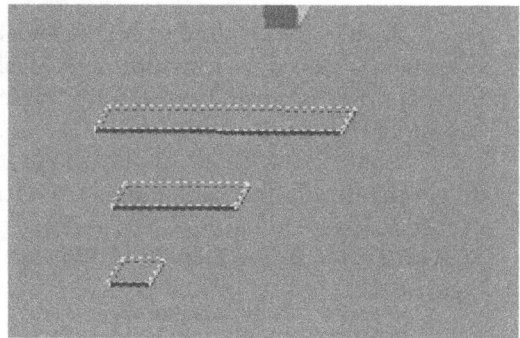

Figure 8-74 Selecting faces

5. Choose the **Generate** tool available in the **Features** toolbar; the selected surfaces on extruded parts are deleted. Also, a green tick mark is placed before **FDelete1** in the **Outline** pane.

Notice there is uniform thickness along all the 4 components of the body. One of the better features of workbench's **DesignModeler** is **Mid-Surface** tool. This tool is used to create a surface model from a solid model. The Mid-Surface feature allows the creation of surface bodies that are midway between existing solid body faces. The resulting surface body/bodies have a thickness property which defines the "thickness" that surface body represents

> **Note**
> *The Element thickness, specified by you, is assigned with half in the +Z direction (element top) and the other half in the -Z direction (element bottom)*

6. Choose **Millimeter** option from the **Units** menu of the Menu bar.

7. Choose the **Mid-Surface** tool from the **Tools** menu; **MidSurf1** is added to the Outline pane. Also, the corresponding **Details View** window is displayed and you are prompted to select a face pairs to create mid-surafce feature.

8. Select **Automatic** from the **Selection Method** drop-down list in the **Details of "MidSurf1"** window, as shown in Figure 8-75; corresponding options are displayed in the **Details View** window, as shown in Figure 8-76.

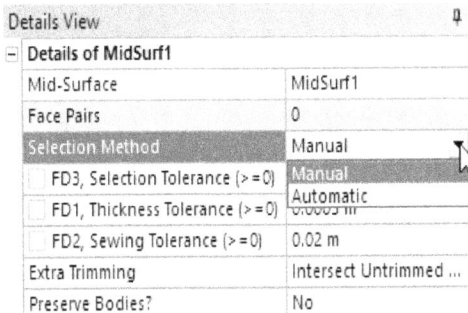

Details View	⎙
− **Details of MidSurf1**	
Mid-Surface	MidSurf1
Face Pairs	0
Selection Method	Manual
FD3, Selection Tolerance (> =0)	Manual / Automatic
FD1, Thickness Tolerance (> =0)	0.0005 m
FD2, Sewing Tolerance (> =0)	0.02 m
Extra Trimming	Intersect Untrimmed ...
Preserve Bodies?	No

Details View	⎙
− **Details of MidSurf1**	
Mid-Surface	MidSurf1
Face Pairs	0
Selection Method	Automatic
Bodies To Search	Visible Bodies
Minimum Threshold	0 m
Maximum Threshold	0 m
Find Face Pairs Now	No
FD3, Selection Tolerance (> =0)	0 m
FD1, Thickness Tolerance (> =0)	0.0005 m
FD2, Sewing Tolerance (> =0)	0.02 m
Extra Trimming	Intersect Untrimmed ...
Preserve Bodies?	No

Figure 8-75 Selection Method drop-down list

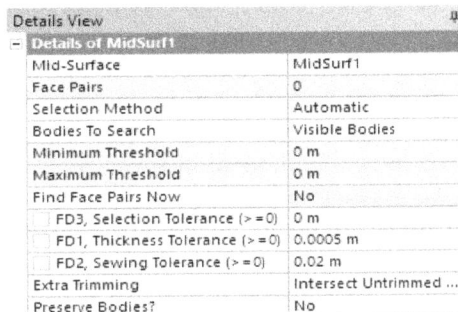

Figure 8-76 Displying options for Automatic selection method

9. Select **All bodies** from the **Bodies To Search** drop-down list in the **Details of MidSurf1** window, if it is not selected by default, refer to Figure 8-77

10. Enter **4** in the **Minimum Threshold** edit box and enter **8** in the **Maximum Threshold** edit box, refer to Figure 8-77.

  The **Minimum Threshold** option is used to set minimum thickness value allowed between the face pairs for automatic selection. Similarly, the **MaximumThreshold** option is used to set maximum thickness value allowed between the face pairs for automatic selection.

11. Select **FD3, Selection Tolerance (>=0)** edit box in the **Details View** window and enter 0.02, refer to Figure 8-77.

  Sometimes, the selected face pairs are not perfectly offset by a fixed distance from one another. To accommodate this offset deviation, you need to provide some tolerance to the offset distance. The **FD3, Selection Tolerance (>=0)** option is used to set deviation value for perfectly offset face pairs.

Note

1. For this model, all the components are having thickness range between 4 mm to 8 mm. It is recommended that you keep a note of the thickness range in the model.

2. Selection tolerance value may vary depending upon the irregularities in the model.

3. In case of manual method, you will create midsurfaces to manually go through and select the face pairs.

12. Select **Yes** from the **Find Face pairs Now** drop-down list in the **Details of "MidSurf1"** window.

You will notice that the **Find Face pairs Now** selection box is reset to **No**, refer to Figure 8-77. In addition to this, **7** is displayed in the **Face Pairs** selection box indicating that seven face pairs have been selected for the **Mid-Surface** operation. The corresponding face pairs of the model in the graphics window is highlighted in two different color.

13. Choose the **Generate** tool from the **Features** toolbar to generate the mid-surface, as shown in Figure 8-78.

If you expand the **4 Parts**, **4 Bodies** node in the **Outline**, four surface bodies are displayed instead of 4 solid bodies.

Details View	⊥
Details of MidSurf3	
Mid-Surface	MidSurf3
Face Pairs	7
Selection Method	Automatic
Bodies To Search	All Bodies
Minimum Threshold	4 mm
Maximum Threshold	8 mm
Find Face Pairs Now	No
FD3, Selection Tolerance (>=0)	0.02 mm
FD1, Thickness Tolerance (>=0)	0.02 mm
FD2, Sewing Tolerance (>=0)	0.02 mm
Extra Trimming	Intersect Untrimmed ...
Preserve Bodies?	No

Figure 8-77 Displaying inputs for Automatic method

Figure 8-78 selecting face pairs

Notice that the gap is generated between the surface bodies because the mid-surface is created at the center of the thickness. These gaps can be filled in different ways. Here two methods are discussed to fill the gap. In the first method, you can extend surface of one body next to the other bodies only if the direction of extrusion is normal to surfaces. In the second method, you can create weld in between the two bodies to join.

14. Choose the **Surface Extension** tool from the **Tools** menu; **SurfaceExt1** is added to the **Outline** pane. Also, the corresponding **Details View** window is displayed and you are prompted to select an edge to perform the operation for extending surfaces.

15. Select the two edges of surface body as shown in Figure 8-79, and then choose the **Apply** button in the **Edges** selection box; **2** is displayed in the **Edges** selection box indicating that two edges have been selected for the surface extension operation.

16. Select **To Next** from the **Extent** drop-down list in the **Details of SurfaceExt1** window.

17. Choose the **Generate** tool from the **Features** toolbar to generate the surface, as shown in Figure 8-80.

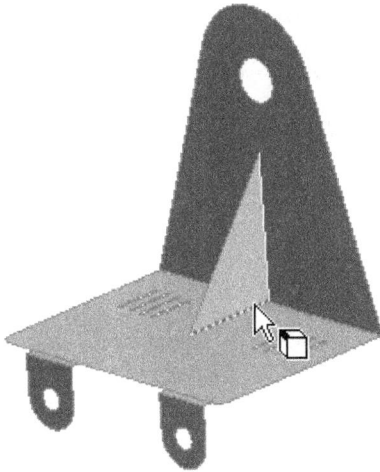

Figure 8-79 Selecting edges for Surface Extension operation

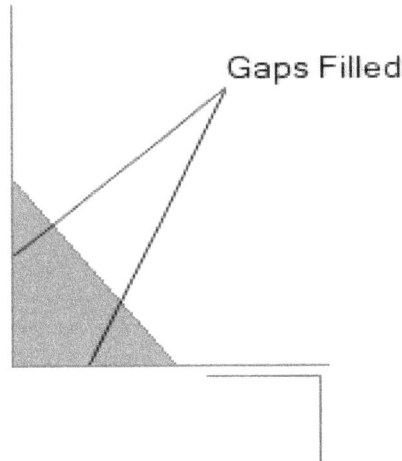

Figure 8-80 Extended surface generated to fill the gaps between bodies

Note
*For this model, only triangular shaped body is eligible for surface extension, refer to Figure 8-79 and Figure 8-80. The remaining gaps will be filled by using **Weld** tool.*

18. Next, choose the **Weld** tool from the **Tools** menu; **Weld1** is added to the **Outline** pane. Also, the corresponding **Details View** window is displayed and you are prompted to select a edges to create weld feature.

19. Select the edges of surface bodies as shown in Figure 8-81, and then choose the **Apply** button in the **Edges** selection box; **8** is displayed in the **Edges** selection box indicating that eight edges have been selected for the **Weld** operation.

20. In the **Details of Weld1** window, select the **Faces** selection box to display the **Apply** and the **Cancel** buttons if they are not already displayed.

21. Select the face of the surface body as shown in Figure 8-82, then choose the **Apply** button in the **Faces** selection box; **1** is displayed in the **Faces** selection box indicating that one face has been selected for the weld operation.

22. Choose the **Generate** tool from the **Features** toolbar to generate the extended surface.

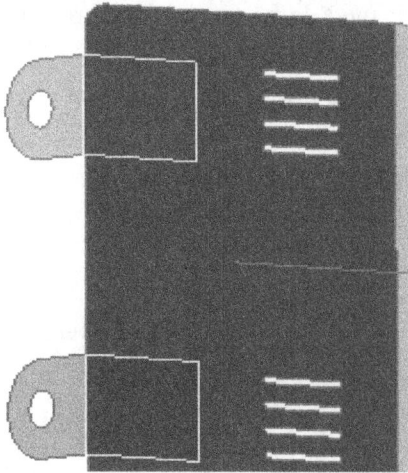

Figure 8-81 *Selecting edges for the weld operation*

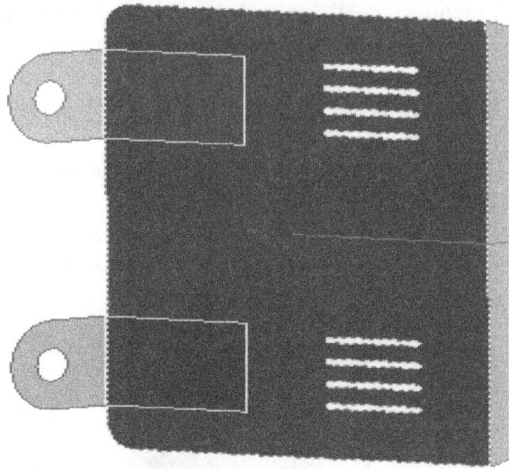

Figure 8-82 *Selecting face for the weld operation.*

23. Choose the **Surface Patch** tool from the **Tools** menu; **SurfPatch1** is added to the **Outline** pane. Also, the corresponding **Details View** window is displayed and you are prompted to select a loop of edges to create the patch.

24. Next, select the **Patch Edges** selection box in the **Details View** window to display the **Apply** and **Cancel** buttons, if they are not already displayed.

25. Select the edges of elongated holes, as shown in Figure 8-83.

Note
*Press and hold the **Ctrl** key and **RMB** and then move the curser over all the edges of the elongated holes to select multiple edges quickly.*

26. Choose the **Apply** button to confirm the selection; the edge turns blue and **40** is displayed in the **Patch Edges** selection box, indicating that forty edges have been selected for applying the patch.

27. Choose the **Generate** tool from the **Features** toolbar to generate the surface patch, as shown in Figure 8-84.

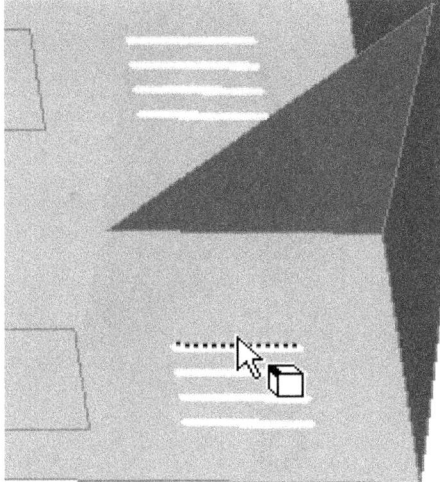

Figure 8-83 Selecting edges for surface patch operation

Figure 8-84 Displaying the generated patched surface

You will Notice that surfaces are created after surface patch operation and these surfaces are not merged with the base surfaces. To do so, use the **Merge** tool from the **Tools** menu.

28. Right-click in the graphics screen and then choose **View > Bottom** view from the shortcut menu displayed to orient the model

29. Choose the **Merge** tool from the **Tools** menu; **Merge1** is added to the **Outline** pane. Also, the corresponding **Details View** window is displayed and you are prompted to select edges to create the merge.

30. Select **Faces** from the **Merge Type** drop-down list in the **Merge1** window; the corresponding options are displayed in the **Details View** window.

31. In the **Details of Merge1** window, select the **Faces** selection box to display the **Apply** and the **Cancel** buttons if they are not already displayed.

32. Select the faces, as shown in Figure 8-85. Choose the **Apply** button to confirm the selection; the faces turns red and **9** is displayed in the **Faces** selection box, indicating that nine faces have been selected for applying the merge.

33. Choose the **Generate** tool from the **Features** toolbar to generate the Surface patch, as shown in Figure 8-86.

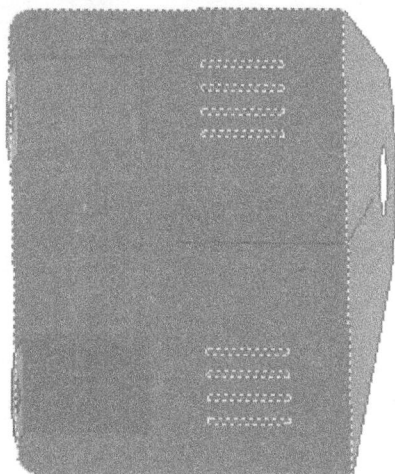

Figure 8-85 *Selecting Surafes for merging*

Figure 8-86 *Displaying the generated merged surface*

34. Exit the **DesignModeler** window to return to the **Workbench** window.

Generating Mesh of the Surface Model

After the model is optimized, you need to generate mesh for the optimized model.

1. In the **Project Schematic** window of the **Workbench** window, right-click on the **Model** cell and then choose the **Refresh** tool from the shortcut menu; the project is updated with the optimized geometry data.

2. In the **Project Schematic** window, double-click on the **Model** cell of the **Mid-Surface_Mesh** analysis system to display the **Mechanical** window.

Note

*1. In the **Mechanical** window, when you expand **Geometry** node, 6 surface bodies are displayed instead of 4 surface bodies. This is because of two surface bodies are generated due to **Weld** operation, which you have used to fill the gape between two surface bodies.*

2. Also, you will notice question mark may have displayed in front of all the surface bodies under the geometry node. Select the surface geometry one by one and assign the Structural Steel material as discussed earlier in this chapter.

3. In the **Outline** pane, right-click on **Mesh**; a shortcut menu is displayed. Choose the **Generate Mesh** option from it to generate the mesh with default settings; the mesh is generated on the surface model, as shown in Figure 8-87.

The element count after generating the mesh with default settings is 2163. Note that the mesh generated in this model will not be uniform.

Note

1. If all the imperfections or minute details are removed, the element and node count in your system will display 2163 and 2612. However, if your model is not optimized accurately like the model shown in Figure 8-87, the element count may vary in your case.

Figure 8-87 Mesh generated on the surafce model

Note that the total number of elements before the optimization is 5,401 which is more than the 2,163. For an effective analysis, you need to visualize the model to reduce the complexities in the model. In this case, you have convert 3D solid model into 2D surface model by using the **Mid-Surface** tool ultimately reducing the element and node count. Also, types of element have changed to 2D element which is 3D element before. This decrease the element count that in turn reduces the computing time. You may further use local and global mesh control options to refine the mesh.

Save the Project and Exit ANSYS Workbench

After generating the mesh, you need to save the project and exit the ANSYS Workbench session.

1. Save the model by choosing the **Save Project** button from the **Standard** toolbar; the model is saved with the name *c08_ansWB_tut04*.

2. Choose **File > Exit** to close the **Workbench** window.

Self-Evaluation Test

Answer the following questions and then compare them to those given at the end of this chapter:

1. As soon as surface body is selected as the geometry to be meshed, the **Automatic Method** is replaced by **Quadrilateral Dominant** in the **Outline** pane. (T/F)

2. In software terms, a surface body is the one which has zero thickness. But, practically such surface bodies do not exist. (T/F)

3. You can either provide a thickness to the surface models in the respective CAD packages or you can provide a thickness to the models in ANSYS Workbench. (T/F)

4. You can set the global mesh control settings in the **Details of "Mesh"** window. (T/F)

5. The procedure to create a finite element model for a surface body is the same as that of a 3D model. (T/F)

Review Questions

1. To insert a method, choose the **Method** tool from the _____ drop-down in the **Mesh** contextual toolbar.

2. The **Look At** tool is available in the _____ toolbar.

3. You may not need to provide a thickness to the model, if it is already given in the CAD package. (T/F)

4. The Tetrahedral method is not available while meshing a very thin model. (T/F)

5. Once the body is selected for generating a mesh, ANSYS Workbench decides if the body is to be treated as a thin model or a 3D model. (T/F)

6. You can modify the names of the components in an assembly while meshing them. (T/F)

7. You can continue to refine the mesh for a component, even if the results obtained from the analysis are same for each refinement done. (T/F)

8. The parameters of a geometry can be viewed in the Status bar when they are selected. (T/F)

9. You have to select face pair manually for **Automatic** method of **Mid-Surface** extraction operation. (T/F)

EXERCISE

Exercise 1

Download the zip file *c08_ansWB_exr01.zip* from *www.cadcim.com*. Extract the file c08_ansWB_*exr01.stp* and then generate a mesh for the bonnet model of a car, as shown in Figure 8-88. Use an appropriate method to generate a localized mesh at the rounds of the model. **(Expected time: 1 hr)**

The complete path for the file is:

Textbooks > CAE Simulation > ANSYS > ANSYS Workbench 2023 R2: A Tutorial Approach > Input Files

Figure 8-88 *The model for Exercise 1*

Answers to Self-Evaluation Test

1. T, **2.** T, **3.** T, **4.** T, **5.** T

Chapter 9

Static Structural Analysis

Learning Objectives

After completing this chapter, you will be able to:

• *Create the static structural analysis system*
• *Perform fatigue failure analysis*
• *Create the eigenvalue buckling analysis system*
• *Apply boundary conditions*
• *Apply different types of constraints*
• *Apply different loads available in ANSYS Workbench*
• *Generate the result of an analysis*
• *Generate report of an analysis*

INTRODUCTION TO STATIC STRUCTURAL ANALYSIS

The Static Structural analysis is one of the important analyses in ANSYS Workbench. It is available as **Static Structural** analysis system under the **Analysis Systems** toolbox in the **Toolbox** window. This system analyses is used for the structural components for displacements (deformation), stresses, strains, and forces under different loading conditions. The loads in this analysis system are assumed not to have damping characteristics (time dependent). Steady loading and response conditions are assumed in this type of analysis system. The types of loading that can be applied in a static analysis include:

*   Externally applied forces and pressures
*   Steady-state inertial forces (such as gravity or rotational velocity)
*   Imposed (nonzero) displacements
*   Temperatures (for thermal strain)

A static structural analysis can be either linear or nonlinear and they are discussed next.

Linear Static Analysis

A linear static analysis is an analysis where the rigidity and corresponding stiffness value of the materials must remain constant. Also, there must be a linear relationship between applied forces and displacements. In general, it applies to structural problems where stresses remain in the linear elastic range of the used material.

Nonlinear Static Analysis

In this type of analysis, all types of nonlinearities, associated with material, geometric, and contact non-linearity of the structure; for example, large deformations, plasticity, stress stiffening, contact (gap) elements, hyperelasticity, and so on are allowed. In practice, the stiffness matrix does not remain constant during the application of load.

This chapter focuses on linear static analyses. When the structure is subjected to repetitive or compressive loading, it may lose its ability to withstand the applied loads and may cause sudden failure or large deformation. Therefore, failure analysis concepts such as fatigue and buckling failure are discussed next.

FATIGUE ANALYSIS

Fatigue is the progressive and localized structural damage that occures when a material is subjected to repeated alternating or cyclic stresses. The nominal maximum stress values are considerably lower than the yield or ultimate strength limit of the material. Although the fracture is of a catastrophic type, it may take some time to propagate, depending on both the intensity and frequency of the stress cycles. It is estimated that 80-90% of all structural failures are caused by fatigue. Catastrophic failure by fatigue often occurs without advanced warning at low-stress levels.

Various terminologies that are used in the design of components subjected to cyclic load are discussed next.

Mean Stress

It is an average of the maximum and minimum stresses in one cycle.

Alternating Stress
It is one-half of the stress range in one cycle.

Endurance Limit
Endurance limit is defined as the maximum value of completely reversed bending stress that a material can withstand without any failure for the infinite number of cycles.

Endurance Strength
Endurance strength is defined as the maximum value of completely reversed bending stress that a material can withstand without any failure for the finite number of cycles.

There are two approaches to fatigue life which are discussed next.

Stress-Life Approach
The stress–life approach is used in a structural problems where induced stress is within the elastic range, and the material has a long cyclic life. In general, the stress–life approach is used in situations where the fatigue life is in excess of 10^3 cycles. This is also referred to as the S-N approach. The S-N approach is suitable for high cycle fatigue, where the material is subject to cyclical stresses that are predominantly within the elastic range. This ensures that no significant plasticity occurs. This is commonly referred to as high-cycle fatigue.

Strain-Life Approach
The strain-life method should be chosen to predict the fatigue life when plastic strain occurs under the given cyclic loading. In general, the strain-life method is used for detailed analysis of plastic deformation at localized regions. This is also referred to as E-N (Strain - Life) approach.

In this analysis, the fatigue tool provides life, damage, and safety factor information by using stress-life or strain life approaches corresponding to the fatigue failure theories.

BUCKLING ANALYSIS
Under compressive loads, slender and thin-walled members such as columns, beams, sheet metal parts and so on, are prone to buckling. For any structural member, generally, bending stiffness is far lower than the axial stiffness. An eigenvalue buckling analysis of an ideal elastic structure can predict the theoretical buckling strength. The buckling failure usually occurs within the elastic range of the material. For these types of failure, the actual compressive stress in the structure, at failure, is often smaller than the yield and ultimate stress of the material. The solving methodology corresponds to the textbook approach to an elastic buckling analysis. Hence, an eigenvalue buckling analysis of a column matches the classical Euler solution. Although, imperfections and nonlinearities prevent most real-world structures from achieving their theoretical elastic buckling strength. Therefore, an eigenvalue buckling analysis often yields quick but non-conservative results. You can predict the buckling failure by using linear structural analysis.

To perform the linear eigenvalue buckling analysis, you need to define loading conditions in the upstream static structural analysis. The results evaluated by the Eigenvalue Buckling analysis are buckling load factors (load multiplier) that scale all loads applied in the Static Structural analysis. For example, if you applied a 100 N compressive load on a structure in the static analysis and if the eigenvalue buckling analysis evaluates a load factor equals to 15, then the predicted buckling load is 100x15 = 1500 N. Therefore, it is always recommended that a unit load is applied to calculate first buckling load.

Euler's formula to calculate the buckling load theoretically is given next.

$$P_{cr} = \frac{\pi^2 EI}{L_e^2}.$$

Where,

Pcr = Critical or Crippling load in N or KN
I   = The moment of inertia for the principal axis about which buckling occurs mm^4
E = Youngs modulus of material in MPa
Le = Effective length of column in mm

Points to Remember

• An eigenvalue buckling analysis must be preceded by static structural analysis. This static analysis can be referred to as the base analysis and it may be either linear or nonlinear.

• A model can have an infinite number of buckling load factors. Each load factor is associated with a different mode shape pattern. Typically the lowest load factor is of interest.

• Based upon the direction of load you apply to a model, load factors can be either positive or negative. The buckling analysis finds load factors from the extreme values negative or positive. The smallest eigenvalue in absolute value is always with respect to the minimum buckling load factor.

• Buckling mode shapes do not depict the actual displacements but help you to visualize the behavior of a model when subjected to buckling.

GENERAL PROCEDURE TO CONDUCT STATIC STRUCTURAL ANALYSIS

To start a new Static Structural analysis system, double-click on **Static Structural** in the **Analysis Systems** toolbox in the **Toolbox** window; the **Static Structural** analysis system will be added to the **Project Schematic** window, as shown in Figure 9-1. To start an analysis, first you need to specify the geometry on which the analysis is to be done. To do so, you can import the geometry from an external CAD package, or you can create the geometry in the ANSYS's **DesignModeler** or **SpaceClaim** software. After the model is specified for an analysis, you need to double-click on the **Model** cell of the **Static Structural** analysis system to open the **Mechanical** window, as shown in Figure 9-2. In this window, you can specify the parameters and run the analysis.

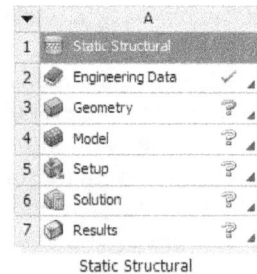

	A	
1	Static Structural	
2	Engineering Data	✓
3	Geometry	?
4	Model	?
5	Setup	?
6	Solution	?
7	Results	?

Static Structural

Figure 9-1 The Static Structural analysis system added to the Project Schematic window

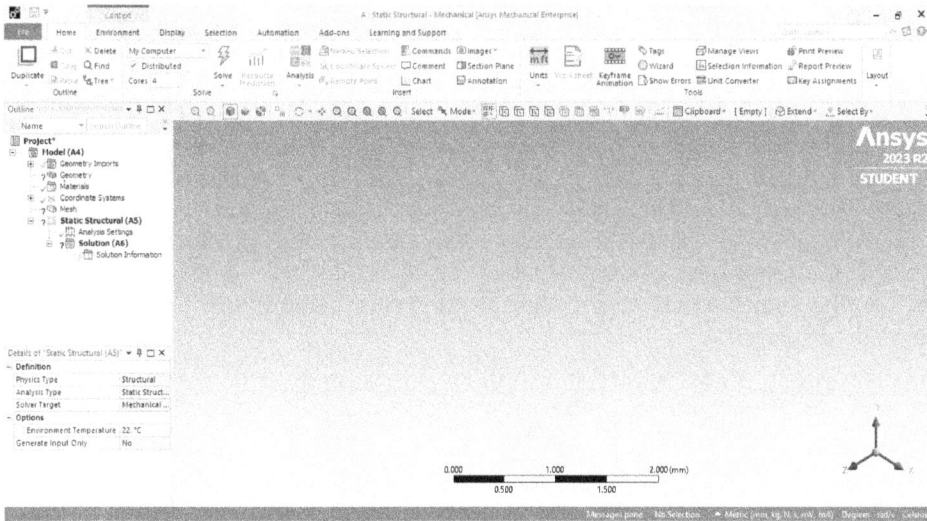

*Figure 9-2 The **Mechanical** window*

As discussed in previous chapters, analysis can be carried out in three major steps: pre-processing, solution, and post-processing. The tools required to carry out these steps are discussed next.

Pre-Processing

The pre-processing of an analysis system involves specifying the material, generating a mesh, and defining boundary conditions.

Note
Tools and options used to specify the material and generate mesh has been already discussed in the previous chapter. In this chapter, you will learn about the tools used for defining boundary condition.

In ANSYS Workbench, the various tools related to boundary conditions are available in the **Environment** contextual tab, which is displayed when you select the **Static Structural** node in the Outline pane, refer to Figure 9-3.

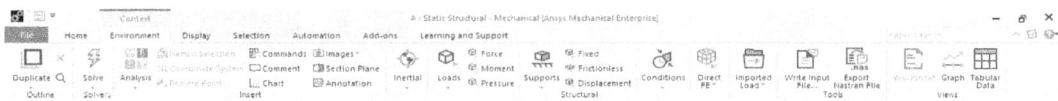

*Figure 9-3 The **Environment** contextual tab*

In order to provide a support to the model, you need to choose the required tool from the **Supports** drop-down. Similarly, to add a load, choose the desired tool from the **Loads** drop-down in the **Environment** contextual tab. Also, when you choose any tool from the **Environment** contextual tab, the corresponding entity is placed under the **Static Structural** node in the Outline pane.

Note

*While performing an analysis, you can display the **Environment** contextual tab by selecting the respective analysis node in the **Mechanical** window.*

The main purpose of an analysis is to evaluate the results. After the boundary condition is set and loads are applied, you need to specify the desired outcomes of the analysis. In ANSYS Workbench, you can analyze various parameters such as deformation, stresses, strains, and so on. To do so, you need to specify the results required and then evaluate them. You can use the tools available in the **Solution** contextual tab to specify results, refer to Figure 9-4. Alternatively, right-click on the **Solution** node in the Outline pane and then use the desired option from the shortcut menu displayed.

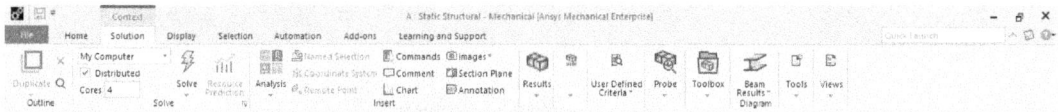

*Figure 9-4 The **Solution** contextual tab*

In order to evaluate deformations, stresses, strains, and so on, choose the desired options from the drop-downs available in the **Solution** contextual tab.

Solution

In an analysis, after pre-processing (meshing, specifying material, and specifying boundary condition) is done, the next step is to solve the analysis. In ANSYS Workbench, you will use the **Solve** tool from the **Standard** tab to run the solver. The solver runs in the background of a software and acquires results of an analysis, based on the specified boundary conditions.

Post-Processing

After the analysis is complete, you need to generate the report in the **Mechanical** window. To do so, choose the **Report Preview** tab from the bottom of the graphics screen, as shown in Figure 9-5, if **Report Preview** tab is not present at the bottom of the graphics screen that choose **Report Preview** and **Print Preview** options from the **Home** tab of **Menu** bar; the **ANSYS Report generation in progress** message is displayed on the screen, as shown in Figure 9-6. After sometime, this message vanishes and the report is generated.

Report generation in progress...

Please wait while the system extracts all necessary project information. During this process, please refrain from all project interaction.

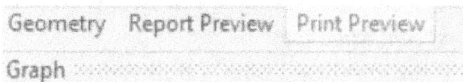

*Figure 9-5 Choosing the **Report Preview** tab from the Graphics Screen*

*Figure 9-6 The **Report generation in progress** message*

TUTORIALS

Tutorial 1

In this tutorial, you will create the model of a cantilever beam, as shown in Figure 9-7. The dimensions to create the model and its boundary and loading conditions are also given in the same Figure. Run a Static Structural analysis on the model and evaluate the total deformation and the directional deformation. Determine directional deformation along the X, Y, and Z axes. After evaluating the results, interpret them.   **(Expected time: 3 hr)**

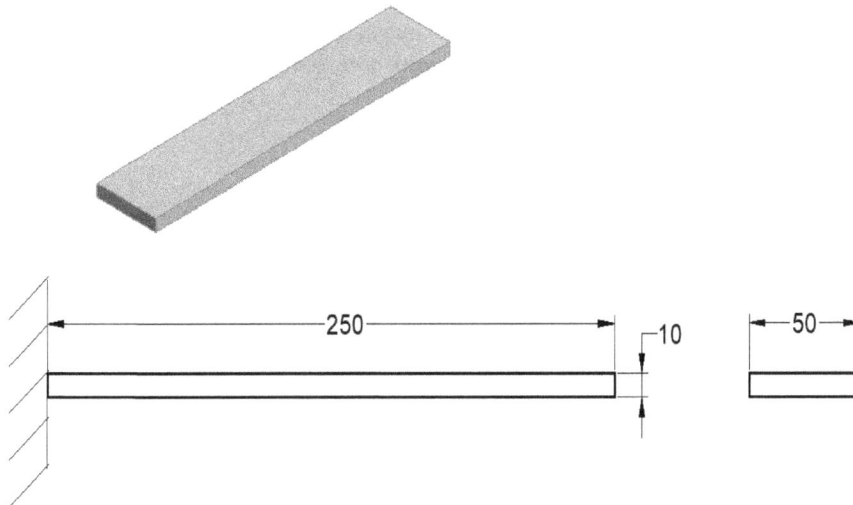

Figure 9-7 The cantilever beam with dimensions and boundary and loading conditions

The following steps are required to complete this tutorial.

a. Start a new project and create the model.
b. Generate the mesh.
c. Set the boundary and loading conditions.
d. Solve the model.
e. Duplicate the existing analysis system.
f. Interpret results.
g. Save the project.

Starting a New Project and Creating the Model

The first step is to start a new project in the **Workbench** window.

1. Start ANSYS Workbench.

2. Choose the **Save Project** button from the **Standard** toolbar; the **Save As** dialog box is displayed.

3. In this dialog box, enter **c09_ansWB_Tut01** in the **File name** field and then save the file in the location: *C:\ANSYS_WB\c09\Tut01*

4. Double-click on **Static Structural** in the **Toolbox** window; the **Static Structural** analysis system is added in the **Project Schematic** window.

5. Rename the **Static Structural** analysis system to **Cantilever**.

6. Right-click on the **Geometry** cell of this component system to display a shortcut menu. Next, choose **New DesignModeler Geometry** from the shortcut menu; the **Starting DesignModeler** message is displayed in the status bar. After sometime, the **DesignModeler** window is displayed.

7. Choose the **Millimeter** option from the **Units** menu of the **Menu** bar. Now, create the sketch on the XY plane and extrude, as shown in Figure 9-8. For dimension of the model, refer to Figure 9-7.

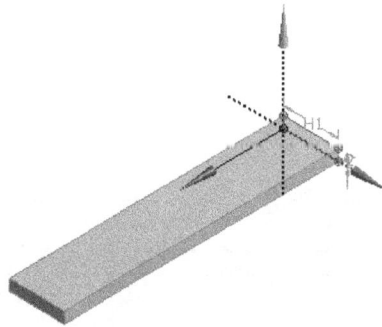

Figure 9-8 *Model created on the XY plane*

8. Exit the **DesignModeler** window to display the **Workbench** window.

Generating the Mesh

After the model is created in the **DesignModeler** window, you need to generate the mesh for the model in the **Mechanical** window.

1. In the **Project Schematic** window, double-click on the **Model** cell in the **Cantilever** analysis system; the **Mechanical** window is displayed.

2. Select **Mesh** in the **Outline** pane to display the **Details of "Mesh"** window.

3. In the **Details of "Mesh"** window, expand the **Sizing** node, if it is not already expanded. Also, notice that **Default** is displayed in the **Element Size** edit box.

    The **Element Size** edit box is used to specify the size of an element. The element size specified in this edit box is according to the size of the geometry. When **Default** is displayed in the

    Element Size edit box, it indicates that a default value, based on the size of the geometry, is already specified by the software.

4. In the Outline pane, right-click on **Mesh**; a shortcut menu is displayed. Choose the **Generate Mesh** option; the mesh is generated with default mesh, as shown in Figure 9-9.

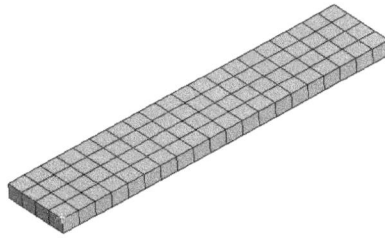

Figure 9-9 *Mesh generated with default mesh controls*

In Figure 9-9, notice that there are 20 elements laid along the length and 4 elements laid along the width of the component. As the component is 250 mm long and 50 mm wide, the size of the elements is 12.5 mm hexahedrals.

5. Expand the **Statistics** node in the **Details of "Mesh"** window to display the total number of elements created. On doing so, you will find that the total number of elements created is 80.

Setting the Boundary and Loading Conditions

After the mesh is generated, you need to set the boundary and loading conditions under which the analysis will be performed.

1. Select the **Static Structural** node in the **Outline** pane; the **Details of "Static Structural (A5)"** window is displayed. Also, the **Environment contextual** tab is displayed, refer to Figure 9-3.

2. In the **Environment contextual** tab, choose the **Fixed** tool from the **Structural** group, refer to Figure 9-10; the **Fixed Support** is attached to the **Outline** pane. Also, the **Details of "Fixed Support"** window is displayed.

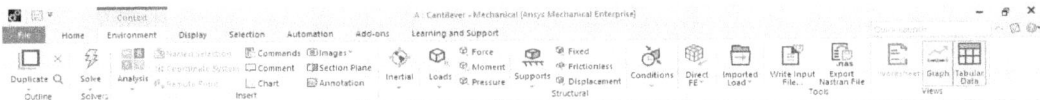

Figure 9-10 *Choosing the **Fixed** tool in the **Environment** contextual tab*

Note
*1. The **Environment** contextual tab is displayed according to the corresponding analysis system node selected in the **Outline** pane.*

*2. The options available in the **Supports** drop-down can also be accessed by using the shortcut menu displayed on right-clicking on the **Static Structural** node.*

3. Select the **Geometry** selection box in the **Scope** node of the **Details of "Fixed Support"** window to display the **Apply** and **Cancel** buttons, if they are not already displayed.

4. Choose the **Face** tool from the **Select** toolbar to select the face to apply fixed support. Next, select the face of the model; the face turns green, as shown in Figure 9-11.

5. Choose the **Apply** button in the **Geometry** selection box to confirm the selection of the face for Fixed support; the color of the face turns violet and a flag is attached to the face, as shown in Figure 9-12.

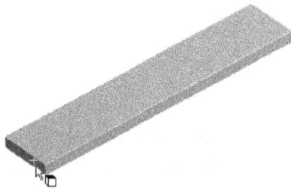

Figure 9-11 *Face selected*

Figure 9-12 *The violet color face of the model displaying the Fixed support*

After the boundary is defined for the model, you need to define the load for which the analysis is to be carried out.

6. Right-click on the **Static Structural** node and then choose **Insert > Force** from the shortcut menu displayed, as shown in Figure 9-13; **Force** is added under the **Static Structural** node in the **Outline** pane and the **Details of "Force"** window is displayed.

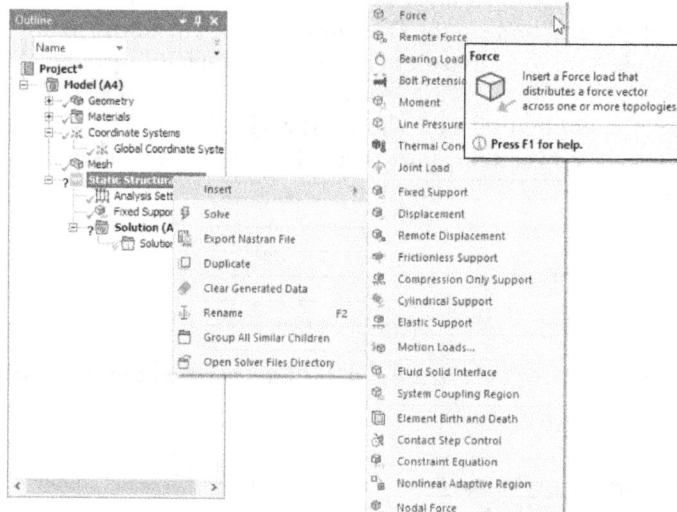

Figure 9-13 *Choosing the **Force** option from the **Static Structural** shortcut menu*

7. In the **Details of "Force"** window, select the **Geometry** selection box to display the **Apply** and **Cancel** buttons, if they are not already displayed.

8. Choose the **Edge** tool from the **Select** toolbar to select an edge from the graphics screen.

9. In the graphics screen, select the edge, just opposite to the face on which you have applied the Fixed support, refer to Figure 9-14.

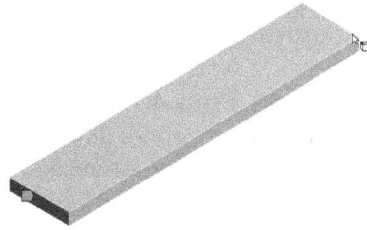

Figure 9-14 Selecting the edge for applying a force load

10. Choose the **Apply** button from the **Geometry** selection box in the **Details of "Force"** window; the edge is selected for applying the Force load.

Note

In this tutorial, the direction of application of load is downward.

11. In the **Magnitude** edit box and then enter **500** as the magnitude of Force load.

The **Magnitude** edit box is used to specify the magnitude of a vector quantity.

12. Select the **Direction** selection box to display the **Apply** and **Cancel** buttons. Next, select the edge of the model, as shown in Figure 9-15.

The **Direction** selection box is used to specify the direction of a vector quantity.

13. Choose the **Flip** toggle button available in the graphics screen, refer to Figure 9-16, to specify the direction of load as downward.

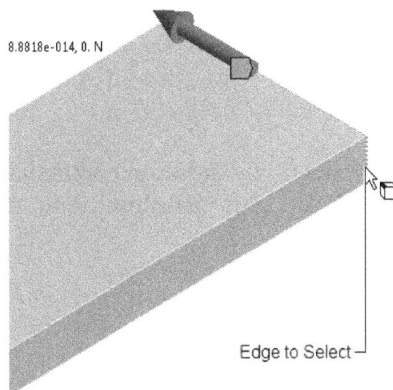

Figure 9-15 Selecting the edge to specify the direction for applying the load

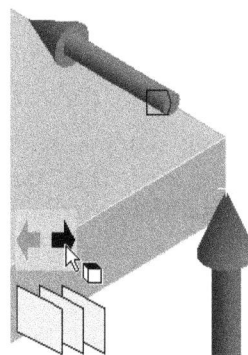

Figure 9-16 Choosing the **Flip** button from the Graphics screen

14. Choose the **Apply** button in the **Direction** selection box in the **Details of "Force"** window.

Now, the preprocessing part is complete. Next, you need to work on the solution part of the analysis.

Solving and Post-Processing the Finite Element Model

After the boundary and loading conditions are specified for the analysis, you need to evaluate the results that are of importance in the case of a particular analysis. You can view the response of the model under the given boundary and loading conditions. The various results that can be evaluated are: Deformation, Stress, Strain, Energy, and Linearized Stress.

To evaluate the results in this analysis, follow the procedure explained next.

1. Select the **Solution** node in the **Outline** pane; the **Solution** contextual tab is displayed, refer to Figure 9-4. Also, the **Details of "Solution"** window is displayed, refer to Figure 9-17.

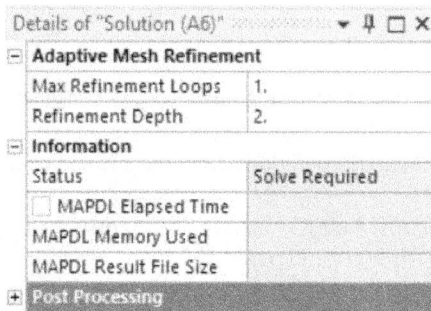

Figure 9-17 The **Details of "Solution"** window

Note
*The tools available in the **Solution** contextual tab can also be accessed by using the **Solution** node. To do so, right-click on the **Solution** node; a shortcut menu will be displayed. Next, choose the **Insert** option from it; a flyout will be displayed showing various options. You can specify the parameters to evaluate by using the corresponding option in this flyout.*

2. Choose the **Total** tool from the **Deformation** drop-down in the **Solution** contextual tab, as shown in Figure 9-18; **Total Deformation** is attached under the **Solution** node. Also, the **Details of "Total Deformation"** window is displayed.

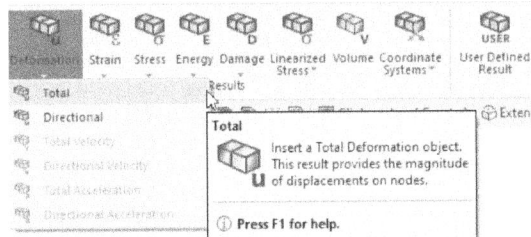

*Figure 9-18 Choosing the **Total** tool from the*
***Deformation** drop-down*

3. Choose the **Directional** tool from the **Deformation** drop-down; **Directional Deformation** is attached to the Outline pane. Also, the **Details of "Directional Deformation"** window is displayed.

A body is called to be deformed if its shape is changed temporarily or permanently. The temporary change of shape is known as elastic deformation and a permanent change of shape is known as plastic deformation. In ANSYS Workbench, you can determine deformation in terms of Total and directional deformations.

Total deformation is the total change of shape in a given working condition. You can view the total deformation induced in any component by using the **Total** tool from the **Deformation** drop-down in the **Solution** contextual tab. Directional deformation is the total change of shape in a particular axis, due to given working conditions. You can view directional deformation by using the **Directional** tool from the **Deformation** drop-down in the **Solution** contextual tab.

Total deformation is the summation of all directional deformations produced in a certain region of the model. The following equation describes the total deformation:

If
Deformation in the X-axis   Ux
Deformation in the Y-axis   Uy
Deformation in the Z-axis   Uz

Then total deformation U will be given as follows:

$$U = (U_x^2 + U_y^2 + U_z^2)^{1/2}$$

4. In the **Details of "Directional Deformation"** window, expand the **Definition** node, if it is not already expanded, refer to Figure 9-19.

Notice that, in the **Orientation** drop-down list, the default selection is X axis, which means the directional deformation shown in the graphics screen is only with respect to X axis. In Finite element modeling where the processing period is small, you can view the directional deformation with respect to the Y and Z axes in the same **Mechanical** window.

5. Select **Y Axis** from the **Orientation** drop-down list, refer to Figure 9-19.

6. Next, choose the **Solve** tool from the **Solution** contextual tab; the directional deformation with respect to the Y axis is displayed in the graphics screen. ⟶ Solve

Figure 9-20 shows the default view of the finite element model with the directional deformation with respect to Y axis.

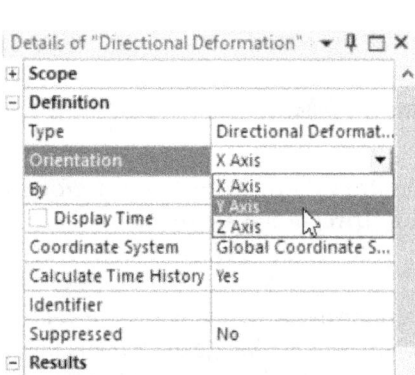

A: Cantilever
Directional Deformation
Type: Directional Deformation(Y Axis)
Unit: mm
Global Coordinate System
Time: 1

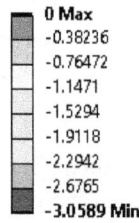

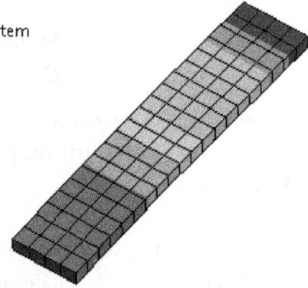

0 Max
-0.38236
-0.76472
-1.1471
-1.5294
-1.9118
-2.2942
-2.6765
-3.0589 Min

Figure 9-19 *Selecting **Y Axis** from the* ***Orientation*** *drop-down list*

Figure 9-20 *Directional deformation with respect to Y axis*

You can change the default scale of the results by selecting the required option from the **Scale** drop-down list that is displayed in the **Result** contextual tab, as shown in Figure 9-21.

7. Select **X Axis** from the **Orientation** drop-down list to evaluate the directional deformation with respect to the X axis only.

8. Next, choose the **Solve** tool from the **Solution** contextual tab to view the directional deformation with respect to the X axis, refer to Figure 9-22.

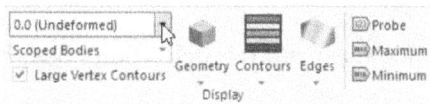

A: Cantilever
Directional Deformation
Type: Directional Deformation(X Axis)
Unit: mm
Global Coordinate System
Time: 1

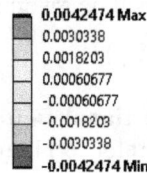

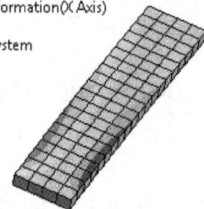

0.0042474 Max
0.0030338
0.0018203
0.00060677
-0.00060677
-0.0018203
-0.0030338
-0.0042474 Min

Figure 9-21 *The **Scale** drop-down list*

Figure 9-22 *Directional deformation with respect to X axis*

Now, if you again have to view the results of the directional deformation along the Y-axis, you have to follow all the steps again, this would result in more processing time as taken earlier. Therefore, it is recommended to create a duplicate of the exiting system, so that you can view the result again at any time without following any steps.

Duplicating the Cantilever Analysis System

Instead of changing the orientation of the axes to find directional deformation, you can duplicate an existing system from the **Workbench** window.

1. Switch from the **Mechanical** window to the **Workbench** window.

2. In the **Project Schematic** window, right-click on **Static Structural** in the **Cantilever** analysis system; a shortcut menu is displayed.

3. In this shortcut menu, choose the **Duplicate** option, as shown in Figure 9-23; the **Copy of Cantilever** analysis system is created in the **Project Schematic** window.

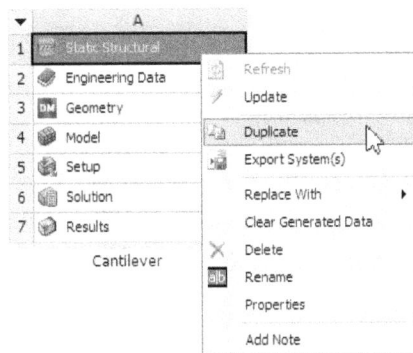

Figure 9-23 *Choosing the **Duplicate** option from the shortcut menu*

4. Rename the **Copy of Cantilever** analysis system to **Cantilever 2**.

5. Double-click on the **Model** cell of the **Cantilever 2** analysis system; another **Mechanical** window is opened.

Note

1. The newly created analysis system has all the characteristics that the original analysis system had.

*2. After the Cantilever analysis system is duplicated, two **Mechanical** windows are opened, namely, **Cantilever - Mechanical** and **Cantilever 2 - Mechanical**.*

6. In the **Mechanical** window of the **Cantilever 2** analysis system, select **Directional Deformation** from the Outline pane to view the **Details of "Directional Deformation"** window.

7. In this window, expand the **Definition** node, if it is not already expanded.

8. Next, select **Y Axis** from the **Orientation** drop-down list, refer to Figure 9-19; a yellow thunderbolt is displayed before the **Solution** node.

9. Choose the **Solve** tool from the **Solution** contextual tab to solve the analysis for the directional deformation along the Y axis; a green tick mark is placed before the components under the **Solution** node in the **Outline** pane, indicating that all the results are evaluated.

10. Select **Directional Deformation** in the **Outline** pane; the corresponding deformation along the Y axis is displayed in the graphics screen, refer to Figure 9-20.

Now, you have two separate windows to analyze directional deformation along the X axis and directional deformation along the Y axis. Similarly, you can create a copy of the existing analysis system and analyze the data with respect to the Z axis.

Figure 9-24 shows the directional deformation along the Z axis.

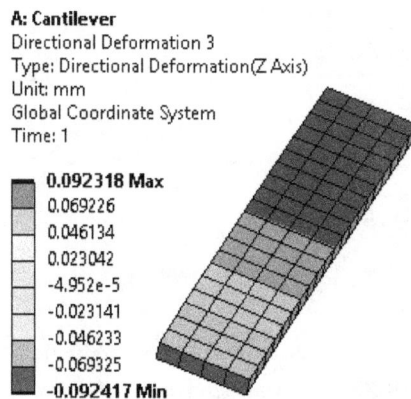

A: Cantilever
Directional Deformation 3
Type: Directional Deformation(Z Axis)
Unit: mm
Global Coordinate System
Time: 1

0.092318 Max
0.069226
0.046134
0.023042
-4.952e-5
-0.023141
-0.046233
-0.069325
-0.092417 Min

Figure 9-24 *Directional deformation along the Z axis*

As discussed earlier, it is convenient to analyze results in separate windows. Therefore, in case of large finite element models, you can validate data for different combinations by opening a new window using the **Duplicate** tool. Figure 9-25 shows the **Project Schematic** window displaying three **Analysis Systems** wherein, **Cantilever 2** and **Cantilever 3** analysis systems have been duplicated from the **Cantilever** analysis system. Note that, **Cantilever 2** and **Cantilever 3** analysis systems are used to determine directional deformations along the Y and Z axes.

Alternatively, you can view the directional deformation with respect to the Y and Z axes in the same **Mechanical** window. To do this, add two **Directional** tools from the **Deformation** drop-down in the **Solution** contextual tab. Notice that the two **Directional Deformation** tools are attached to the **Outline** pane with name **Directional Deformation 2** and **Directional Deformation 3**. Next, you can select **Y Axis** from the **Orientation** drop-down list from the **Details of "Directional Deformation 2"** window. Similarly, select **Z Axis** from the **Orientation** drop-down list from the **Details of "Directional Deformation 3"** window. Now, choose the **Solve** tool from the **Solution** contextual tab to solve the analysis for the directional deformations along the Y and Z axes.

Tip

*In ANSYS Workbench, you can specify multiple instances of the same result. As you created multiple analysis systems for directional deformation along the Y and Z axes, you can also insert multiple instances of directional deformation and set the environment for directional deformation in Y and Z axes in the **Mechanical** window.*

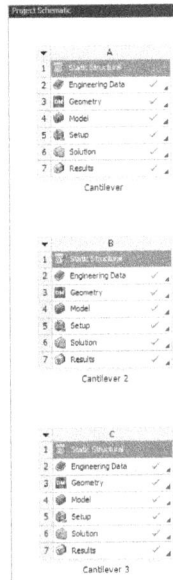

*Figure 9-25 The **Project Schematic** window with the copied analysis systems*

Interpreting the Results

After the analysis is finished, the next important step is to understand the evaluated results. In this tutorial, you have evaluated deformation so far. Now, it is required to check if the evaluated results fall under the permissible limit or not.

1. Select **Total Deformation** under the **Solution** node in the **Outline** pane, as shown in Figure 9-26; the total deformation of the cantilever is displayed in the graphics screen, as shown in Figure 9-27. Also, the corresponding legend is displayed in the graphics screen.

*Figure 9-26 Selecting the **Total Deformation** node*

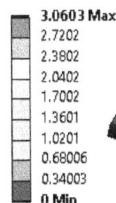

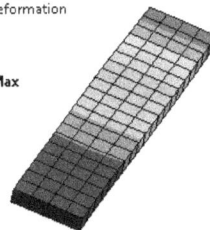

Figure 9-27 Total deformation displayed in the graphics screen

The legend has colors arranged in a band from top to bottom. Depending upon the type of analysis and the parameters evaluated, each color will indicate a different value. Figure 9-28 shows a typical legend displayed when total deformation is selected from the outline pane.

The blue color in the legend indicates the minimum value of total deformation. In this case, It displays **0** which means there is no deformation at that region, as shown in Figure 9-29.

A: Cantilever
Total Deformation
Type: Total Deformation
Unit: mm
Time: 1

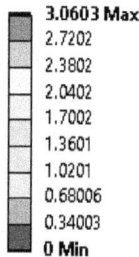

3.0603 Max
2.7202
2.3802
2.0402
1.7002
1.3601
1.0201
0.68006
0.34003
0 Min

Maximum
Total
Deformation

Minimum Total Deformation

Figure 9-28 The legend   *Figure 9-29 Color contours displaying the result*

The value that is displayed next to each color is the total deformation in the region which is depicted by that particular color in the model. The blue color in the model represents the lowest value of the total deformation, whereas the red color denotes the maximum value of the total deformation.

Note
To view the color contours in the model, you need to select the desired node in the Outline pane.

2.   Select **Total Deformation** in the Outline pane, if it is not already selected.

3.   Right-click on the graphics screen and then choose **View > Right** from the shortcut menu displayed to orient the model, as shown in Figure 9-30.

Maximum Total
Deformation Region

Minimum Total
Deformation Region

*Figure 9-30 The view of the model on choosing the
Right option from the shortcut menu*

4.   Select the **Show Undeformed WireFrame** option from the **Edges** drop-down list in the **Result** contextual tab to visualize the extent of the deformed shape, as shown in Figure 9-31; the undeformed wireframe model is displayed along with the deformed model in the graphics screen, as shown in Figure 9-32.

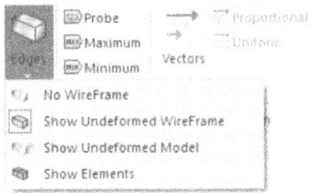

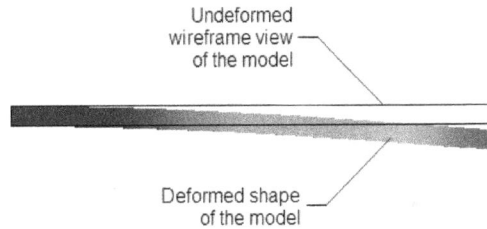

Figure 9-31 *Selecting **Show Undeformed** **WireFrame** from the **Edges** drop-down list*

Figure 9-32 *Deformed and undeformed views of the model*

To view the transparent mode of the model, select the **Show Undeformed Model** option from the **Edges** drop-down list, refer to Figure 9-31. Similarly to show the elements that are created after meshing in the deformed shape of the model, select the **Show Elements** option from the **Edges** drop-down list, refer to Figure 9-31. Figure 9-33 shows the undeformed model when the **Show Undeformed Model** option is selected. Figure 9-34 displays the elements when the **Show Elements** option is selected.

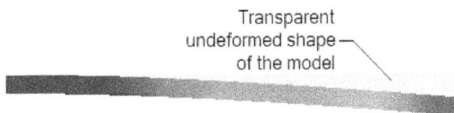

Figure 9-33 *Displaying the undeformed model when the **Show Undeformed Model** option is selected*

Figure 9-34 *Elements displayed when the **Show Elements** option is selected*

In this tutorial, you may be able to represent a 3-D solid model to 1-D Line/Beam model. To do so, you need to draw a line body and assign the cross-section, as shown in Figure 9-7, and discussed in Chapter 4. This 1-D beam model incorporates the properties of 3-D solid model and effectively reduces the computational efforts, and ultimately optimizes the analysis results.

Note
*A **Static Structural** analysis can be followed by a "pre-stressed" analysis such as modal or linear (eigenvalue) buckling analysis. In this subsequent analysis, the effect of stress on stiffness of the structure (stress-stiffness effect) is taken into account.*

5. Exit the **Cantilever - Mechanical** window.

Saving the File

After all the actions are performed and the desired solution is achieved, you need to save the model.

1. In the **Workbench** window, choose the **Save Project** button from the toolbar to save ⊞
    the project.

2. Exit the **Workbench** window to end the session.

Tutorial 2

In this tutorial, you will create two holes on the back planar face of the model created in Tutorial 2
of Chapter 3. Figures 9-35 and 9-36 show the dimensions of model and the holes, respectively.
The material to be applied on the model is structural steel. Next, you will run the analysis under
two conditions and evaluate the total deformation, directional deformation, equivalent stress,
maximum principal stress, and minimum principal stress.   **(Expected time: 1 hr 30 min)**

Case I
Provide fixed support at the circular holes and apply a force load of 30N at the end of the model,
as shown in Figure 9-37.

Case II
Provide displacement support along one of the axes of the holes and allow displacement axially
only. Movements along any other directions must be restricted. The applied Force in this case
is 30 N.

Case III
Create a surface model from the solid model and provide same analysis conditions as stated for
case I.

The following steps are required to complete this tutorial:

a. Open the existing project and save it with a different name.
b. Edit the model and add an analysis system.
c. Assign the material and generate the mesh.
d. Specify the boundary conditions.
e. Solve the analysis and analyze the results.
f. Duplicate the model and set the boundary condition for Case II.
g. Solve the model.
h. Create the surface model and generate the mesh.
i. Specify the boundary conditions for Case III.
j. Solve the model and analyze results.
k. Interpret the results for all cases.
l. Save the project.

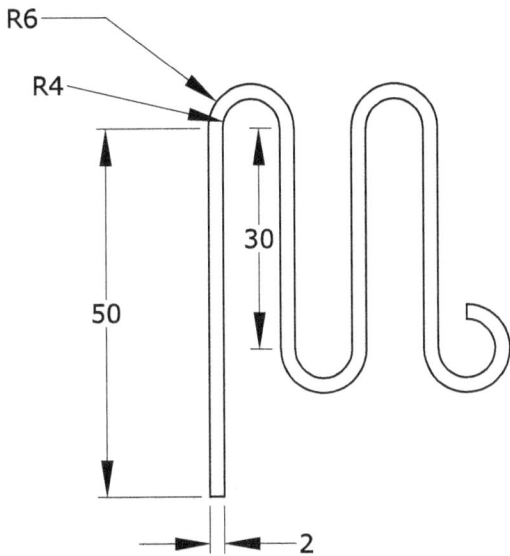

Figure 9-35 *Side view of the model showing the dimensions*

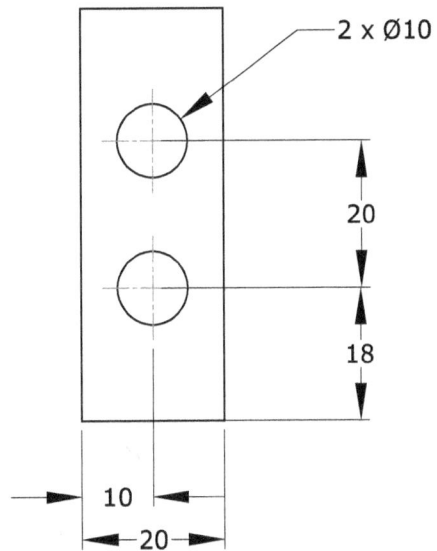

Figure 9-36 *Back view of the model showing dimensions of the holes*

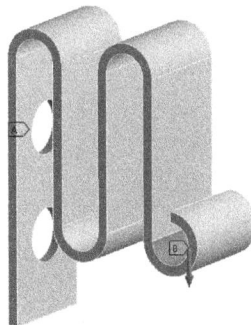

Figure 9-37 *Applied boundary conditions*

Opening the Existing Project

In this tutorial, you will use the model created in Tutorial 2 of Chapter 3. To use the model, first you need to open it in the **Workbench** window and then save it with a different name.

1. Choose the **Open** button from the toolbar; the **Open** dialog box is displayed.

2. Browse to the location *C:\ANSYS_WB\c03\Tut02*.

3. Double-click on **c03_ansWB_Tut02** to open it.

4. Choose the **Save Project As** button and then browse to the location *C:\ANSYS_WB\c09*.

5. Create a new folder with the name **Tut02** and save the project file with the name **c09_ansWB_Tut02**.

Editing the Model and Adding an Analysis System

As per the requirement of the tutorial, you need to create two holes at the back face of the model to provide space for fasteners, and then add an analysis system to it.

1. Double-click on the **Geometry** cell of the **Spring Plate** component system to start the **DesignModeler** window.

2. Right-click in the graphics screen to display a shortcut menu and then choose **View > Left View** from it to display the left face of the model where holes are to be created.

3. Choose the **Faces** button from the **Select** toolbar and then select the left face of the model, refer to Figure 9-38.

4. Create a new plane on the selected face by using the **New Plane** tool from the **Active Plane/Sketch** toolbar and choose the **Generate** tool from the **Features** toolbar; the new plane is created as shown in Figure 9-38.

5. Draw two circles each of diameter 10. For other dimensions, refer to Figures 9-36.

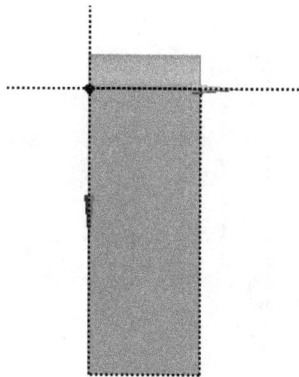

Figure 9-38 New plane created on the left face of the model

6. Create two holes from the circles by using the **Extrude** tool.

7. Now, exit the **DesignModeler** window to display the **Workbench** window.

8. In the **Workbench** window, add a new **Static Structural** analysis system in the **Project Schematic** window.

9. Drag the **Geometry** cell from the **Spring Plate** component system and drop in the **Geometry** cell of the **Static Structural** analysis system to share the created geometry, refer to Figure 9-39.

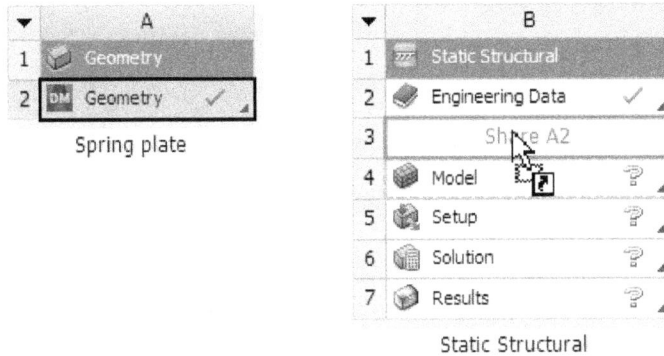

Figure 9-39 *Dragging the* **Geometry** *cell to the* **Static Structural** *analysis system*

10. In the **Project Schematic** window, rename the **Static Structural** analysis system to **Spring Plate**.

Specifying a Material and Generating Mesh

The next step is to specify the material and generate a mesh.

1. In the **Spring Plate** analysis system, double-click on the **Model** cell to open the **Mechanical** window.

2. In the **Mechanical** window, expand the **Geometry** node in the **Outline** pane.

3. Select the **Solid** node under the **Geometry** node to display the **Details of "Solid"** window.

4. In the **Details of "Solid"** window, expand the **Material** node to display the **Assignment** drop-down list.

   You will notice in the **Assignment** drop-down list, refer to Figure 9-40, the structural steel material is applied by default.

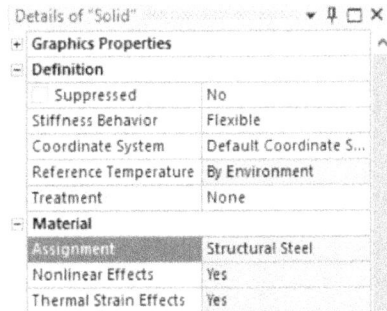

Figure 9-40 *Displaying the default* **Structural Steel** *material option*

5. Right-click on the **Project** node in the **Outline** pane and then choose the **Rename** option from the shortcut menu displayed, refer to Figure 9-41; the **Project** node is highlighted in the **Outline** pane.

6. Enter **Spring Plate** as the new name of the node; the **Project** node is renamed as **Spring Plate**, as shown in Figure 9-42.

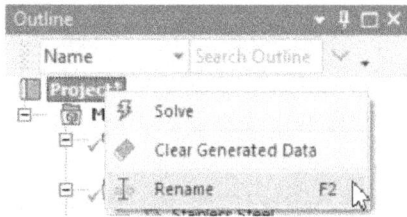

Figure 9-41 *Choosing the* **Rename** *option from the shortcut menu*

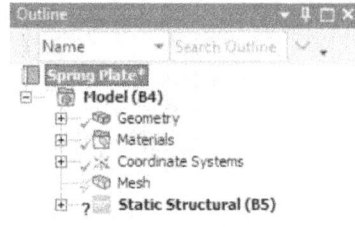

Figure 9-42 *The* **Project** *node renamed as* **Spring Plate**

7.   Select **Mesh** in the **Outline** pane to display the **Details of "Mesh"** window.

8.   In the **Details of "Mesh"** window, expand the **Defaults** node, if it is not already expanded. Enter **1.2** in the **Element Size** edit box.

9.   In the **Sizing** node, select the **Yes** option from the **Use Adaptive Sizing** drop-down list and enter **3** in the **Resolution** edit box. Then, select **Yes** from the **Mesh Defeaturing** drop-down list and keep the other option default, as shown in Figure 9-43; the **Details of "Mesh"** window is modified.

10. In the **Outline** pane, right-click on **Mesh**; a shortcut menu is displayed. Choose the **Generate Mesh** option; the mesh is generated with all other options left as default, as shown in Figure 9-44.

Note
For case I, the element and node count in your system will display 16,985 and 31,375.

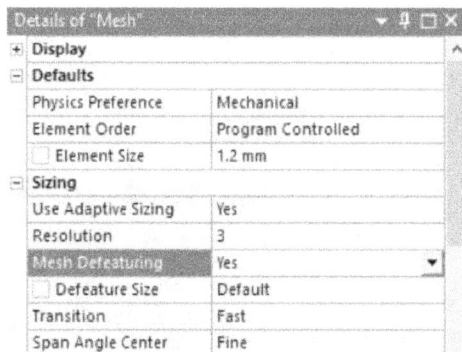

Figure 9-43 *Selecting the* **Yes** *option from the* **Use Adaptive Sizing** *and* **Yes** *option from* **Mesh Defeaturing** *drop-down list*

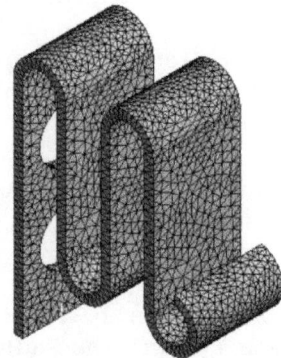

Figure 9-44 *The mesh generated*

Specifying the Boundary Conditions

After you mesh the model, it is required to specify the boundary and loading conditions.

1. In the **Mechanical** window, select the **Static Structural** node from the **Outline** pane; the **Details of "Static Structural"** window is displayed along with the **Environment** contextual tab.

2. Choose the **Fixed** tool from the **Structural** group of the **Environment** contextual tab; **Fixed Support** is added under the **Static Structural** node. Also, the **Details of " Static Structural"** window is displayed.

3. Click on the **Geometry** selection box to display the **Apply** and **Cancel** buttons.

4. Choose the **Face** tool from the **Select** toolbar to enable selection of faces.

5. Select two circular faces of the holes on the model, as shown in Figure 9-45. Next, choose the **Apply** button in the **Geometry** selection box; the selected faces turn purple indicating that fixed support is applied, refer to Figure 9-46.

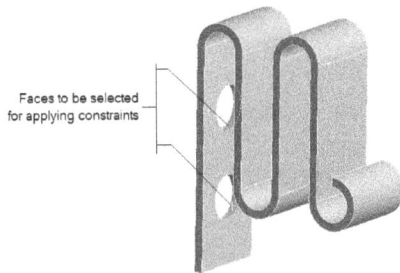

Figure 9-45 Faces to be selected for applying constraints

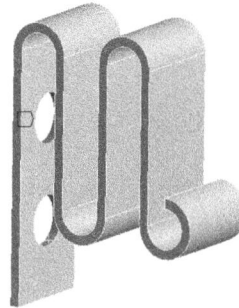

Figure 9-46 Model after the boundary conditions are applied

6. Choose the **Force** tool from the **Structural** group of the **Environment** contextual tab; **Force** is added under the **Static Structural** node in the **Outline** pane. Also, the **Details of "Force"** window is displayed.

7. Click on the **Geometry** selection box to display the **Apply** and **Cancel** buttons, if they are not already displayed.

8. Next, select the circular face on the right of the model, as shown in Figure 9-47.

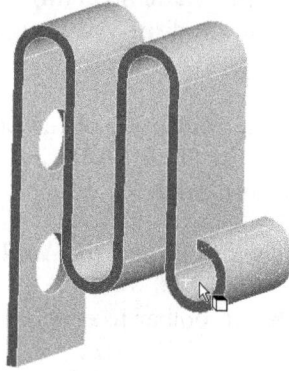

Figure 9-47 *Selecting the circular face of the model*

9. Choose the **Apply** button from the **Geometry** selection box; the cylindrical face turns red indicating that the force load is applied.

10. In the **Details of "Force"** window, expand the **Definition** node, if it is not already expanded.

11. Select **Vector** from the **Define By** drop-down list, if it is not already selected.

12. In the **Details of "Force"** window, click on the right arrow next to the **Magnitude** edit box; a flyout is displayed.

13. Choose **Constant** from the flyout, if it is not already chosen, as shown in Figure 9-48.

    The **Constant** option is chosen when the force applied remains constant with respect to time.

14. In the **Magnitude** edit box, enter **30**.

15. Click on the **Direction** selection box to display the **Apply** and **Cancel** buttons.

    As the application of force under consideration is vertically downward, you need to define the direction by selecting edges for the force vector.

16. Select any vertical edge on the model, as shown in Figure 9-49, to specify the direction of force application.

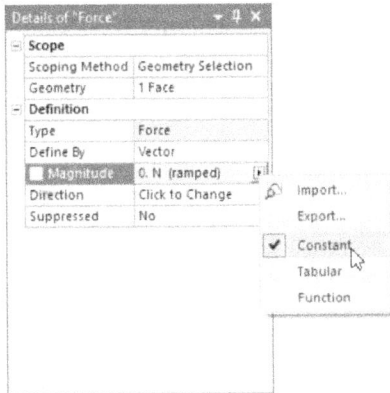

Figure 9-48 *Choosing the **Constant** option*

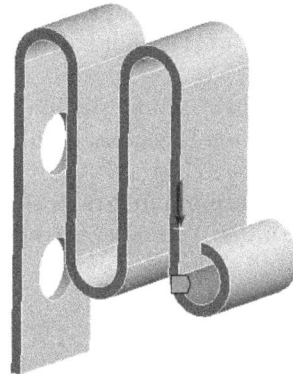

Figure 9-49 *Selecting the edge to specify the direction*

17. Next, choose the **Apply** button from the **Direction** selection box; a downward force is specified for the analysis.

After you have selected the edge for specifying the direction, you can flip the direction specified by choosing the **Flip** button available in the graphics screen.

Solving the FE Model and Analyzing the Results

After the boundary and load conditions are specified for the model, you need to solve the analysis. After solving, you will get the total and directional deformations due to the given condition. Also, you will get equivalent Stress, maximum principal, and minimum principal stresses.

1. Select the **Solution** node in the **Outline** pane; the **Solution** contextual tab is displayed. Also, the **Details of "Solution (B6)"** window is displayed.

2. Choose the **Total** tool from the **Deformation** drop-down of the **Solution** contextual tab; **Total Deformation** is added under the **Solution** node.

3. Now, choose the **Directional** tool from the **Deformation** drop-down; **Directional Deformation** is added under the **Solution** node.

Note

1. In this tutorial, the directional deformation has been calculated with respect to the X axis.

*2. To find directional deformation along the other two axes, you need to insert more instances of directional deformation by using the **Deformation** drop-down in the **Solution** contextual tab. Next, you need to change the axes by selecting the required option from the **Orientation** drop-down in the **Details of "Directional Deformation"** window.*

4. Choose the **Equivalent (von-Mises)** tool from the **Stress** drop-down in the **Solution** contextual tab.

The equivalent or von-mises stress is the criteria by which the effect of all the directional stresses acting at a point is considered. This helps in finding out whether the model will fail or bear the stress at that particular point.

5. Choose the **Maximum Principal** option from the **Stress** drop-down in the **Solution** contextual tab; **Maximum Principal** is added under the **Solution** node.

6. Choose the **Minimum Principal** option from the **Stress** drop-down in the **Solution** contextual tab; **Minimum Principal Stress** is added under the **Solution** node.

7. Choose the **Solve** tool from the **Solution** contextual tab; the parameters are evaluated.

8. In the **Outline** pane, select **Total Deformation** to visualize the results; the deformed model is shown in the graphics screen, as shown in Figure 9-50.

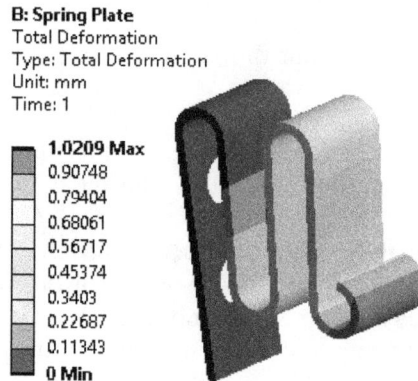

B: Spring Plate
Total Deformation
Type: Total Deformation
Unit: mm
Time: 1

1.0209 Max
0.90748
0.79404
0.68061
0.56717
0.45374
0.3403
0.22687
0.11343
0 Min

Figure 9-50 *Total deformation in the model*

9. In the **Details of "Total Deformation"** window, expand the **Results** node, if it is not already expanded. Note that the maximum and minimum deformations displayed are **1.0209 mm** and **0 mm**, respectively.

The maximum and minimum values of total deformation can also be obtained from the Legend displaying the color bands in the graphics screen.

10. Select all other parameters from the **Solution** node; the respective view is displayed in the graphics screen. The table given next lists all the results obtained from the analysis. Also, Figures 9-51 to 9-54 show the corresponding graphical representation of values obtained.

Parameter	Max. Value	Min. Value
Total Deformation	1.0209 mm	0 mm
Directional Deformation	0.47268 mm	-0.70794 mm
Equivalent Stress (von-mises)	109.63 MPa	1.3897e-005 MPa
Max. Principal Stress	165.19 MPa	-51.161 MPa
Min. Principal Stress	59.971 MPa	-161.59 MPa

B: Spring Plate
Directional Deformation
Type: Directional Deformation(X Axis)
Unit: mm
Global Coordinate System
Time: 1

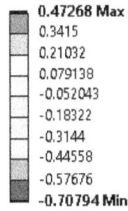

0.47268 Max
0.3415
0.21032
0.079138
-0.052043
-0.18322
-0.3144
-0.44558
-0.57676
-0.70794 Min

Figure 9-51 *Directional deformation along the X axis*

B: Spring Plate
Equivalent Stress
Type: Equivalent (von-Mises) Stress
Unit: MPa
Time: 1

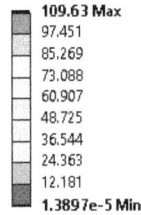

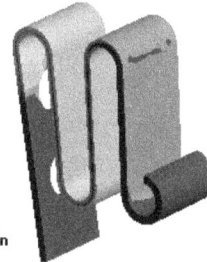

109.63 Max
97.451
85.269
73.088
60.907
48.725
36.544
24.363
12.181
1.3897e-5 Min

Figure 9-52 *Equivalent stress*

B: Spring Plate
Maximum Principal Stress
Type: Maximum Principal Stress
Unit: MPa
Time: 1

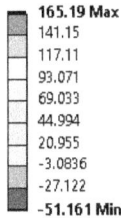

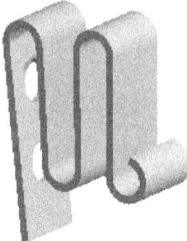

165.19 Max
141.15
117.11
93.071
69.033
44.994
20.955
-3.0836
-27.122
-51.161 Min

Figure 9-53 *Maximum principal stress*

B: Spring Plate
Minimum Principal Stress
Type: Minimum Principal Stress
Unit: MPa
Time: 1

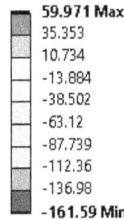

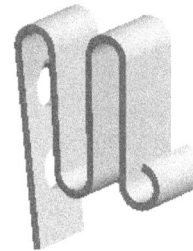

59.971 Max
35.353
10.734
-13.884
-38.502
-63.12
-87.739
-112.36
-136.98
-161.59 Min

Figure 9-54 *Minimum principal stress*

11. Close the existing **Mechanical** window; the **Workbench** window is displayed.

Setting the Boundary Condition for Case II Analysis System

After you retrieve the results of the case I analysis system, you now need to work for case II.

1. In the **Spring Plate** analysis system of the **Project Schematic** window, right-click on **Static Structural**; a shortcut menu is displayed.

2. From this shortcut menu, choose the **Duplicate** option, as shown in Figure 9-55; the **Copy of Spring Plate** analysis system is created in the **Project Schematic** window.

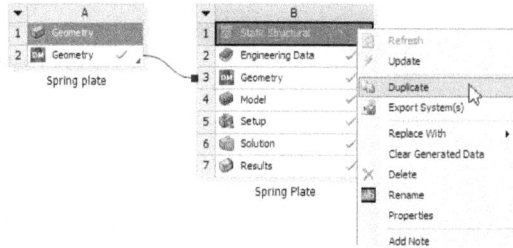

Figure 9-55 *Choosing the **Duplicate** option*

3. Double-click on the name field of the **Copy of Spring Plate** analysis system and then specify **Case II** as the name of this analysis system; it is renamed as **Case II**.

4. Double-click on the **Model** cell of the **Case II** analysis system; **Mechanical** window of this analysis system is displayed.

> **Note**
> *As the **Case II** analysis system is an exact copy of the **Spring Plate** analysis system, all the contents of the **Mechanical** window of the **Case II** analysis system are exactly the same as that of the **Spring Plate** analysis system.*

5. Right-click on **Fixed Support** in the **Outline** pane; a shortcut menu is displayed.

6. Choose the **Suppress** option from the shortcut menu displayed, as shown in Figure 9-56; a cross icon (×⚫) is placed on the left of **Fixed Support** in the **Outline** pane, indicating that this support is not available for analysis anymore.

7. Select the **Static Structural** node in the **Outline** pane to display the **Environment** contextual tab.

8. Choose the **Displacement** tool from the **Structural** group of the **Environment** contextual tab, as shown in Figure 9-57; **Displacement** is added under the **Static Structural** node in the **Outline** pane. Also, the **Details of "Displacement"** window is displayed.

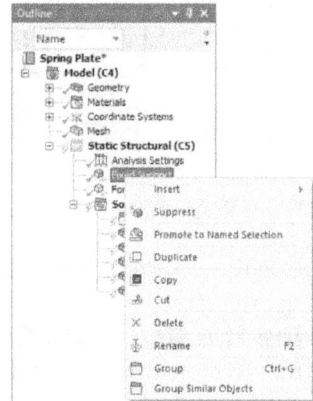

Figure 9-56 *Choosing the **Suppress** option from the shortcut menu*

Displacement support is similar to the fixed support but has partially restrained degrees of freedom.

Figure 9-57 *Choosing the **Displacement** tool from the **Structural** group*

9. In the **Details of "Displacement"** window, select the **Geometry** selection box to display the **Apply** and the **Cancel** buttons, if they are not displayed already.

10. Choose the **Face** tool from the **Select** toolbar.

11. Select both the circular faces, refer to Figure 9-58.

12. Next, choose the **Apply** button from the **Geometry** selection box in the **Details of "Displacement"** window; **2 Faces** is displayed in the **Geometry** selection box.

   Notice that, with the inclusion of displacement support in the analysis system, a local coordinate system icon is placed in the model, as shown in Figure 9-59. This coordinate system will help you to control the displacement of the support along a certain axis.

   In this tutorial, you will restrict the movement of the component along the Y and Z axes, whereas a displacement of 2 mm is allowed along the X axis.

13. In the **Details of "Displacement"** window, expand the **Definition** node, if it is not already expanded.

14. Enter **2** in the **X Component** edit box; an arrow is placed in the model, refer to Figure 9-59, indicating that the displacement is possible only along positive X axis.

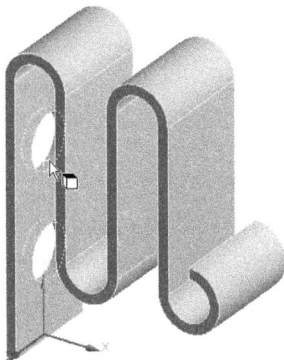

Figure 9-58 *Selecting the circular faces*   *Figure 9-59* *Coordinate system placed in the model*

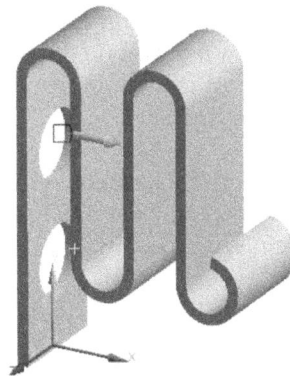

   There are three edit boxes displayed when the **Components** option is selected in the **Define By** drop-down list. To allow movement of the component along a particular axis, specify a value in the corresponding edit box.

15. Similarly, enter **0** in both the **Y Component** and **Z Component** edit boxes.

   Note
   The direction of the arrow is a resultant of all the values specified in the X, Y, and Z axes.

Solving the Analysis

After you finish setting the boundary conditions, you now need to solve it in order to get results. The parameters specified under the **Solution** node of this system are same as that of **Spring Plate** analysis system.

1. Choose the **Solve** tool from the **Solution** contextual tab toolbar; green tick mark is placed before the respective results under the **Solution** node in the **Outline** pane.

2. Exit the **Mechanical** window for the *Case II* analysis system.

Creating the Surface Model and Generating the Mesh for Surface Model for Case III Analysis System

After you retrieve the results of the case I and case II analysis system, you now need to work for case III.

1. In the **Spring Plate** Geometry component system of the **Project Schematic** window, right-click on **Geometry**; a shortcut menu is displayed.

2. From this shortcut menu, choose the **Duplicate** option, as shown in Figure 9-60; the **Copy of Spring Plate** component system is created in the **Project Schematic** window.

3. Double-click on the name field of the **Copy of Spring Plate** component system and then specify **Spring Plate_Surface** as the name of this component system; it is renamed as **Spring Plate_Surface**.

4. Double-click on the **Geometry** cell of the **Spring Plate_Surface** component system; the **DesignModeler** window of this component system is displayed.

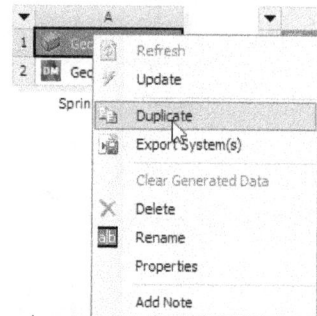

Figure 9-60 *Choosing the* ***Duplicate*** *option*

> **Tip**
> *Surface bodies can be created from planar 2D sketches, or by revolving, extruding, or sweeping lines or curves. In many cases, they can also be created from solid models by using the* ***Mid-Surface*** *tool provided in Workbench as discussed in earlier chapter.*

5. Choose the **Mid-Surface** tool from the **Tools** menu; **MidSurf1** is added to the **Outline** pane. Also, the corresponding **Details View** window is displayed and you are prompted to select a face pairs to create the mid-surface feature.

6. In the **Details of MidSurf1** window, choose **Automatic** from the **Selection Method** drop-down.

7. In the **Minimum Threshold** edit box of the **Details View** window, enter **1**, refer to Figure 9-61. In the **Maximum Threshold** edit box of the **Details View** window, enter **3**, refer to Figure 9-61.

Note that the spring plate has a thickness of 2 mm.

8. In the **FD3, Selection Tolerance (>=0)** edit box of the **Details View** window, enter **0.01**, refer to Figure 9-61. Select **Yes** from the **Find Face pairs Now** drop-down list in the **Details of MidSurf1** window.

You will notice that the **Find Face pairs Now** selection box is reset to **No**, refer to Figure 9-61. In addition to this, **8** is displayed in the **Face Pairs** selection box indicating that eight face pairs have been selected for the mid-surface operation.

9. Choose the **Generate** tool from the **Features** toolbar to generate the Mid-Surface, as shown in Figure 9-62.

Note that when you expand **1 Part, 1 Body** node in the **Tree Outline**, **Solid** is replaced by **Surface Body**.

Details View	🔼
Details of MidSurf1	
Mid-Surface	MidSurf1
Face Pairs	8
Selection Method	Automatic
Minimum Threshold	1 mm
Maximum Threshold	3 mm
FD3, Selection Tolerance (>=0)	0.01 mm
FD1, Thickness Tolerance (>=0)	0.01 mm
FD2, Sewing Tolerance (>=0)	0.02 mm
Extra Trimming	Intersect Untrimmed ...
Preserve Bodies?	No

Figure 9-61 Displaying inputs for the Automatic method

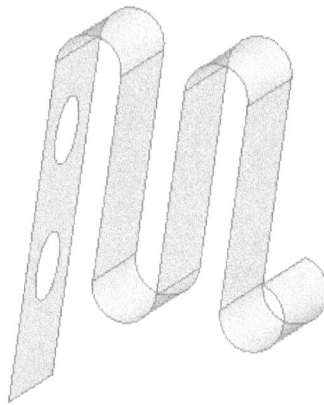

Figure 9-62 Generating the mid-surface feature

10. Exit the **DesignModeler** window to return to the **Workbench** window.

11. In the **Workbench** window, add a new **Static Structural** analysis system in the **Project Schematic** window.

12. Drag the **Geometry** cell from the **Spring Plate_Surface** component system and drop in the **Geometry** cell of the **Static Structural** analysis system to share the created surface geometry.

13. In the **Project Schematic** window, rename the **Static Structural** analysis system to **Case III**.

14. In the **Case III** analysis system, double-click on the **Model** cell to open the **Mechanical** window.

15. In the **Mechanical** window, expand the **Geometry** node in the **Outline** pane.

16. Select the **Surface Body** node under the **Geometry** node to display the **Details of "Surface Body"** window.

17. In the **Details of "Surface Body"** window, expand the **Material** node to display the **Assignment** drop-down list.

You will notice in the **Assignment** drop-down list, refer to Figure 9-40, the structural steel material is applied by default.

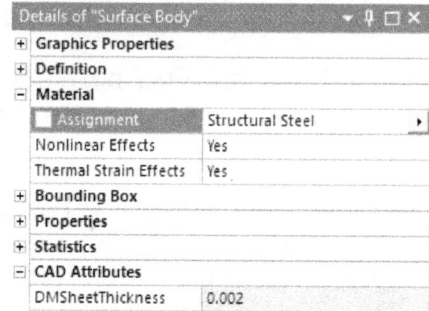

Figure 9-63 Displaying the default Structural Steel material option

18. Select **Mesh** in the **Outline** pane to display the **Details of "Mesh"** window.

19. In the **Details of "Mesh"** window, expand the **Defaults** node, if it is not already expanded. Enter **1.2** in the **Element Size** edit box.

20. In the **Sizing** node, select the **Yes** option from the **Use Adaptive Sizing** drop-down list and enter **3** in the **Resolution** edit box. Then, select **Yes** from the **Mesh Defeaturing** drop-down list and keep the other option default, as shown in Figure 9-64; the **Details of "Mesh"** window is modified.

21. In the **Outline** pane, right-click on **Mesh**; a shortcut menu is displayed. Choose the **Generate Mesh** option; the mesh is generated with all other options left as default, as shown in Figure 9-65.   🖫 Generate Mesh

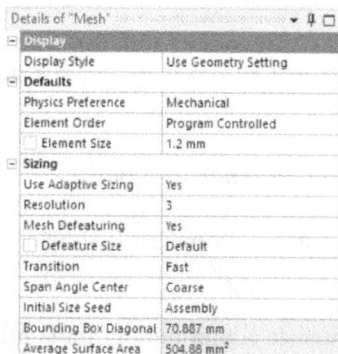

*Figure 9-64 Selecting the **Yes** option in the* **Use Adaptive Sizing** *and* **Mesh Defeaturing** *drop-down lists*

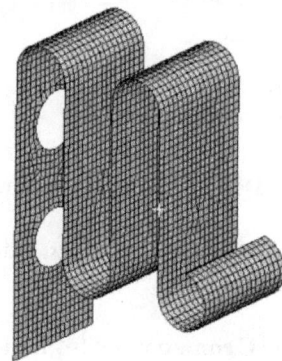

*Figure 9-65 The **mesh** generated*

Note
For this case III, the element and node count in your system will display 2932 and 3149.

Specifying the Boundary Conditions for Case III Analysis System

After you mesh the model, it is required to specify the boundary and loading conditions.

1. In the **Mechanical** window, select the **Static Structural** node from the **Outline** pane. Choose the **Fixed** tool from the **Structural** group of the **Environment** contextual tab; **Fixed Support** is added under the **Static Structural** node. Also, the **Details of " Fixed Support"** window is displayed.   Fixed

2. Choose the **Edge** tool from the **Select** toolbar to enable selection of edges.

3. Select two circular edges of the holes on the model, as shown in Figure 9-66. Next, choose the **Apply** button in the **Geometry** selection box; the selected edges turn purple indicating that fixed support is applied, refer to Figure 9-67.

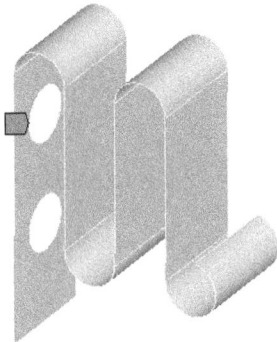

Figure 9-66 Edges to be selected for applying constraints

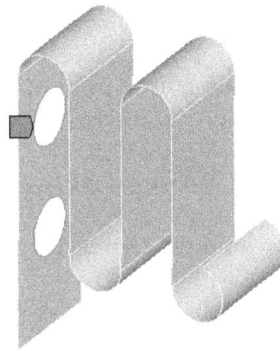

Figure 9-67 Model after the boundary conditions are applied

Note

*The display of mesh has been turned off to see the surface model clearly, as shown in Figure 9-66 and Figure 9-67. To do so, choose the **Show Mesh** tool from the **Style** group of the **Display** contextual tab.*

4. Choose the **Force** tool from the **Structural** group of the **Environment** contextual tab; **Force** is added under the **Static Structural** node in the **Outline** pane. Also, the **Details of "Force"** window is displayed.

5. Choose the **Face** tool from the **Select** toolbar to enable selection of faces.

6. Next, select the circular face on the right of the model, as shown in Figure 9-68.

Figure 9-68 *Selecting the circular face of the model*

7. Choose the **Apply** button from the **Geometry** selection box; the cylindrical face turns red indicating that the force load is applied.

8. In the **Details of "Force"** window, expand the **Definition** node, if it is not already expanded.

9. Select **Vector** from the **Define By** drop-down list, if it is not already selected.

10. In the **Magnitude** edit box, enter **30**.

11. Click on the **Direction** selection box to display the **Apply** and **Cancel** buttons. Select any vertical edge on the model, as shown in Figure 9-69, to specify the direction of force application in downward direction.

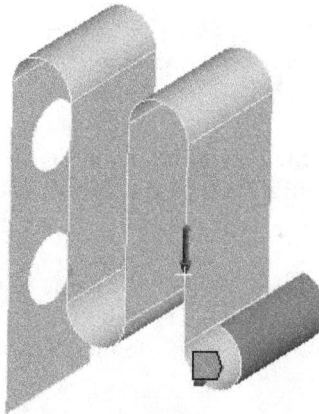

Figure 9-69 *Selecting the face to specify the direction*

12. Next, choose the **Apply** button from the **Direction** selection box; a downward force is specified for the analysis.

Solving the FE Model and Analyzing the Results for Case III

After the boundary and load conditions are specified for the surface model, you need to solve the analysis.

1. Select the **Solution** node in the **Outline** pane; the **Solution** contextual tab is displayed. Also, the **Details of "Solution"** window is displayed.

2. Choose the **Total** tool from the **Deformation** drop-down of the **Solution** contextual tab; **Total Deformation** is added under the **Solution** node.

3. Now, choose the **Directional** tool from the **Deformation** drop-down; **Directional Deformation** is added under the **Solution** node. Next, select **X Axis** from the **Orientation** drop-down list in the **Details of "Direction Deformation"** window, if not selected by default.

4. Choose the **Equivalent (von-Mises)** tool from the **Stress** drop-down in the **Solution** contextual tab.

5. Choose the **Maximum Principal** option from the **Stress** drop-down in the **Solution** contextual tab; **Maximum Principal Stress** is added under the **Solution** node.

6. Choose the **Minimum Principal** option from the **Stress** drop-down in the **Solution** contextual tab; **Minimum Principal Stress** is added under the **Solution** node.

7. Choose the **Solve** tool from the **Solution** contextual tab; the parameters are evaluated.

8. Exit the **Mechanical** window for the case III analysis system.

Interpreting the Results for All the Cases

After solving the three analysis systems, it is required to understand the data.

1. In the **Project Schematic** window, double-click on the **Model** cell in the **Spring Plate** analysis system; the **Mechanical** window is displayed.

2. In the **Mechanical** window, select the **Total Deformation** node available under the **Solution** node in the **Outline** pane; the respective legend is displayed in the graphics screen. Also, the deformed shape of the model is displayed in the graphics screen, refer to Figures 9-70.

3. In the **Outline** pane, select **Directional Deformation**; the corresponding legend is displayed in the graphics screen.

4. Similarly, select all other nodes available under the **Solution** node in the **Outline** pane to display their corresponding legends and effects on the model.

    Figures 9-70 and 9-71 display the total deformation and equivalent stress for the **Spring Plate** analysis system.

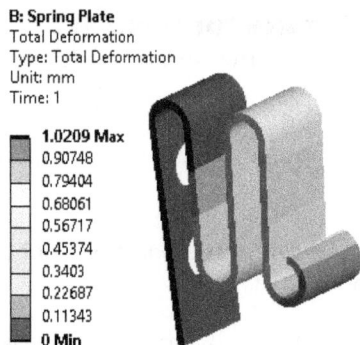

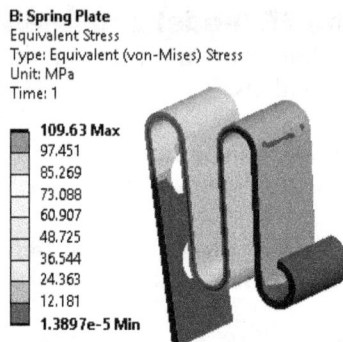

Figure 9-70 *Total deformation for the* ***Spring*** ***Plate*** *analysis system*

Figure 9-71 *Equivalent stress for the* ***Spring*** ***Plate*** *analysis system*

5. Exit the **Mechanical** window corresponding to the **Spring Plate** analysis system; the **Workbench** window is displayed.

6. In the case II analysis system, double-click on the **Model** cell to display the **Mechanical** window.

7. In the **Outline** pane, select the **Total Deformation** node; the corresponding legend is displayed.

    Figures 9-72 and 9-73 show the total deformation and equivalent stress for the case II analysis system.

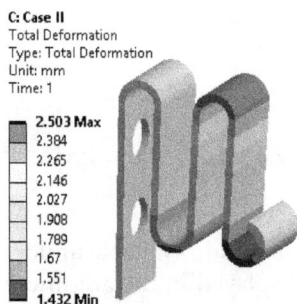

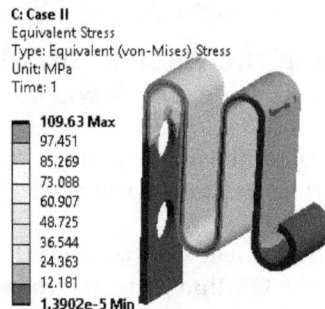

Figure 9-72 *Total deformation for case II analysis system*

Figure 9-73 *Equivalent stress for case II analysis system*

8. Exit the **Mechanical** window corresponding to the case II analysis system; the **Workbench** window is displayed.

9. In the case III analysis system, double-click on the **Model** cell to start the **Mechanical** window.

10. In the **Outline** pane, select the **Total Deformation** node; the corresponding legend is displayed.

Figures 9-74 and 9-75 show the total deformation and equivalent stress for the case III analysis system.

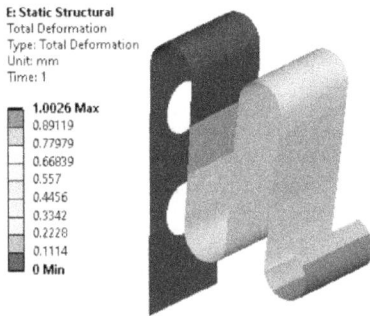

Figure 9-74 *Total deformation for case III analysis system*

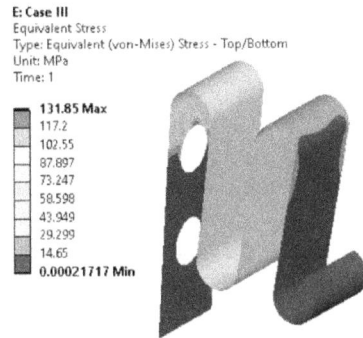

Figure 9-75 *Equivalent stress for case III analysis system*

11. Similarly, select all other nodes available under the **Solution** node in the **Outline** pane to display their corresponding legends and effects on the model.

The results obtained from the three analysis systems are shown in the table given next.

Description	Spring Plate Analysis System	Case II Analysis System	Case III Analysis System
Total Deformation	Max: 1.0209 mm Min: 0 mm	Max: 2.503 mm Min: 1.432 mm	Max: 1.0026 mm Min: 0 mm
Directional Deformation (Along X axis)	Max: 0.472689 mm Min: -0.70794 mm	Max: 2.4727 mm Min: 1.2921 mm	Max: 0.45082 mm Min: -0.68599 mm
Equivalent Stress	Max: 109.63 MPa Min: 1.3897e-006 MPa	Max: 109.63 MPa Min: 1.3902e-005 MPa	Max: 131.85 MPa Min: 2.1717e-004 MPa
Max. Principal Stress	Max: 165.19 MPa Min: -51.161 MPa	Max: 165.19 MPa Min: -51.161 MPa	Max: 145.37 MPa Min: -1.1316 MPa
Min. Principal Stress	Max: 59.971 MPa Min: -161.59 MPa	Max: 59.971 MPa Min: -161.59 MPa	Max: 1.117 MPa Min: -148.34 MPa

Note
The tabular format is very helpful in jotting down the data of various iterations of the same analysis with different settings. With the help of tabular data generated from various iterations in ANSYS Workbench, you can also save the preprocessing time by manipulating the desired mesh option, refer to Chapter 7 for details.

The above table shows that the maximum total deformation of the model when it is provided fixed support is 1.0212 mm and 1.0026 mm, whereas the maximum total deformation when displacement support is provided is 2.503 mm. In all the three analyses, the load applied is same; the only difference is the way the support is applied. However, the more displacement is observed in the case II analysis system as the cylindrical faces can move to a distance of 2 mm along the X axis.

The maximum equivalent (von-mises) stress developed is 131.85 MPa in the last analysis system. However, the less maximum equivalent (von-mises) stress is observed in the case I analysis system.

The minimum equivalent (von-mises) stress developed in the Spring Plate, Case II, and Case III analysis systems are 1.3897e-006 MPa, 1.3902e-005 MPa, and 2.1717e-004 MPa, respectively.

In this tutorial, the main aspects of the shell elements used for analyzing shell structures are discussed. Shells can be regarded as the extensions of the beam elements from 1-D line elements to 2-D surface elements. The elements and nodes count has also reduced drastically. In addition to this, analysis results have improved effectively.

Tip
To evaluate directional deformation along the Y axis, select Y Axis from the Orientation drop-down list in the Details of "Directional Deformation" window and then choose the Solve tool from the Solution contextual tab.

11. Exit the **Mechanical** window to display the **Workbench** window.

Saving the Project

You have already saved the project with the name *c09_ansWB_Tut02*, therefore now you just need to save your work.

1. Choose the **Save Project** button from the **Standard** toolbar to save the project.

2. Next, exit the **Workbench** window to close the session.

Tutorial 3   Fatigue Analysis

In this tutorial, you will first download the zip file *c09_ansWB_tut03.zip* from *www.cadcim.com* and then extract it to the specified folder. After extracting, add a **Static Structural** analysis system to the **Project Schematic** window and import the file into ANSYS Workbench. Next, you will apply bearing load at the cylindrical faces of the model and fixed support at the top planar face of the model. Figure 9-76 and Figure 9-77 show the model with the cylindrical surfaces annotated for applying bearing load and fixed support, respectively. You will also apply the gray cast iron material to the model. Next, you will solve the analysis to evaluate total deformation, directional deformation along Y axis, equivalent shear stress, and equivalent elastic strain. If fully reversed cyclic pressure load with a magnitude of 100 MPa is applied normally at the cylindrical faces, find the fatigue life, damage, and safety factor of the model. Also, determine whether or not fatigue failure occurs in the model assuming a design life of 10^6 cycles. You will apply structural steel material to the model.   **(Expected time: 60 min)**

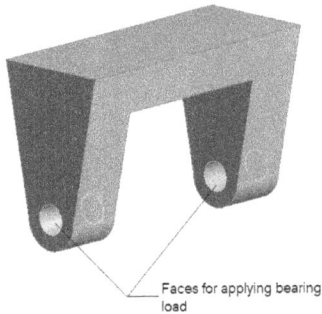

Figure 9-76 Component with surfaces selected for applying bearing load

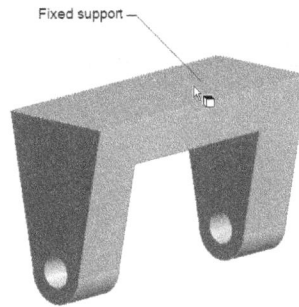

Figure 9-77 The top face of the model selected for applying Fixed support

The following steps are required to complete this tutorial:

a. Download the part file and import it into workbench.
b. Generate a mesh and apply boundary and loading conditions.
c. Edit the material and start the **Mechanical** window.
d. Solve the analysis and compare the results.
e. Apply the boundary conditions for fatigue analysis.
f. Retrieve and analyze the results for fatigue analysis.
g. Save the project and exit Workbench.

Downloading and Importing the File into Workbench

Before you start the tutorial, you need to download the *c09_ansWB_tut03.zip* file from *www.cadcim.com* and then import it into ANSYS Workbench. Next, you need to import this file to ANSYS Workbench.

1. Create a folder with the name **Tut03** at the location *C:\ANSYS_WB\C09*.

2. Download the zip file *c09_ansWB_tut03.zip* from the *www.cadcim.com*. The complete path of the file is :

    Textbooks > CAE Simulation > ANSYS > ANSYS Workbench 2023 R2: A Tutorial Approach > Input Files

    After downloading, extract the zip file and save the *c09_ansWB_tut03.stp* file at the location *C:\ANSYS_WB\C09\Tut03*.

3. Start the ANSYS Workbench session.

4. In the **Workbench** window, add the **Static Structural** analysis system to the **Project Schematic** window.

5. Right-click on the **Geometry** cell of the **Static Structural** analysis system to display a shortcut menu.

6. Choose **Import Geometry > Browse** from the shortcut menu; the **Open** dialog box is displayed.

7. In the **Open** dialog box, browse to the location *C:\ANSYS_WB\c09\Tut03* and then select **c09_ansWB_tut03.stp**. Next, choose the **Open** button from the **Open** dialog box; the file is imported into the **Workbench** window. Also, a green tick mark is placed corresponding to the **Geometry** cell in the **Mesh** component system.

8. Choose the **Save Project** button from the **Standard** toolbar; the **Save As** dialog box is displayed.

9. In this dialog box, browse to the location *C:\ANSYS_WB\c09\Tut03* and then save the project with the name **c09_ansWB_tut03**.

Assigning the Material, Generating Mesh, and Specifying the Boundary condition

Once the file is imported, you now need to assign a material to it and then generate a mesh. To do so, follow the procedure explained next.

1. In the Engineering Data workspace, add Gray Cast Iron material to the Engineering Data and return to the **Project Schematic** window.

2. In the **Project Schematic** window, double-click on the **Model** cell of the **Static Structural** analysis system; the **Mechanical** window is displayed.

3. In the **Mechanical** window, assign the Gray Cast Iron material to the model.

4. In the Outline pane, right-click on **Mesh**; a shortcut menu is displayed. ⚡ Generate Mesh Choose the **Generate Mesh** option.

Figure 9-78 shows the **Details of "Mesh"** window with the default global mesh control settings and Figure 9-79 shows the mesh generated for the model by using the default global mesh control settings.

Notice that the total elements created are 975 with the default global mesh control settings.

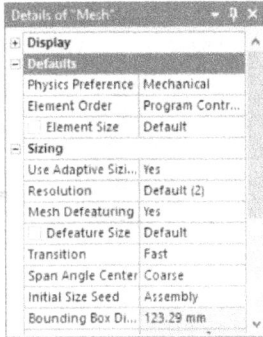

Figure 9-78 The Details of "Mesh" window

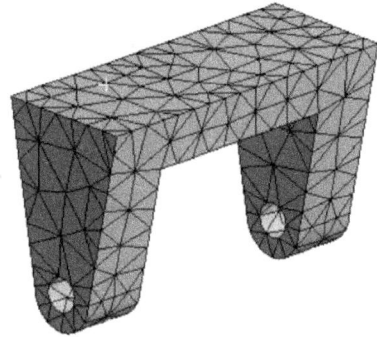

Figure 9-79 Mesh generated with default global mesh control settings

5. Select the **Static Structural** node in the **Outline** pane to display the **Environment** contextual tab.

6. Choose the **Fixed** tool from the **Structural** group of the **Environment** contextual tab; **Fixed Support**, with a question symbol (₂⊚), is attached to the **Static Structural** node in the Outline pane.

7. Choose the **Face** tool from the **Select** toolbar to select a face for applying Fixed support.

8. Select the **Geometry** selection box in the **Details of "Fixed Support"** window to display the **Apply** and **Cancel** buttons, if they are not already displayed.

9. Select the top face of the model in the graphics screen to apply fixed support, refer to Figure 9-80.

10. Choose the **Apply** button in the **Geometry** selection box; the fixed support is applied on the selected face of the component.

11. In the **Environment** contextual tab, choose the **Bearing Load** tool from the **Loads** drop-down; **Bearing Load**, with a question symbol (₂○), is attached to the **Static Structural** node in the **Outline** pane. Also, the **Details of "Bearing Load"** window is displayed.

    Bearing load is applied when it is required to apply variable load on a cylindrical surface.

    Notice that in the **Details of "Bearing Load"** window, the **Geometry** selection box is selected by default. Also the **Apply** and **Cancel** buttons are displayed.

12. In the graphics screen, select the cylindrical face of the model, as shown in Figure 9-81.

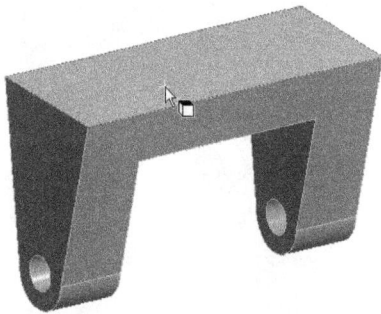

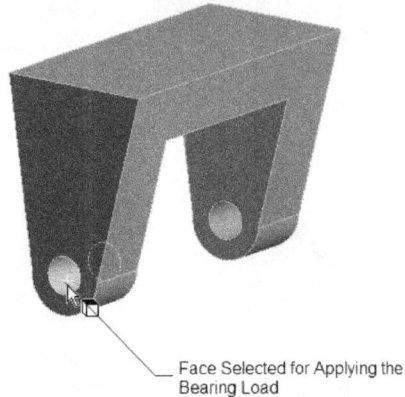

Face Selected for Applying the
Bearing Load

Figure 9-80 *Face selected for applying the fixed constraint*

Figure 9-81 *Face selected for applying the bearing load*

13. Choose the **Apply** button from the **Geometry** selection box; the color of the selected face turns red indicating that the face has been selected for applying the bearing load.

    After the face is selected for applying the load, the magnitude of the load needs to be specified.

    Next step is to specify the direction of the load. In the case of bearing loads, note that the direction of the load should always be radial to the cylindrical surface. The most common practice while specifying the direction of any vector quantity is to select an edge in the model in the graphics screen. In case, there is no such edge available in the model along which a direction can be specified, you need to specify a vector quantity in the graphics screen by providing data for its X, Y, and Z components. For example, in this tutorial, you need to specify the direction of the 10N bearing load along negative Y axis (- Y). But as you can notice, there are no edges that are along the Y axis. Therefore, in order to specify a direction in this model, you need to specify the X, Y, and Z components.

14. In the **Details of "Bearing Load"** window, select the **Components** option from the **Define By** drop-down list, refer to Figure 9-82; the components of **Details of "Bearing Load"** window are modified and **Coordinate System** drop-down list is displayed in it along with the **X Component**, **Y Component**, and **Z Component** edit boxes, refer to Figure 9-83.

15. Enter **0** in the **X Component, -10** in the **Y Component** and **0** in the **Z Component** edit boxes; a red arrow is displayed on the selected surface, as shown in Figure 9-84. Also, a green tick mark is placed before **Bearing Load** in the **Outline** pane.

    Notice that the direction of the load is now radial and hence it is correct to define the direction by selecting the **Components** option from the **Define By** drop-down list.

    The magnitude of load applied on each cylindrical surface is 10 N.

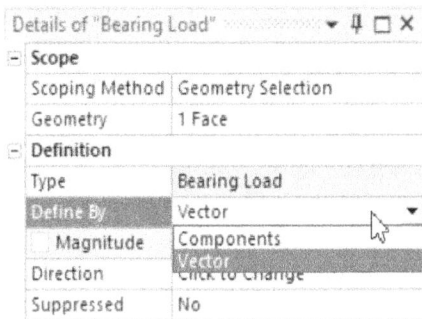

Figure 9-82 *Options in the* **Define By** *drop-down list*

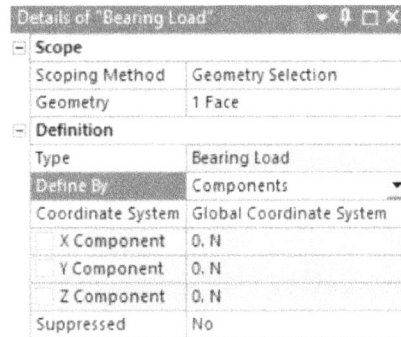

Figure 9-83 *The options in the* **Details of** *"Bearing Load" window*

16. Again, choose the **Bearing Load** tool from the **Loads** drop-down in the **Environment** contextual tab; the **Bearing Load 2** node is attached to the **Static Structural** node in the Outline pane.

17. Select the **Geometry** selection box in the **Details of "Bearing Load 2"** window to display the **Apply** and **Cancel** buttons.

18. Select the second cylindrical surface of the component and then choose the **Apply** button to specify this cylindrical face as the face to apply bearing load.

19. In the **Details of "Bearing Load 2"** window, select the **Components** option in the **Define By** drop-down list; the components of this window are modified and the **Coordinate System** drop-down list is displayed along with the **X Component**, **Y Component**, and **Z Component** edit boxes.

20. In the **Y Component** edit box, specify **-10** as the component of the bearing load vector. The direction and magnitude of the load is specified, refer to Figure 9-84.

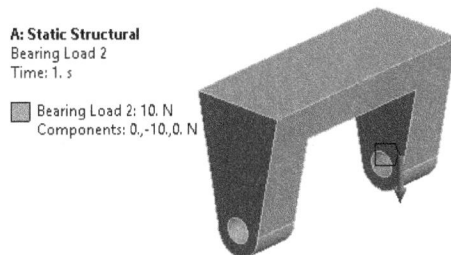

Figure 9-84 *Resulting direction of the loads*

After the mesh is generated and the loading and boundary conditions are specified, you need to specify the parameters to evaluate and then solve the analysis.

Analyzing the Results

Now, you will analyze the component for total and directional deformation and maximum von-mises stress and strain.

1. Select the **Solution** node in the **Outline** pane; the **Solution** contextual tab is displayed.

2. Choose the **Total** tool from the **Deformation** drop-down in the **Solution** contextual tab; the **Total Deformation** node is attached to the **Solution** node in the **Outline** pane.

3. Similarly, using the **Directional** deformation, **Equivalent (von-Mises) Stress** and **Equivalent (von-Mises) Strain** tools from the drop-downs available in the **Solution** contextual tab, you can insert the corresponding results. The **Outline** pane after the results added is shown in Figure 9-85.

Notice that there are yellow thunderbolt symbols attached to the results in the **Solution** node in the **Outline** pane. Also, there is a thunderbolt attached to the **Static Structural** node. These yellow thunderbolts indicate that the solution is incomplete and you need to solve the model in order to achieve results.

Figure 9-85 *Outline pane after the results are added*

4. Choose the **Solve** tool from the **Solution** contextual tab; the **ANSYS Workbench Solution Status** window is displayed. Wait for sometime, the results are evaluated against the given boundary and loading conditions.

Figure 9-86 shows the model when the **Total Deformation** node is selected in the **Outline** pane and Figure 9-87 shows the model when the **Equivalent Stress** node is selected in the **Outline** pane.

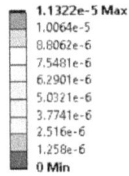

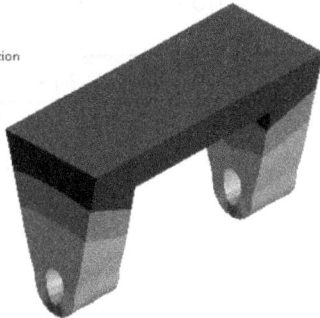

Figure 9-86 Total deformation of the model

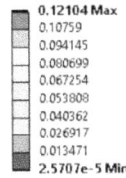

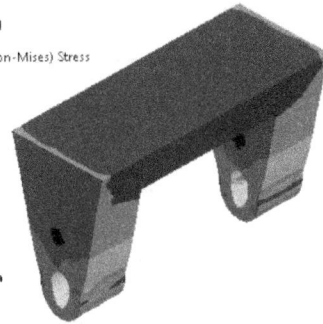

Figure 9-87 Equivalent stress of the model

5. Select the **Directional Deformation** node in the Outline pane; the respective directional deformation of the model is displayed in the graphics screen. Also, the **Details of "Directional Deformation"** window is displayed.

6. In the **Details of "Directional Deformation"** window, expand the **Definition** node, if it is not already expanded.

7. In the **Definition** node, the default selection in the **Orientation** drop-down is **X axis**. Select the **Y axis** option from this drop-down list; a yellow thunderbolt is displayed before **Directional Deformation** in the Outline pane, indicating that the solution has changed and you need to update the results.

8. Right-click on the **Solution** node in the **Outline** pane; a shortcut menu is displayed.

9. Choose **Evaluate All Results** from the shortcut menu displayed; a green tick mark is displayed before the **Directional Deformation** node in the Outline pane, indicating that the result files are updated.

10. Select the **Directional Deformation** node in the Outline pane to display the directional deformation along the Y axis, as shown in Figure 9-88.

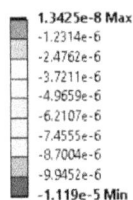

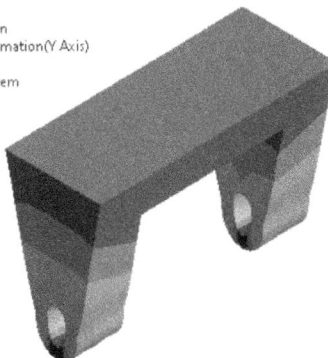

Figure 9-88 Directional deformation along the Y axis

The table displayed next displays all the values obtained from the analysis.

Parameter	Max. Value	Min. Value
Total Deformation	1.1322E-005 mm	0 mm
Directional Deformation (Y axis)	1.3425E-8 mm	-1.119E-5 mm
Equivalent Stress	0.12104 MPa	2.5707E-005 MPa
Equivalent Elastic Strain	6.1251E-007 mm/mm	2.8556 E-10 mm/mm

Specifying the Boundary Condition for Fatigue Life

Now, you need to specify boundary conditions for fatigue analysis.

1. In the **Mechanical** window, right-click on **Bearing Load** in the **Outline** pane; a shortcut menu is displayed. Choose the **Suppress** option from the shortcut menu displayed. Similarly, right-click on **Bearing Load 2** in the **Outline** pane; a shortcut menu is displayed. Choose the **Suppress** option from the shortcut menu displayed.

2. Click on the **Static Structural** node in the **Outline** pane to display the **Environment** contextual tab.

3. Choose the **Pressure** tool from the **Structural** group of the **Environment** contextual tab; **Pressure** is added under the **Static Structural** node in the **Outline** pane. Also, the **Details of "Pressure"** window is displayed.

4. Next, select the circular face on the right of the model, as shown in Figure 9-89.

5. Choose the **Apply** button from the **Geometry** selection box; the cylindrical face turns red indicating that the Pressure load is applied.

6. In the **Details of "Pressure"** window, click on the **Magnitude** edit box and then enter **100**, as shown in Figure 9-90.

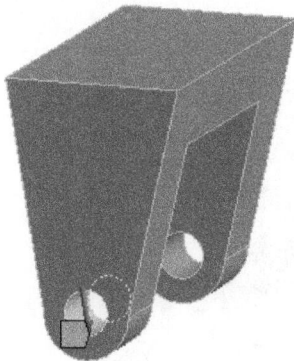

Figure 9-89 *Selecting the circular face of the model*

Figure 9-90 *Displaying magnitude of pressure*

7. Again, choose the **Pressure** tool from the **Structural** group of the **Environment** contextual tab; the **Pressure 2** node is attached to the **Static Structural** node in the **Outline** pane.

8. Next, select the circular face on the left of the model.

9. Choose the **Apply** button from the **Geometry** selection box; the cylindrical face turns red indicating that the pressure load is applied, as shown in Figure 9-91.

10. In the **Details of "Pressure 2"** window, click on the **Magnitude** edit box and then enter **100**.

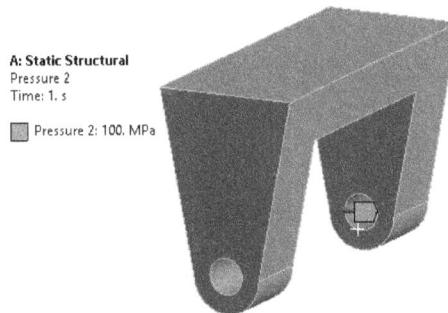

A: Static Structural
Pressure 2
Time: 1. s

Pressure 2: 100. MPa

Figure 9-91 *Resulting direction of the loads*

Retrieving and Analyzing the Results for Fatigue Life

Now, you will analyze the component for Life, Damage, and Safety Factor.

1. Select the **Solution** node in the **Outline** pane; the **Solution** contextual tab is displayed.

2. Choose the **Fatigue Tool** from the **Toolbox** drop-down in the **Solution** contextual tab; the **Fatigue Tool** node is attached to the **Solution** node in the **Outline** pane. Also, the **Details of "Fatigue Tool"** window is displayed.

   You will notice that the **Geometry** window has switched to the **Worksheet** window, as shown in Figure 9-92. The **Worksheet** window displays the graphs for **Constant Amplitude Load Fully reversed** and **Mean Stress Correction Theory** as they are the default options set for the **Fatigue Tool**, refer to Figure 9-92.

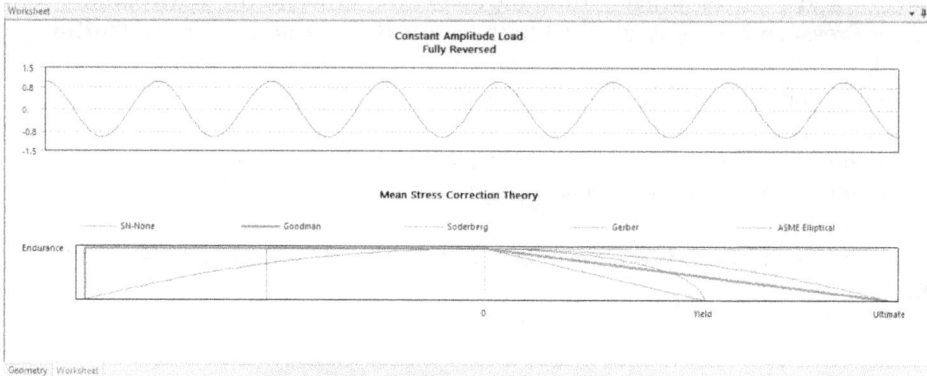

*Figure 9-92 The **Worksheet** window*

3. Select **Goodman** from **Mean Stress Theory** drop-down list under the **Options** node in the **Details of "Fatigue Tool"** window, as shown in Figure 9-93.

Note

*In the **Details of "Fatigue Tool"** window, the **Fully Reversed** option from the **Type** drop-down under the **Loading** node is selected by default. Similarly, the **Stress Life** option is chosen in the **Analysis Type** drop-down under the **Options** node by default, refer to Figure 9-94.*

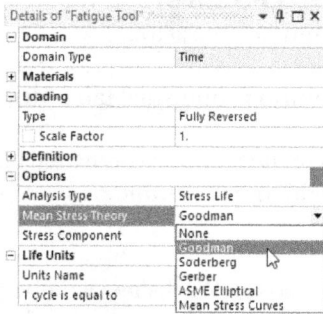

*Figure 9-93 Choosing **Goodman** from the **Mean Stree Theory** node*

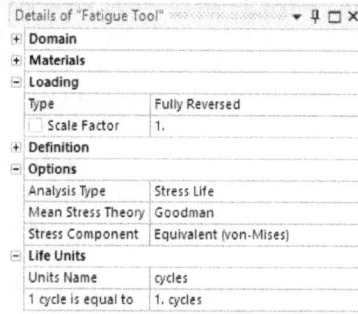

Figure 9-94 Displaying default setting in Details of "Fatigue Tool" window

4. In the **Fatigue Tool contextual** tab, choose the **Life** tool from the **Contour** group, refer to Figure 9-95; the **Life** is attached to the **Outline** pane. Also, the **Details of "Life"** window is displayed.

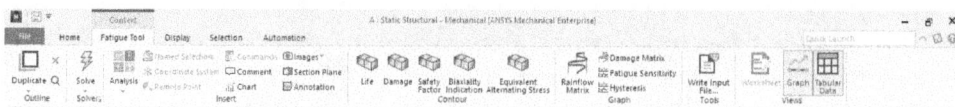

*Figure 9-95 The **Fatigue Tool** contextual tab*

5. Choose the **Damage** tool from the **Contour** group, refer to Figure 9-95; the **Damage** is attached to the **Outline** pane. Also, the **Details of "Damage"** window is displayed.

6. In this window, expand the **Definition** node if it is not already expanded. Next, enter **10⁶** in the **Design Life** edit box, as shown in Figure 9-96.

7. Choose the **Safety Factor** tool from the **Contour** group, refer to Figure 9-95; the **Safety Factor** is attached to the **Outline** pane. Also, the **Details of "Safety Factor"** window is displayed.

8. In this window, expand the **Definition** node, if it is not already expanded. Next, enter **10⁶** in the **Design Life** edit box, as shown in Figure 9-97.

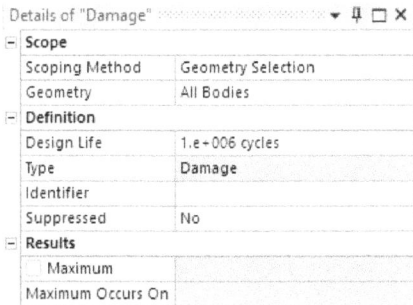

Details of "Damage" ▾ ♯ □ ✕	
− **Scope**	
Scoping Method	Geometry Selection
Geometry	All Bodies
− **Definition**	
Design Life	1.e+006 cycles
Type	Damage
Identifier	
Suppressed	No
− **Results**	
Maximum	
Maximum Occurs On	

Details of "Safety Factor" ▾ ♯ □ ✕	
− **Scope**	
Scoping Method	Geometry Selection
Geometry	All Bodies
− **Definition**	
Design Life	1.e+006 cycles
Type	Safety Factor
Identifier	
Suppressed	No
− **Results**	
Minimum	
Minimum Occurs On	

*Figure 9-96 Setting the design life for the **Damage** tool*

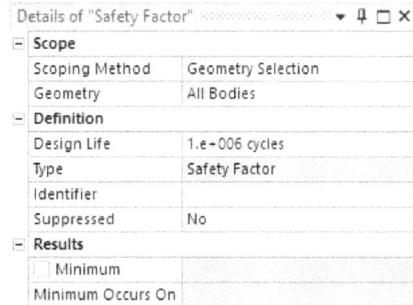

*Figure 9-97 Setting the design life for the **Safety Factor** tool*

9. Choose the **Solve** tool from the **Solution** contextual tab; the parameters are evaluated. Wait for sometime, the results are evaluated against the given boundary and loading conditions.

Figure 9-98 shows the model when the **Life** node is selected in the **Outline** pane. Figure 9-99 shows the model when the **Damage** node is selected in the **Outline** pane and Figure 9-100 shows the model when the **Safety factor** node is selected in the **Outline** pane.

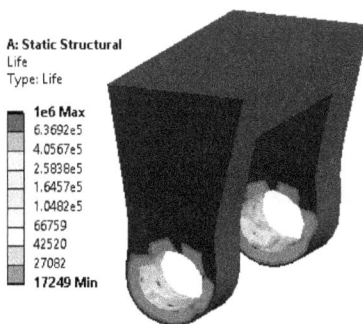

A: Static Structural
Life
Type: Life

1e6 Max
6.3692e5
4.0567e5
2.5838e5
1.6457e5
1.0482e5
66759
42520
27082
17249 Min

Figure 9-98 Life of the model

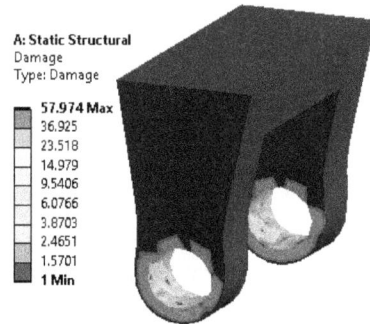

A: Static Structural
Damage
Type: Damage

57.974 Max
36.925
23.518
14.979
9.5406
6.0766
3.8703
2.4651
1.5701
1 Min

Figure 9-99 Damage to the model

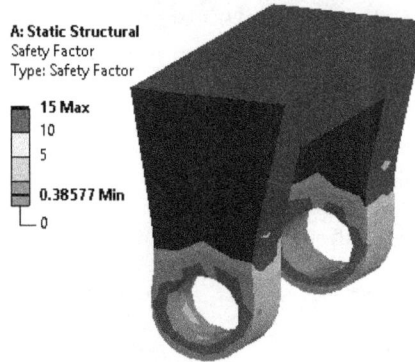

Figure 9-100   Safety factor of the model

Above figures show the minimum life of the model as 17,249 cycles. Again, maximum damage of the model for the same boundary condition is 57.974. Also, minimum safety factor of the model is 0.38577. For the design life of 10^6 cycles, the model will not survive the fatigue testing.

Note
Damage is the ratio of design life to minimum life of the model. To avoid the damage to the model the maximum damage should be less than 1.

10. Exit the **Mechanical** window; the **Workbench** window is displayed.

Saving the Project and Exiting Workbench

After you have evaluated all the results, you now need to save the file before you exit the ANSYS Workbench session. To do so, follow the procedure explained next.

1. Choose the **Save Project** button from the **Standard** toolbar to save the project.

2. Exit the **Workbench** window.

Tutorial 4   Eigenvalue Buckling Analysis

In this tutorial, you will create the model for a structural steel column that has a thickness of 2 mm, as shown in Figure 9-101 and Figure 9-102. The dimensions to create the model and its boundary and loading conditions are mentioned in those figures. The bottom face of the column is fixed and the static pressure of 40 MPa is applied at the top face of the column. Perform a Static Structural analysis on the model and evaluate the total deformation and the equivalent stress in the model. Also, determine the critical buckling load under the given static pressure load, and obtain the first three buckling mode shapes. After evaluating the results, interpret them.   **(Expected time: 3 hr)**

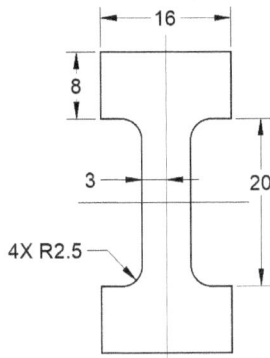

Figure 9-101 *Dimensions of the model for Tutorial 4*

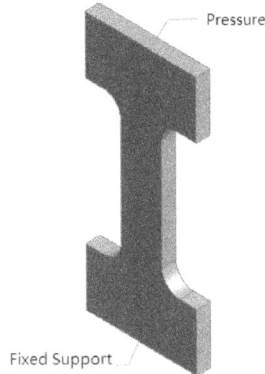

Figure 9-102 *Fixed support and pressure applied to the model*

The following steps are required to complete this tutorial.

a. Start a new project and create the model.
b. Generate the mesh.
c. Set the boundary and loading conditions.
d. Solve the model.
e. Create a Eigenvalue Buckling Analysis System.
f. Solve the model and Interpret results.
g. Animate the results.
h. Save the project.

Starting a New Project and Creating the Model
The first step is to start a new project in the **Workbench** window.

1. Start ANSYS Workbench.

2. Choose the **Save Project** button from the **Main** toolbar; the **Save As** dialog box is displayed.

3. In this dialog box, enter **c09_ansWB_Tut04** in the **File name** field and then save the file at the location: *C:\ANSYS_WB\c09\Tut04*

4. Double-click on **Static Structural** in the **Toolbox** window; the **Static Structural** analysis system is added to the **Project Schematic** window.

5. Right-click on the **Geometry** cell of this component system to display a shortcut menu. Next, choose **New DesignModeler Geometry** from the shortcut menu; the **Starting DesignModeler** message is displayed in the status bar. After sometime, the **DesignModeler** window is displayed.

6. Choose the **Millimeter** option from the **Units** menu of the **Menu** bar. Now, create the sketch on the XY plane and extrude, as shown in Figure 9-103. For dimension of the model, refer to Figure 9-101.

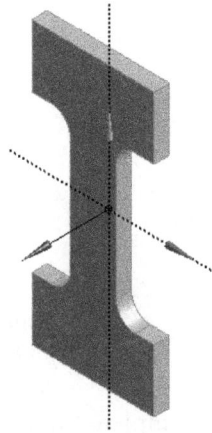

Figure 9-103 Model created
on the XY plane

7. Exit the **DesignModeler** window to display the **Workbench** window.

Generating the Mesh

After the model is created in the **DesignModeler** window, you need to generate the mesh for the model in the **Mechanical** window.

1. In the **Project Schematic** window, double-click on the **Model** cell in the **Static Structural** analysis system; the **Mechanical** window is displayed.

2. Select **Mesh** in the **Outline** pane to display the **Details of "Mesh"** window.

3. In the **Details of "Mesh"** window, expand the **Sizing** node if it is not already expanded. Also, notice that **Default** is displayed in the **Element Size** edit box. Enter **0.5** in the **Element Size** edit box.

4. In the **Outline** pane, right-click on **Mesh**; a shortcut menu is displayed. Choose the **Generate Mesh** option; the mesh is generated with default mesh, as shown in Figure 9-104.

Figure 9-104 Mesh generated
with 0.5 mm element size

5. Expand the **Statistics** node in the **Details of "Mesh"** window to display the total number of elements created. On doing so, you will find that the total number of elements created is 6160.

Setting the Boundary and Loading Conditions

After the mesh is generated, you need to set the boundary and loading conditions under which the analysis will be performed.

1. Select the **Static Structural** node in the **Outline** pane; the **Details of "Static Structural"** window is displayed. Also, the **Environment** contextual tab is displayed.

2. In the **Environment** contextual tab, choose the **Fixed** tool from the **Structural** group, refer to Figure 9-105; the **Fixed Support** is attached to the **Outline** pane. Also, the **Details of "Fixed Support"** window is displayed.

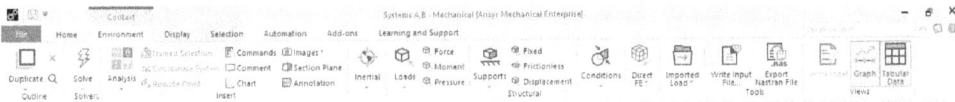

Figure 9-105 Choosing the Fixed tool in the Environment contextual tab

3. Select the **Geometry** selection box in the **Scope** node of the **Details of "Fixed Support"** window to display the **Apply** and **Cancel** buttons, if they are not already displayed.

4. Choose the **Face** tool from the **Select** toolbar to select the face to apply fixed support. Next, select the face of the model; the face turns green, as shown in Figure 9-106.

5. Choose the **Apply** button in the **Geometry** selection box to confirm the selection of the face for fixed support; the color of the face turns violet and a flag is attached to the face, as shown in Figure 9-107.

Figure 9-106 Face selected

Figure 9-107 The violet color face of the model displaying the fixed support

After the boundary is defined for the model, you need to define the load for which the analysis is to be carried out.

6. Right-click on the **Static Structural** node and then choose **Insert > Pressure** from the shortcut menu displayed; **Pressure** is added under the **Static Structural** node in the **Outline** pane and the **Details of "Pressure"** window is displayed.

7. In the **Details of "Pressure"** window, select the **Geometry** selection box to display the **Apply** and **Cancel** buttons, if they are not already displayed.

8. Choose the **Face** tool from the **Select** toolbar to select a face from the graphics screen.

9. In the graphics screen, select the face, just opposite to the face on which you have applied the Fixed support, refer to Figure 9-108.

10. Choose the **Apply** button from the **Geometry** selection box in the **Details of "Pressure"** window; the face is selected for applying the pressure load.

Figure 9-108 Selecting the face for applying a pressure load

> **Note**
> *In this tutorial, the direction of application of pressure is downward.*

11. Select the **Magnitude** edit box and then enter **40** as the magnitude of pressure load.

   Note that the directon of pressure is always normal to surface, therefore it is downward by default.

Solving and Post-Processing the Finite Element Model

After the boundary and loading conditions are specified for the analysis, you need to evaluate the results that are of importance in the case of a particular analysis.

   To evaluate the results in this analysis, follow the procedure explained next.

1. Select the **Solution** node in the **Outline** pane; the **Solution** contextual tab is displayed. Also, the **Details of "Solution"** window is displayed.

2. Choose the **Total** tool from the **Deformation** drop-down in the **Solution** contextual tab, as shown in Figure 9-109; **Total Deformation** is added under the **Solution** node. Also, the **Details of "Total Deformation"** window is displayed.

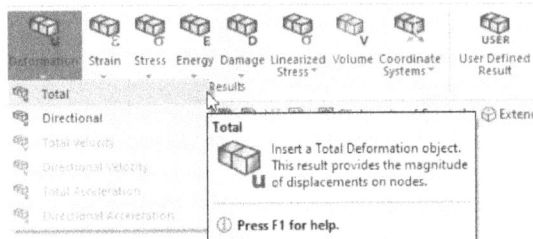

Figure 9-109 *Choosing the* **Total** *tool from the* **Deformation** *drop-down*

3. Choose the **Equivalent (von-Mises)** tool from the **Stress** drop-down; **Equivalent Stress** is added to the **Outline** pane. Also, the **Details of "Equivalent Stress"** window is displayed.

4. Choose the **Stress Tool** from the **Toolbox** drop-down; **Stress Tool** is attached to the **Outline** pane. Also, the **Details of "Stress Tool"** window is displayed, as shown in Figure 9-110. Next, expand the **Stress Tool** node, you will notice **Safety Factor** with a yellow thunderbolt added under the **Stress Tool** node in the **Outline** pane.

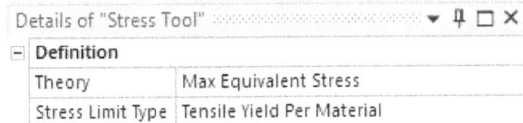

Details of "Stress Tool"	▼ ⏻ ☐ ✕
⊟ **Definition**	
Theory	Max Equivalent Stress
Stress Limit Type	Tensile Yield Per Material

Figure 9-110 *Displaying the* **Details of** *"Stress Tool" window*

5. Choose the **Solve** tool from the **Solution** contextual tab; the parameters are evaluated.

Figure 9-111 shows the model when the **Total Deformation** node is selected in the **Outline** pane and Figure 9-112 shows the model when the **Equivalent Stress** node is selected in the **Outline** pane. Similarly, Figure 9-113 shows the model when the **Safety Factor** node is selected in the **Outline** pane.

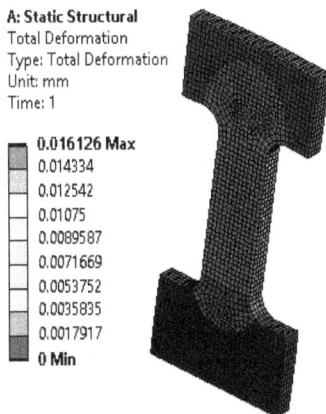

Figure 9-111 *Total deformation of the model*

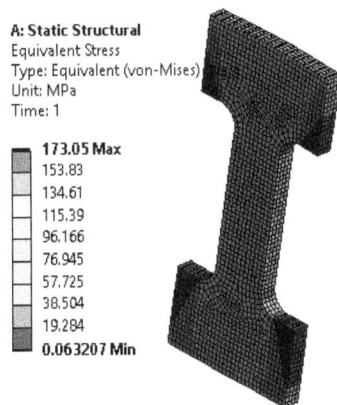

Figure 9-112 *Equivalent stress of the model*

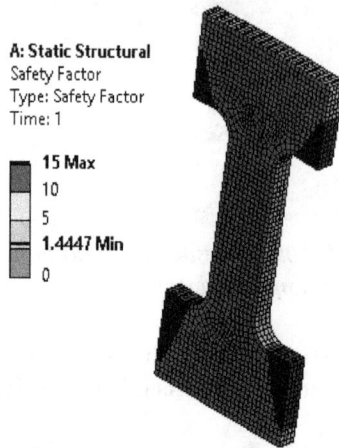

A: Static Structural
Safety Factor
Type: Safety Factor
Time: 1

15 Max
10
5
1.4447 Min
0

Figure 9-113 Safety factor of the model

The table given next lists all the results obtained from the analysis.

Parameter	Max. Value	Min. Value
Total Deformation	0.016126 mm	0 mm
Equivalent Stress (von-mises)	173.05 MPa	0.063207 MPa
Safety Factor	15	1.4447

In this table, maximum equivalent stress value is 173.05 MPa which is smaller than the yield strength of the model material. Therefore, the minimum safety factor value, 1.4447, is greater than unity. Based on the results, model stability can be evaluated for the given loading condition.

Note
*In this tutorial, the **Stress Tool** is added in the **Solution** node to calculate safety factor by comparing the maximum equivalent von-mises stress in the model with the tensile yield strength of the model material.*

6.   Exit the **Mechanical** window; the **Workbench** window is displayed.

Creating and Setting a Eigenvalue Buckling Analysis

After performing the Static Structural analysis, you need to create and set a Eigenvalue Buckling analysis system.

1.   In **Project Schematic** window, drag the **Eigenvalue Buckling** analysis system and drop in the **Solution** cell of the **Static Structural** analysis system, as shown in Figure 9-114. On doing so, the **Project Schematic** window will will be displayed showing that the **Static Structural** analysis system has transferred the data to the **Eigenvalue Buckling** analysis system, as shown in Figure 9-115.

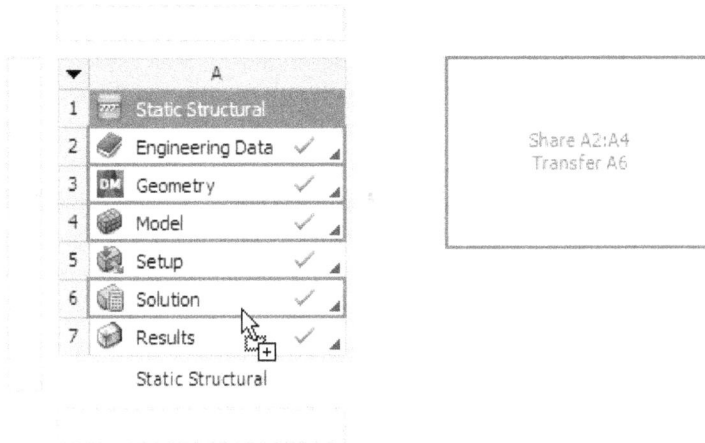

Figure 9-114 Dragging the **Eigenvalue Buckling** analysis system into the **Solution** cell of the **Static Structural** analysis system

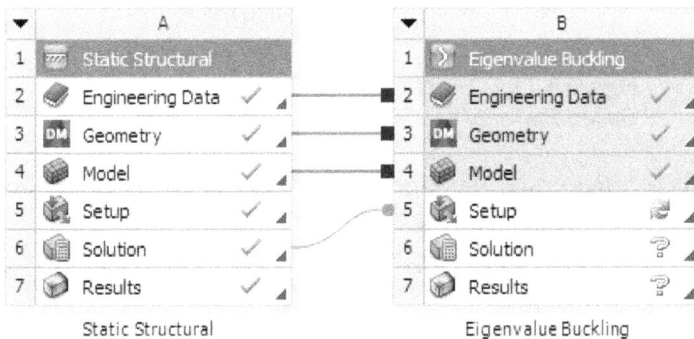

Figure 9-115 The **Static Structural** analysis system showing the transfer of data to the **Eigenvalue Buckling** analysis system

Notice that **Setup**, **Solution**, and **Results** cells in the **Eigenvalue Buckling** analysis system has refresh and unfulfilled attached to them indicating that they need to be updated.

2. Right-click on the **Setup** cell of the **Eigenvalue Buckling** analysis system in the **Project Schematic** window; a shortcut menu is displayed. Choose **Update** from this shortcut menu to update the analysis system.

3. Double-click on the **Setup** cell of the **Eigenvalue Buckling** analysis system in the **Project Schematic** window; the **Mechanical** window for **Eigenvalue Buckling** analysis system is displayed.

Notice that the **Static Structural (A5)** and **Solution (A6)** nodes in the **Outline** pane has green tick mark indicating that they are updated and imported from initial analysis system, as shown in Figure 9-116.

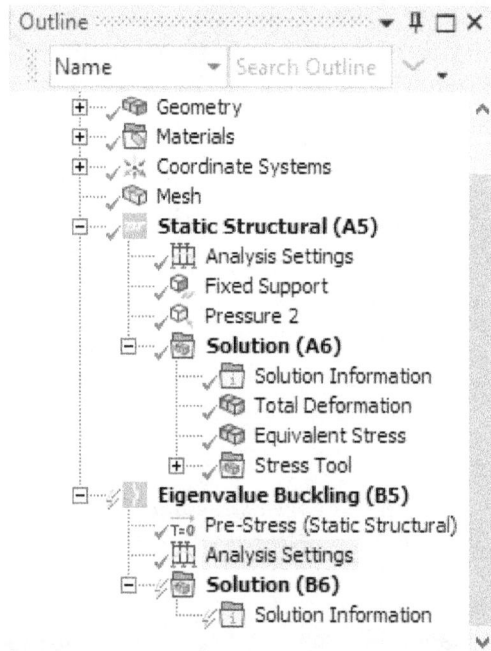

Figure 9-116 The **Outline** pane displaying data imported from the initial analysis system

4. Select **Analysis Settings** under the **Eigenvalue Buckling (B5)** node in tree of the **Outline** pane; the **Details of "Analysis Settings"** window is displayed.

5. In the **Details of "Analysis Settings"** window, expand the **Options** node if not already expanded.

6. Enter **3** in the **Max Modes to Find** edit box if not already specified by default, as shown in Figure 9-117.

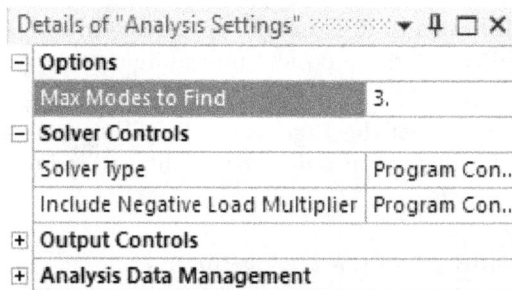

Figure 9-117 The **Details of "Analysis Settings"** window

Solving the FE Model and Analyzing the Results

After the analysis settings are specified for the model, you need to perform the analysis. On doing so, you will get the total deformations and buckling modes for the given condition.

1. Select the **Solution** node in the **Outline** pane; the **Solution** contextual tab is displayed. Also, the **Details of "Solution (B6)"** window is displayed.

2. Choose the **Total** tool from the **Deformation** drop-down of the **Solution** contextual tab; **Total Deformation** is added under the **Solution** node.

3. In the **Details of "Total Deformation"** window, expand the **Definition** node if it is not already expanded, refer to Figure 9-118.

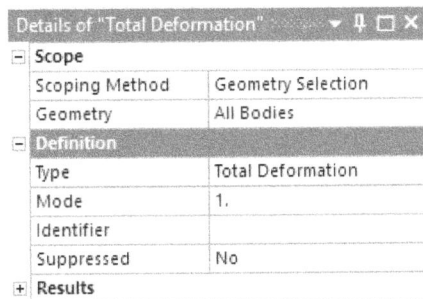

Figure 9-118 The Details of "Total Deformation" window

Notice that, in the **Mode** edit box, the default selection is 1, which means the total deformation shown in the graphics screen is only with respect to Mode 1.

4. Similarly, choose the **Total** tool from the **Deformation** drop-down of the **Solution** contextual tab; **Total Deformation 2** is added under the **Solution** node. Next, in the **Mode** edit box of the **Details of "Total Deformation 2"** window, enter **2**, which means the total deformation 2 shown in the graphics screen is only with respect to Mode 2.

5. Again, choose the **Total** tool from the **Deformation** drop-down of the **Solution** contextual tab; **Total Deformation 3** is added under the **Solution** node. Next, in the **Mode** edit box of the **Details of "Total Deformation 3"** window, enter **3**, which means the total deformation 3 shown in the graphics screen is only with respect to Mode 3.

6. Choose the **Solve** tool from the **Solution** contextual tab; the parameters are evaluated.

7. Select the **Solution** node in the **Outline** pane; the **Graph** and **Tabular Data** windows are displayed, refer to Figure 9-119.

B: Eigenvalue Buckling
Solution
Time: 1, s

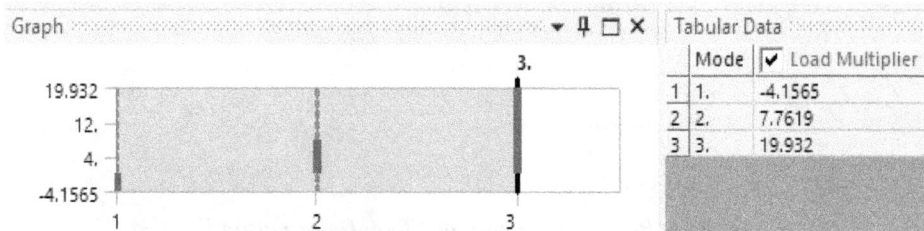

Figure 9-119 Partial view of the **Mechanical** window displaying the **Graph** and **Tabular Data** windows

8. Select **Total Deformation** under the **Solution** node in the **Outline** pane; 1st mode is displayed in the graphics screen with load multiplier -4.1565, as shown in Figure 9-120.

Note that the negative load multiplier for Mode 1 indicating that the model will buckle if the load is applied in the opposite direction.

9. Select **Total Deformation 2** under the **Solution** node; 2nd mode shape is displayed in the graphics screen with load multiplier 7.7619, as shown in Figure 9-121.

10. Select **Total Deformation 3** under the **Solution** node; 3rd mode shape is displayed in the graphics screen with load multiplier 19.932, as shown in Figure 9-122.

The corresponding legends of the mode shapes are also displayed in figures.

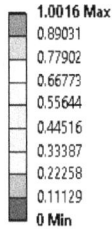

Figure 9-120 *Total deformation for the Mode 1 of the model*

Figure 9-121 *Total deformation for the Mode 2 of the model*

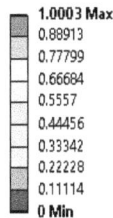

Figure 9-122 *Total deformation for the Mode 3 of the model*

Note that, in case of the 1st buckling mode, load multiplier is -4.1565. To find the buckling load of the model, multiply the applied load by the load multiplier. For example, the first buckling load will be 166.26 MPa (-4.1565 × 40 MPa) but the load must be applied in opposite direction. As a result, the applied pressure of 40 MPa will not cause the model to buckle.

Tip

In ANSYS Workbench, you can specify multiple defined mode shapes for corresponding total deformation results. As you added multiple total deformation tools for total deformations with respect to mode shapes 1, 2, and 3. You can also retrieve the mode shapes results using the **Graph** *and* **Tabular Data** *window. To do this, right-click in the* **Graph** *window; a shortcut menu is displayed, refer to Figure 9-119. Choose* **Select All** *from this shortcut menu to select all the data available in the* **Graph** *window. After the columns in the* **Graph** *window are selected, right-click again to display a shortcut menu. Choose the* **Create Mode Shape Results** *option from the shortcut menu displayed. Total deformation results are added under the* **Solution** *node in the* **Outline** *pane as:* **Total Deformation, Total Deformation 2, Total Deformation 3***. Also, you will notice that there are yellow thunderbolts attached to each one of them indicating that these results need to be evaluated.*

Playing the Animation

Now you need to animate the behavior of the model.

1. Select **Total Deformation** under the **Solution** node from the **Outline** pane in the **Mechanical** window; first mode shape is displayed in the graphics screen, refer to Figure 9-120. Also, the **Graph** and **Tabular Data** windows corresponding to the first mode are displayed.

2. Choose the **Play** button available in the **Animation** area in the **Graph** window; the animation corresponding to the first mode shape is played in the graphics screen.

3. Choose the **Export Video File** button from the **Animation** area of the **Graph** window; the **Save As** dialog box is displayed.

4. Browse to the location *C:\ANSYS_WB\c09\Tut04*

5. Enter **Mode 1** in the **File name** edit box of the **Save As** dialog box and then choose the **Save** button to save the video of the animation in the specified location.

6. Next, select **Total Deformation 2** from the **Outline** pane and play animation for the second mode shape.

7. Save the animations for second and third mode shapes named as **Mode 2** and **Mode 3** respectively in the same folder.

8. Exit the **Mechanical** window; the **Workbench** window is displayed.

Saving the Project and Exiting Workbench

After you have evaluated all the results, you now need to save the file before you exit the ANSYS Workbench session.

1. Choose the **Save Project** button from the **Standard** toolbar to save the project.

2. Exit the **Workbench** window.

Self-Evaluation Test

Answer the following questions and then compare them to those given at the end of this chapter:

1. The _____ tool is used to determine the Equivalent Stress.

2. When the **Equivalent (von-mises)** tool is chosen from the **Stress** drop-down in the **Solution** contextual tab, _____ is added under the **Solution** node in the **Outline** pane.

3. You can specify the exact direction of application of the Bearing Load along an axis by selecting the _____option from the **Define By** drop-down list in the **Details of "Bearing Load"** window.

4. The Legend is a band of _____ to differentiate different areas of results in an analysis system.

5. You cannot apply Fixed support on cylindrical surfaces. (T/F)

6. In ANSYS Workbench, you can apply support on multiple surfaces. (T/F)

7. You can apply a support only by using the options in the **Environment** contextual tab. (T/F)

8. You can rename a load in the **Outline** pane. (T/F)

9. The Von-mises Stress is the resultant of stresses along all the axes. (T/F)

10. The Bearing Load applies the same load throughout the cylindrical surface on which it is applied. (T/F)

Review Questions

Answer the following questions:

1. Which of the following parameters can be evaluated by using the **Deformation** drop-down in the **Environment** contextual tab?

    (a) **Directional**   (b) **Method**
    (c) **Force**   (d) **Bearing Load**

2. The options in the_____ toolbar are used to apply supports and loads in any analysis.

3. You can clear all the analysis settings and results by choosing the _____ option from the **File** menu in the **Mechanical** window.

4. You can apply the Fixed support to a component by using the **Fixed** tool from the **Supports** drop-down in the **Environment** contextual tab. (T/F)

5. By applying Fixed support to a part, you can restrict its few degrees of freedom. (T/F)

6. The **Environment** contextual tab is displayed when the analysis node is selected in the Outline pane. (T/F)

7. In a Static Structural analysis, the deformation can only be achieved along the X axis. (T/F)

EXERCISES

Exercise 1

Download the zip file *c09_ansWB_exr01.zip* file from *www.cadcim.com*. After downloading, extract the zip file to save the stp part file in the project. The model displayed in the **Mechanical** window, after it is downloaded and imported into ANSYS Workbench. Next, apply the Stainless Steel material to the model. Evaluate the Equivalent Stress, total deformation and the directional deformation for the given material and boundary conditions. Evaluate the same results by changing the material to Aluminium Alloy. The side faces of the plate are fixed and a downward force of 450 N is applied on the edge of the circular cutout, refer to Figure 9-123. Figure 9-124 shows the schematic representation of the boundary and loading conditions.   **(Expected time: 1 hr)**

The complete path for downloading the file is :

> *Textbooks > CAE Simulation > ANSYS > ANSYS Workbench 2023 R2: A Tutorial Approach > Input Files*

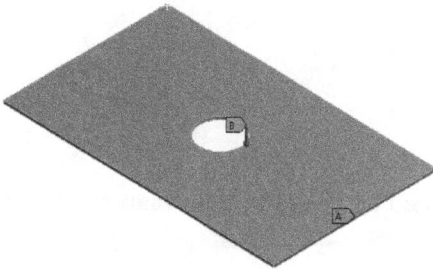

Figure 9-123 Model with the Fixed constraint and the Force load applied on it

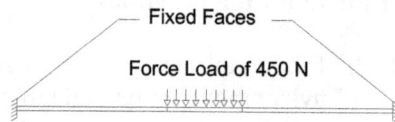

Figure 9-124   Schematic representation of boundary and loading conditions

Exercise 2

Download the *c09_ansWB_exr02.zip* file from *www.cadcim.com*. After downloading, extract it to save the stp part file at the desired location. Figure 9-125 shows the model displayed in the **Mechanical** window. After it is downloaded and imported into ANSYS Workbench, apply the Aluminium Alloy material to it. Next, evaluate total deformation, directional deformation, and Equivalent Stress for the given boundary conditions. Figure 9-126 shows the model with boundary and loading conditions. The magnitude of the Force load applied at B is 250 N and model is fixed at position A, refer to Figure 9-126 .   **(Expected time: 40 min)**

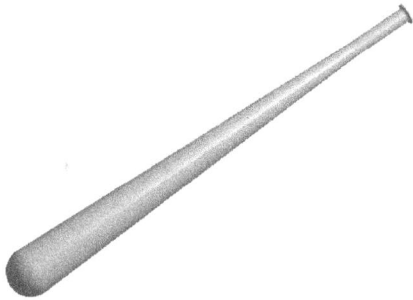

Figure 9-125 *Model for Exercise 2*

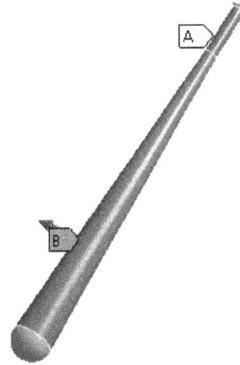

Figure 9-126 *Boundary conditions on the model*

Answers to Self-Evaluation Test

1. Equivalent (von-Mises), 2. Equivalent Stress, 3. Components, 4. colors, 5. F, 6. T, 7. F, 8. T, 9. T, 10. F

Chapter *10*

Vibration Analysis

Learning Objectives

After completing this chapter, you will be able to:

- *Understand the Modal analysis system*
- *Understand the Harmonic analysis system*
- *Set the analysis parameters*
- *Analyze the model for optimization*
- *Understand modes and mode shapes*
- *Generate mode shapes*

INTRODUCTION TO VIBRATION ANALYSIS

Vibration can be defined as the cyclic and oscillating motion of a machine or machine components. Vibration analysis plays an important role in machine design. A structure may fail if it is designed inappropriately for its dynamic integrity. Sometimes, it leads to massive engineering failures such as collapse of bridge, building and can cost human lives. The finite element analysis is used widely to calculate vibrational characteristics of the structures during the application of forces. In this chapter, Modal and Harmonic analyses are discussed.

Modal Analysis

The modal analysis is used to calculate the vibration characteristics such as natural frequency and mode shape (deformed shapes) of a structure or a machine component. The output of the modal analysis can be further used as an input for the harmonic and transient analyses.

For example, Figure 10-1 shows a cantilever beam, attached to a system vibrating at a certain frequency. It is important for the designer to find out whether the beam will sustain the vibrations induced by the machine to which it is connected.

Figure 10-1 Cantilever beam model

When the cantilever vibrates, various shapes are attained at certain frequencies. The shape of the component corresponding to a frequency is known as mode shape. The mode shape is a graphical representations of the deformation attained due to vibration. The main aim of the modal analysis is to find whether the natural frequency of the component is closer to the vibrations induced in the component. In this example, with this cantilever, the maximum number of modes found is six. Figures 10-2 through 10-5 display the various mode shapes of 1st, 2nd, 3rd, and 5th modes, respectively.

Note
The magnitude of deformation provides an approximate idea about the deformation in the model.

Figure 10-2 *Mode shape for the 1st mode*   ***Figure 10-3*** *Mode shape for the 2nd mode*

Figure 10-4 *Mode shape for the 3rd mode*   ***Figure 10-5*** *Mode shape for the 5th mode*

If the natural frequency of a system is very close to the excitation frequency, the component can get into resonance and fail. Therefore, to avoid the resonance, you need to strengthen the component on the basis of the mode shape. However, sometimes strengthening the component may not be possible due to the design limitations. Also, in actual practice, the displacement produced at resonance may not be infinite due to the presence of damping. Therefore, you need to calculate the response of a system under the time/frequency based loads. If the stress/strain/displacement response is less than the permissible limit, the component will not be required to strengthen or redesign.

Harmonic Analysis

Harmonic analysis is the steady-state response of a system under a sustained cyclic load. Harmonic analysis system is used to analyze a system working under periodic or sinusoidal loads. This analysis helps in determining whether a particular structure will be able to withstand resonance, fatigue, and other effects of forced vibration.

In ANSYS Workbench, harmonic analysis is a linear type of analysis where all types of nonlinearities are ignored, even if they are defined. All the loads and displacements vary sinusoidally at the known corresponding frequency.

Solution methods for harmonic analysis are discussed next.

Mode Superposition

For the Harmonic Response analysis, the response is obtained for linear structures by performing a Modal analysis to the given loading conditions. Mode Superposition is a powerful technique in which the modal analysis is selected directly, thus reducing the computation time.

Full

This method gives direct solution of the simultaneous equations of motion to obtain the harmonic response of the structure.

PERFORMING THE MODAL ANALYSIS

The Modal analysis is performed to find out the natural frequencies of a model. You can find out more than one natural frequency of a model depending upon the degrees of freedom available.

The following steps are involved to perform a Modal analysis:

a. Set the analysis preference.
b. Create or import the geometry into ANSYS Workbench.
c. Define element attributes (element types, real constants, and material properties).
d. Define meshing attributes.
e. Generate a mesh for the model.
f. Specify the analysis type and analysis options.
g. Obtain the solution.
h. Review the results.

Most of these steps have already been discussed in previous chapters.

Adding Modal Analysis System to ANSYS Workbench

To perform a Modal analysis in ANSYS Workbench, you need to add the **Modal** analysis system from the **Analysis Systems** toolbox in the **Toolbox** window to the **Project Schematic** window, refer to Figure 10-6.

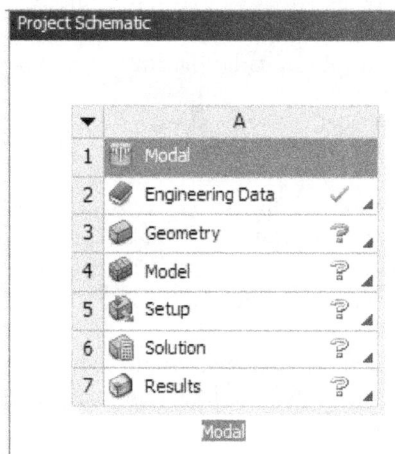

Figure 10-6 *Partial view of the **Project Schematic** window with the **Modal** analysis system added to it*

Starting the Mechanical Window

To start the analysis, double-click on the **Model** cell of the **Modal** analysis system to display the **Mechanical** window, refer to Figure 10-7. The components of the **Mechanical** window displayed by using the **Model** cell of the **Modal** analysis system are similar to the components of the **Mechanical** window displayed by using the **Static Structural** analysis system.

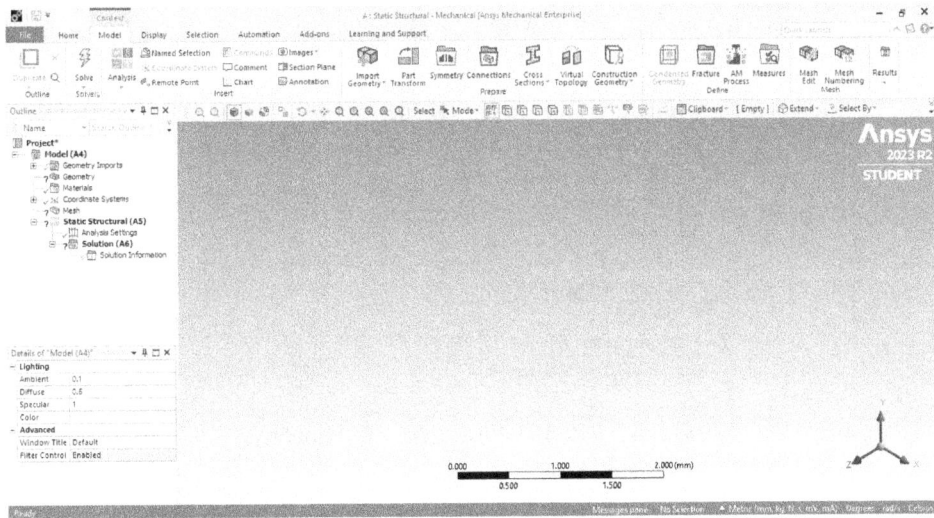

Figure 10-7 *The **Mechanical** window with the **Modal** node displayed in the **Outline** pane*

In the **Mechanical** window, you can set the number of modes or natural frequencies you need to find. Figure 10-8 shows the **Outline** window pane of the **Mechanical** window.

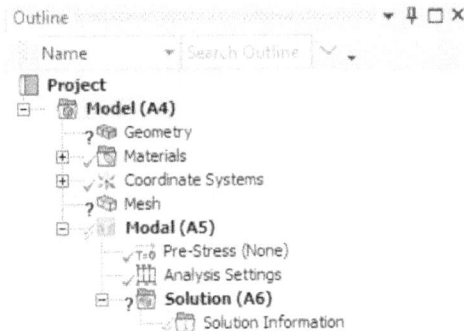

Figure 10-8 *The **Outline** window pane*

Specifying Analysis Settings

After generating mesh for the model, it is required to specify the settings needed to run the Modal analysis. To do so, select **Analysis Settings** displayed under the **Modal** node in the **Outline** pane, refer to Figure 10-8; the **Details of "Analysis Settings"** window will be displayed. In this window, specify a value in the **Max Modes to Find** edit box to display the various mode shapes. A limit can be assigned to the search of mode shape display by selecting **Yes** from the **Limit Search to Range** drop-down list. On doing so, the **Range Minimum** and **Range Maximum** edit boxes will be displayed. Specify the values for the minimum and maximum frequencies in these edit boxes to find mode shapes of the model within that specified range. Select the **Yes** option from the **Damped** drop-down list in the **Details of "Analysis Settings"** window to apply damping on a model, refer to Figure 10-9. The default selection in the **Damped** drop-down list is **No.** As a result, ANSYS Workbench will not consider the system to be damped. In such cases, the analysis is known as undamped modal analysis where the mode shapes and natural frequencies are complex.

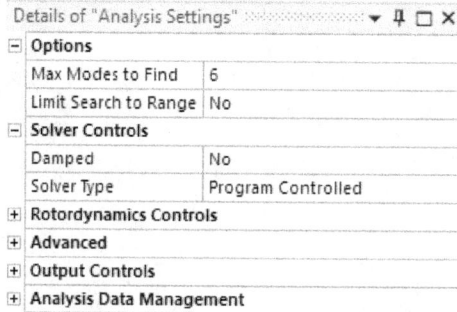

Details of "Analysis Settings"	▼ ♯ ☐ ✕
⊟ **Options**	
Max Modes to Find	6
Limit Search to Range	No
⊟ **Solver Controls**	
Damped	No
Solver Type	Program Controlled
⊞ **Rotordynamics Controls**	
⊞ **Advanced**	
⊞ **Output Controls**	
⊞ **Analysis Data Management**	

Figure 10-9   The ***Details of "Analysis Settings"*** *window*

After the analysis setup is done, you need to solve the model. You can do so by choosing the **Solve** tool from the **Environmental** contextual tab. After the model is solved, you need to plot the mode shapes. The procedure to plot the mode shapes is explained next.

Plotting the Deformed Shape (mode shape)

You can plot the mode shape (deformed shape) at each mode. However, before plotting the deformed shape, you need to specify the mode in the **Graph** window. To create the mode shapes, select the **Solution** node in the **Outline** pane; the **Graph** and **Tabular Data** windows will be displayed in the graphics screen, as shown in Figure 10-10 and 10-11.

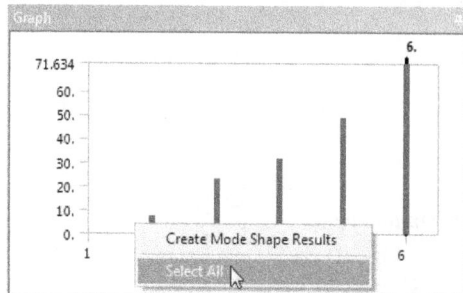

Figure 10-10 The ***Graph*** *window*

	Mode	✔ Frequency [Hz]
1	1.	2.0711e-002
2	2.	7.1546
3	3.	23.214
4	4.	31.236
5	5.	48.527
6	6.	71.634

Figure 10-11 The ***Tabular Data*** *window*

Now, right-click in the **Graph** window to display a shortcut menu and then choose **Select All** from it. Right-click again in the **Graph** window and then choose the **Create Mode Shape Results** option from the shortcut menu; the modes are added under the **Solution** node. Based on the number specified in the **Max Modes to Find** edit box, the number of modes are created with the names **Total Deformation, Total Deformation 2, . . . Total Deformation 6**. Select the required mode from the **Solution** node to visualize the corresponding mode shape in the graphics screen.

TUTORIALS

Tutorial 1   Modal Analysis

In this tutorial, you will create the model of a cantilever, as shown in Figure 10-12. The dimensions of the model are given in Figure 10-13. You will generate the mesh with default global mesh control settings and find six natural frequencies and their respective mode shapes. The material used is Structural Steel.   **(Expected time: 30 min)**

The following steps are required to complete this tutorial:

a. Create a new project.
b. Create the model.
c. Generate the mesh.
d. Specify the boundary conditions.
e. Solve the analysis.
f. Retrieve the analysis results.
g. Play the animation.
h. Save the model.

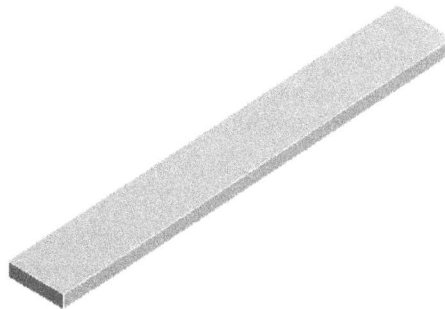

Figure 10-12 *Model for Tutorial 1*

Figure 10-13 *Dimensions of the cantilever beam*

Creating a New Project

Before starting the tutorial, it is important to create a new project and save it.

1. Start ANSYS Workbench session and then add the **Modal** analysis system to the **Project Schematic** window.

2. In the **Project Schematic** window, rename the **Modal** analysis system as **Cantilever**.

3.   Choose the **Save Project** button from the **Standard** toolbar; the **Save As** dialog box is displayed.

4.   Create a new folder with the name **c10** at the location *C:\ANSYS_WB*. Open the *c10* folder and then create another folder in it with the name **Tut01**.

5.   In this folder, save the project with the name **c10_ansWB_tut01**.

Creating the Model

After creating the project, you now need to work in the **DesignModeler** to create the model.

1.   Right click on the **Geometry** cell and then select **New DesignModeler Geometry**; the **DesignModeler** window is displayed.

2.   Select the **Millimeter** option from the **Units** menu of **Menu** bar.

3.   In the **DesignModeler** window, select **XY Plane** from the **Outline** pane to specify it as the sketching plane. Next, orient the view normal to the viewing direction.

4.   Invoke the **Sketching** mode. Next, create a rectangle and then dimension it, as shown in Figure 10-14. For dimensions of the model, refer to Figure 10-13.

5.   Change H1 to **20** and V2 to **5** in the **Details View** window.

6.   Switch to the **Modeling** mode and then change the view to Isometric, refer to Figure 10-12.

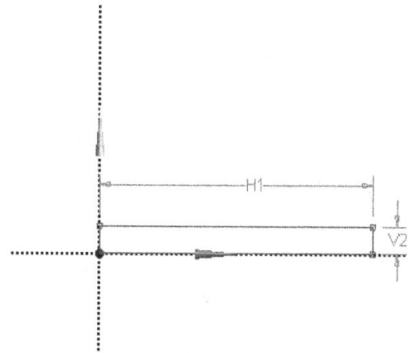

Figure 10-14 Rectangle created on the XY plane

7.   Extrude the sketch upto a depth of 150 mm. Press **F5** button to generate the model; the model after extrusion is shown in Figure 10-12.

8.   Close the **DesignModeler** window.

Generating Mesh on the Model

Now, you need to generate mesh of the model.

1.   Double-click on the **Model** cell in the **Cantilever** analysis system and wait for sometime; the **Mechanical** window is displayed. Now, you will notice that in the **Outline** pane, the **Mesh** node is displayed in the **Outline** pane with a yellow thunderbolt attached to it.

2.   Click on **Mesh** in the **Outline** pane; the **Details of "Mesh"** window is displayed.

3.   In the **Details of "Mesh"** window, expand the **Sizing** node if not already expanded.

4. In the **Defaults** node in the **Details of "Mesh"** window, enter **2.5** in the **Element Size** edit box.

5. Right-click on **Mesh** in the **Outline** pane and then choose the **Preview > Surface Mesh** from the shortcut menu displayed; preview of the mesh on the model is displayed.

6. Press **F5** button to generate the mesh; the mesh is generated, as shown in Figure 10-15.

Setting the Boundary Conditions

After the mesh is generated, you need to set the boundary conditions under which the analysis is to be performed.

1. Right-click on **Modal** node in the **Outline** pane and then choose **Insert > Fixed Support** from the shortcut menu displayed; **Fixed Support** with a question symbol is added under the **Modal** node in the **Outline** pane. Also, the **Details of "Fixed Support"** window is displayed.

2. In the **Details of "Fixed Support"** window, click on the **Geometry** cell to display the **Apply** and **Cancel** buttons if not already displayed.

3. Select the side face of the model, as shown in Figure 10-16.

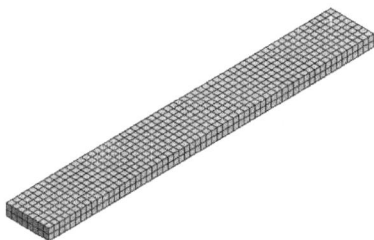

Figure 10-15 Mesh generated on the model

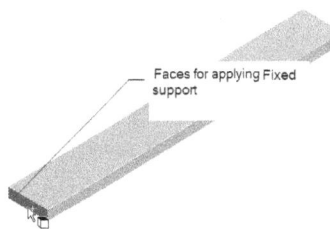

Figure 10-16 Face selected for applying Fixed support

4. Next, choose the **Apply** button from the **Geometry** selection box in the **Details of "Fixed Support"** window; Fixed support is applied to the selected face.

Solving the Modal Analysis

After specifying the boundary conditions in the **Mechanical** window, you need to set the variables to define the results and solve the analysis.

1. Select **Analysis Settings** under the **Modal** node in tree of the **Outline** pane; the **Details of "Analysis Settings"** window is displayed.

2. In the **Details of "Analysis Settings"** window, expand the **Options** node if not already expanded.

3. Enter **6** in the **Max Modes to Find** edit box if not already specified by default. Also make sure that **No** is selected in the **Limit Search to Range** drop-down list, refer to Figure 10-17.

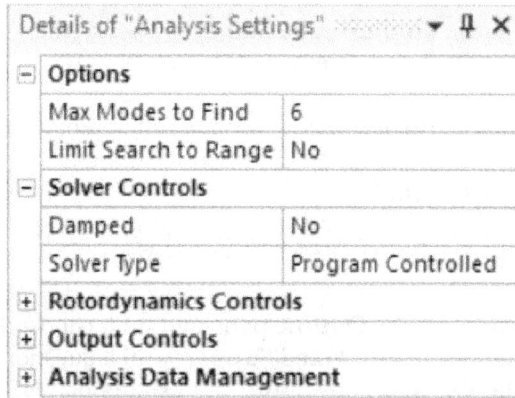

Details of "Analysis Settings" ▼ ♦ ✕	
⊟ **Options**	
Max Modes to Find	6
Limit Search to Range	No
⊟ **Solver Controls**	
Damped	No
Solver Type	Program Controlled
⊞ **Rotordynamics Controls**	
⊞ **Output Controls**	
⊞ **Analysis Data Management**	

Figure 10-17 The Details of "Analysis Settings" window

4. Expand the **Solver Controls** node in the **Details of "Analysis Settings"** window if not already expanded.

5. In the **Damped** drop-down list, select the **No** option if not already selected.

6. Right-click on the **Solution** node in the **Outline** pane and then choose the **Solve**   ≡∮ Solve option from the shortcut menu displayed; the analysis is solved.

7. Select the **Solution** node in the **Outline** pane; the **Graph** and **Tabular Data** windows are displayed, refer to Figure 10-18.

> **Note**
> *If the **Tabular Data** and **Graph** windows are not displayed automatically, then click on the **Manage** drop-down list in the **Home tab** and choose the **Tabular Data** option; the **Tabular Data** window will be displayed on the left corner of the graphics screen, Similarly, choose the **Graph** option from the **Manage** drop-down list to display the **Graph** window.*

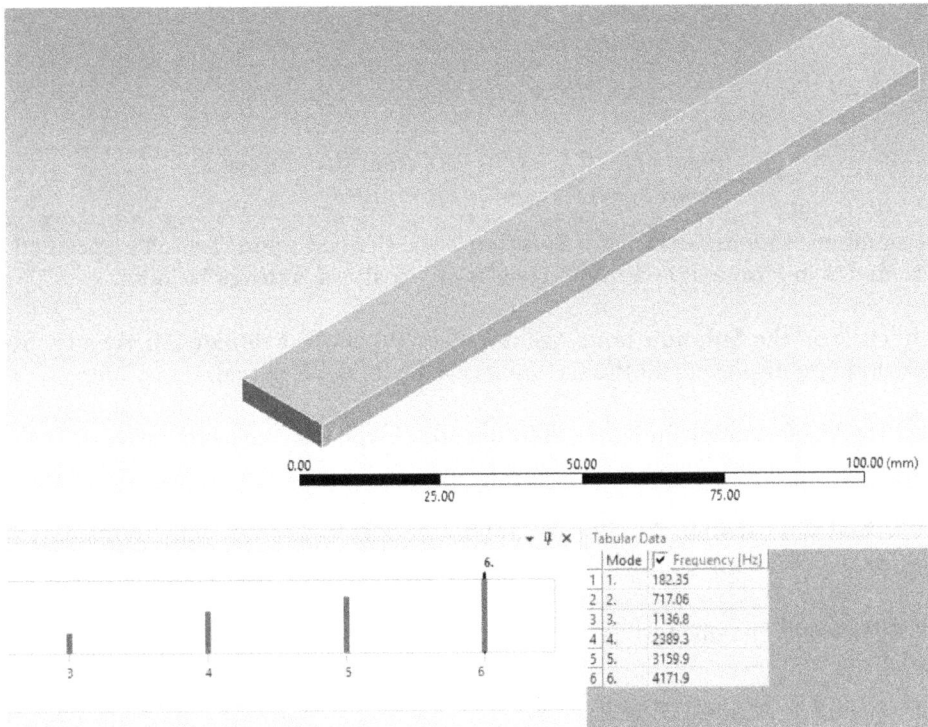

Figure 10-18 *Partial view of the* ***Mechanical*** *window displaying the* ***Graph*** *and* ***Tabular Data*** *windows*

Retrieving Analysis Results

After the analysis is solved, you need to find the mode shapes.

1. Right-click in the **Graph** window, a shortcut menu is displayed.

2. Choose **Select All** from this shortcut menu to select all the data available in the **Graph** window, as shown in Figure 10-19.

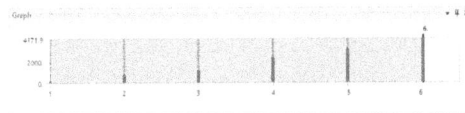

Figure 10-19 *The* ***Graph*** *window*

3. After the columns in the **Graph** window are selected, right-click again to display a shortcut menu.

4. Choose the **Create Mode Shape Results** option from the shortcut menu displayed, refer to Figure 10-20; Total deformation results are added under the **Solution** node in the **Outline** pane as: **Total Deformation**, **Total Deformation 2**, - - - **Total Deformation** 6. Also, you will notice that there are yellow thunderbolts attached to each one of them indicating that these results need to be evaluated.

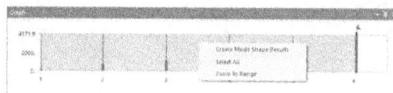

Figure 10-20 *Choosing the* *Create Mode Shape*
Results *from the shortcut menu displayed*

The number of modes under the **Solution** node depend upon the value specified in the
Max Modes to Find edit box in the **Details of "Analysis Settings"** window.

5. Right-click on the **Solution** node again and then choose **Evaluate All Results** from the
    shortcut menu displayed; all the six results are ready to be viewed.

6. Select **Total Deformation** under the **Solution** node in the **Outline** pane; 1st mode is displayed
    in the graphics screen, as shown in Figure 10-21.

7. Select **Total Deformation 2** under the **Solution** node; 2nd mode shape is displayed in the
    graphics screen, as shown in Figure 10-22.

    The corresponding legends of the mode shapes are also displayed in the figures.

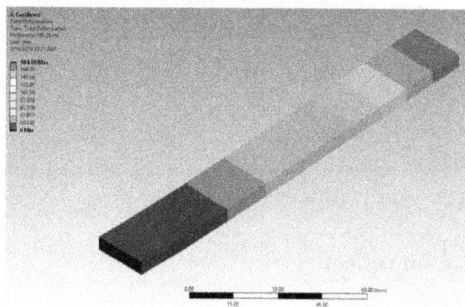

Figure 10-21 *1st mode shape*

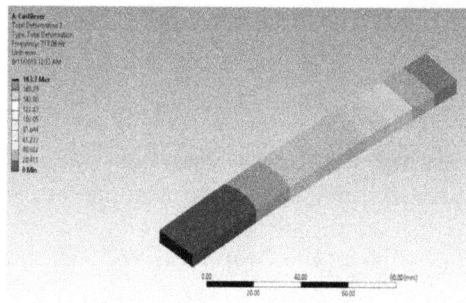

Figure 10-22 *2nd mode shape*

8. Similarly, select other results from the **Solution** node to view the corresponding mode shape
    in the graphics screen.

    Figure 10-23 shows the **Tabular Data** window. The three columns in this window display the
    serial number, mode, and frequency of the model.

	Mode	✔ Frequency [Hz]
1	1.	182.35
2	2.	717.06
3	3.	1136.8
4	4.	2389.3
5	5.	3159.9
6	6.	4171.9

*Figure 10-23 The **Tabular Data** window*

Playing and Saving the Animation

After retrieving the results, you will now play the animation.

1. Select **Total Deformation** in the **Solution** node in the **Outline** pane; the **Graph** window is displayed.

2. Choose the **Play** button available in the **Animation** area of the **Graph** window; the animation with respect to the first mode shape is played in the graphics screen.

    Note
    *You can set the duration of the animation by specifying a value in the **Time** edit box in the **Animation** area. Also, you can set the frame rate by specifying a value in the **Frames** edit box.*

3. Similarly, select **Total Deformation 2** from the **Outline** pane and play animation for the second mode shape.

4. Choose the **Export Video File** button from the **Animation** area of the **Graph** window; the **Save As** dialog box is displayed.

5. Browse to the location *C:\ANSYS_WB\c10\Tut01*.

6. Enter **mode 1** in the **File name** edit box in the **Save As** dialog box and then choose the **Save** button to save the video.

7. Similarly, export the animated video files of all the remaining modes to the desired location.

8. Exit the **Mechanical** window; the **Workbench** window is displayed.

9. Choose the **Save Project** button from the **Standard** toolbar to save the project with the name *c10_ansWB_tut01*.

10. Close the ANSYS Workbench session.

Tutorial 2 — Modal Analysis

Download the file *c10_ansWB_tut02.zip* from *www.cadcim.com* and import it into ANSYS Workbench, refer to Figure 10-24. After importing the file, apply Aluminium Alloy material to the model. Next, you will run the modal analysis under two conditions and evaluate the six natural frequencies and their respective mode shapes. **(Expected time: 60 min)**

Case I
Provide Fixed support at the cylindrical portion of the model, as shown in Figure 10-25.

Case II
Provide Displacement support along one of the axes of the cylindrical portion and allow displacement axially only. Movements along any other directions must be restricted.

The path of the file is as follows:

Textbooks > CAE Simulation > ANSYS > ANSYS Workbench 2023 R2: A Tutorial Approach > Input Files

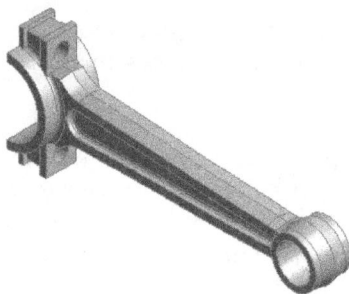

Figure 10-24 Model for Tutorial 2

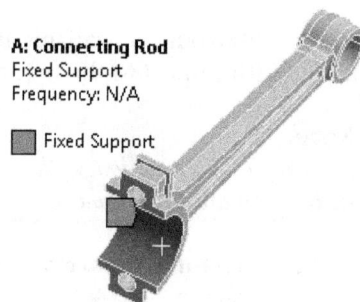

A: Connecting Rod
Fixed Support
Frequency: N/A

☐ Fixed Support

Figure 10-25 Fixed support applied to the model

The following steps are required to complete this tutorial:

a. Create a new project.
b. Download and import the file.
c. Apply the material and generate mesh for the model.
d. Apply boundary condition.
e. Solve the analysis.
f. Retrieve and analyze the results.
g. Play the animation and save the video file.
h. Duplicate the model and set the boundary condition for Case II.
i. Solve the model and interpret the results.
j. Save the project and close the ANSYS Workbench session.

Creating a New Project

Before you start the tutorial, create a new project and save it.

1. Start an ANSYS Workbench session and then add the **Modal** analysis system to the **Project Schematic** window.

2. Choose the **Save Project** button from the **Standard** toolbar; the **Save As** dialog box is displayed.

3. Browse to the location *C:\ANSYS_WB\c10* and then create a new folder with the name **Tut02** in it.

4. In the *Tut02* folder, save the project with the name **C10_ansWB_tut02** at the location *C:\ANSYS_WB\c10*.

5. In the **Project Schematic** window, rename the **Modal** analysis system as **Connecting Rod**, refer to Figure 10-26.

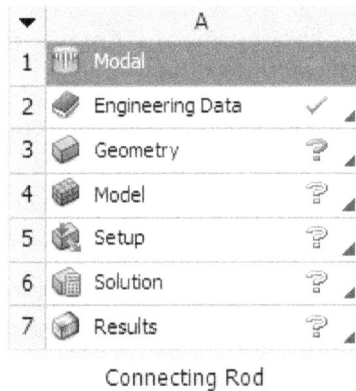

	A
1	Modal
2	Engineering Data ✓
3	Geometry ?
4	Model ?
5	Setup ?
6	Solution ?
7	Results ?

Connecting Rod

Figure 10-26 The Connecting Rod analysis system

Downloading and Importing the File

To run an analysis for the Connecting Rod, you need to download the zip file and then extract the zip file to the specified folder.

1. Download the file *c10_ansWB_tut02.zip* from *www.cadcim.com*. The path for the file is:

    Textbooks > CAE Simulation > ANSYS > ANSYS Workbench 2023 R2: A Tutorial Approach > Input Files

    After downloading the zip file, extract it to save the *c10_ansWB_tut02.igs* file at the location *C:\ANSYS_WB\c10\Tut02*.

2. In the **Connecting Rod** analysis system, right-click on the **Geometry** cell and then choose **Import Geometry > Browse** from the shortcut menu displayed; the **Open** dialog box is displayed.

3. In the **Open** dialog box, browse to the location *C:\ANSYS_WB\c10\Tut02* and open the file *c10_ansWB_tut02.igs*; a green tick mark is displayed before the **Geometry** cell in the **Connecting Rod** analysis system indicating that the geometry is satisfied for the analysis.

Applying Material and Generating a Mesh

After the model is imported into the **Connecting Rod** analysis system, you will apply material to it and then generate mesh for the model.

1. In the **Project Schematic** window, double-click on the **Engineering Data** cell of the **Connecting Rod** analysis system to display the Engineering Data workspace.

2. Choose the **Engineering Data Sources** toggle button from the **Standard** toolbar; the **Engineering Data Sources** window is added to the Engineering Data workspace.

3. Select the **General Materials** library from the **Engineering Data Sources** window; the **Outline of General Materials** window is displayed.

4. Choose the plus ( ) symbol displayed next to the **Aluminium Alloy** material in the **Outline of General Materials** window; the material is added to the Engineering Data.

5. Choose the **Project** button from the **Standard** toolbar; the **Project Schematic** window is displayed.

6. Choose the **Update Project** button from the **Standard** toolbar to update the project.

    After the material is added to the Engineering Data workspace, you need to go to the **Mechanical** window and then apply the material to the model. After applying material, generate mesh for the model. To do so, follow the procedure explained next.

7. Double-click on the **Model** cell of the **Connecting Rod** analysis system in the **Project Schematic** window; the **Mechanical** window is displayed.

    Figure 10-27 shows the **Outline** pane in the **Mechanical** window. Notice that green tick marks are placed before all the components except **Mesh** indicating that mesh has to be generated.

8. Expand the **Geometry** node in the **Outline** pane to display **Part 1**. Next, select **Part 1** under the **Geometry** node; the **Details of "Part 1"** window is displayed.

9. In the **Details of "Part 1"** window, expand the **Material** node and then choose **Aluminium Alloy** from the **Assignment** flyout; Aluminium Alloy material is applied to the model.

10. Right-click on **Mesh** in the **Outline** pane and choose the **Generate Mesh** option from the shortcut menu displayed; mesh is generated with default global mesh control settings, as shown in Figure 10-28. Notice that the **Details of "Mesh"** window is displayed below the **Outline** pane.

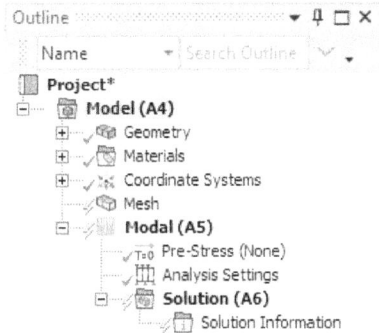

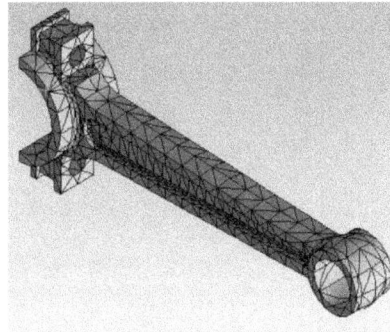

Figure 10-27 *The* **Outline** *pane displayed in the* **Mechanical** *window*

Figure 10-28 *The meshed model with default global mesh control settings*

Notice that after generating the mesh of the model, the total element count is 3872. You can also notice that the quality of the mesh is not fine around the circular regions in the model. The next step is to discretize the model in such a manner that a smooth transition will be achieved around the circular edges.

11. Expand the **Sizing** node in the **Details of "Mesh"** window if not already expanded.

12. In the **Details of "Mesh"** window, select the **No** option from the **Use Adaptive Sizing** drop-down list and then choose the **Yes** option from the **Capture Proximity** drop-down list.

13. Enter **2.5 mm** in the **Proximity Min Size** edit box and 2 in the **Num Cells Across Gap** edit box in the **Sizing** node in the **Details of "Mesh"** window.

14. Choose the **Update** tool from the **Mesh** contextual tab; the **ANSYS Workbench Update Model Status** message box is displayed. After sometime, this message box is closed and the mesh is generated with changed settings, as shown in Figure 10-29.

The total element count with the optimized model is approximately 18,424 which is a lot more than the previous count of 3872. The specified global mesh control settings are justified due to the complexity of the geometry and the required accuracy of the results.

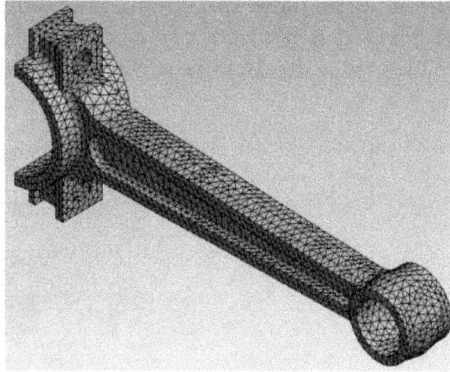

Figure 10-29 Mesh generated with the changed settings

> **Note**
> *Depending upon the complexity and type of model, you can insert more local mesh control settings to generate a fine mesh.*

Applying the Boundary Conditions

After the model is meshed, you need to apply the boundary conditions to the model.

1. Right-click on the **Modal** node in the **Outline** pane and then choose **Insert > Fixed Support** from the shortcut menu displayed; **Fixed Support** with a question symbol is added under the **Modal** node in the **Outline** pane. Also, the **Details of "Fixed Support"** window is displayed.

2. In the **Details of "Fixed Support"** window, select the **Geometry** selection box to display the **Apply** and **Cancel** buttons if not already displayed.

3. Now, choose the **Face** tool from the **Select** toolbar and then select the cylindrical face of the model, as shown in Figure 10-30.

4. Choose the **Apply** button in the **Geometry** selection box; **1 Face** is displayed in the **Geometry** selection box. Also, a green tick mark is placed before **Fixed Support** under the **Modal** node in the **Outline** pane indicating that Fixed support is applied on the model.

Solving the Modal Analysis

After the boundary conditions are specified in the **Mechanical** window, you need to define the results to be evaluated and solve the finite element model.

1. Select **Analysis Settings** in the **Modal** node from the **Outline** pane; the **Details of "Analysis Settings"** window is displayed.

Figure 10-30 *Cylindrical face selected for applying Fixed support*

2. In the **Details of "Analysis Settings"** window, enter **6** in the **Max Modes to Find** edit box if not already specified.

3. From the **Limit Search to Range** drop-down list, select the **Yes** option; the **Range Maximum** and **Minimum** edit boxes are displayed.

4. Enter **500 Hz** and **100000 Hz** in the **Range Minimum** and **Range Maximum** edit boxes, respectively.

5. Next, choose the **Solve** tool from the **Solve** panel of the **Home** tab; the analysis is solved. Also, the **Graph** and **Tabular Data** windows are displayed.

Retrieving and Analyzing the Results

After the model is solved, you need to find the mode shapes.

1. Select the **Solution** node in the **Outline** pane; the **Graph** and the **Tabular Data** windows are displayed, refer to Figure 10-31.

2. Right-click in the **Graph** window, a shortcut menu is displayed.

3. Choose **Select All** from this shortcut menu to select the data available in the **Graph** window.

4. Next, right-click again to display the shortcut menu.

5. Choose the **Create Mode Shape Results** option from the shortcut menu; the Total Deformation results are added under the **Solution** node in the

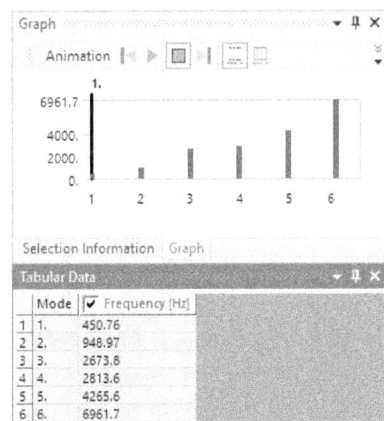

Figure 10-31 *The Graph and Tabular Data windows*

Outline pane as **Total Deformation**, **Total Deformation 2**, - - - **Total Deformation 6**. Also, there are yellow thunderbolt symbols attached to each one of them indicating that you need to evaluate them.

6.   Right-click on the **Solution** node to display a shortcut menu.

7.   Next, choose the **Evaluate All Results** option from the shortcut menu; the **ANSYS Workbench Solution Status** message box is displayed for sometime and the results are evaluated. Also, green tick marks are placed before total deformation results under the **Solution** node in the **Outline** pane.

8.   Select **Total Deformation** under the **Solution** node in the **Outline** pane; first mode shape is displayed in the graphics screen, as shown in Figure 10-32.

9.   Similarly, select **Total Deformation 2** from the **Outline** pane; second mode shape is displayed in the graphics screen, as shown in Figure 10-33.

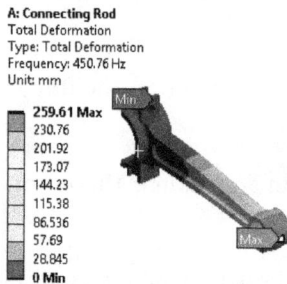

Figure 10-32 1st mode shape

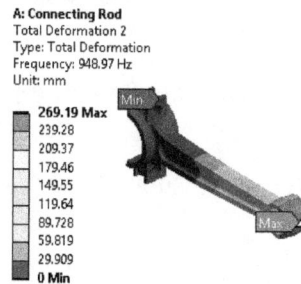

Figure 10-33 2nd mode shape

Figure 10-34 shows the **Tabular Data** window. Table given next displays the maximum and minimum values of the deformation in mode shapes and corresponding natural frequencies.

	Mode	✔ Frequency [Hz]
1	1.	450.76
2	2.	948.97
3	3.	2673.8
4	4.	2813.6
5	5.	4265.6
6	6.	6961.7

*Figure 10-34 The **Tabular Data** window*

S.N.	Mode Shape	Natural Frequency (Hz)	Minimum Deformation (mm)	Maximum Deformation (mm)
1	First mode shape	450.76	0	259.61
2	Second mode shape	948.97	0	269.19
3	Third mode shape	2673.8	0	258.26
4	Fourth mode shape	2813.6	0	390.25
5	Fifth mode shape	4265.6	0	283.16
6	Sixth mode shape	6961.7	0	298.19

Note that minimum deformation in all the cases is 0 because the model is constrained with a fixed support at one end and there is no deformation near the constrained face of the model.

There are six mode shapes for this model and the natural frequency result of the model ranges between 450.76 Hz and 6961.7 Hz.

10. Select **Total Deformation** from the **Outline** pane to display the first mode shape in the graphics screen, as shown in Figure 10-35.

11. Similarly, select **Total Deformation 2**, **Total Deformation 4**, and **Total Deformation 6** to display the corresponding mode shapes.

Figures 10-36, 10-37, and 10-38 display the second, fourth, and sixth mode shapes, respectively.

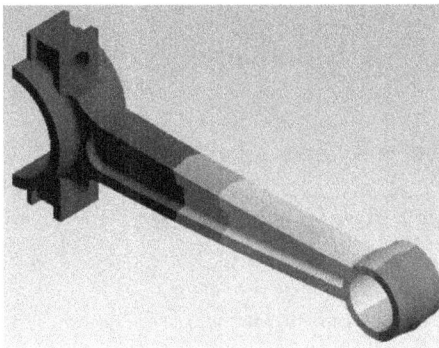

Figure 10-35 First mode shape

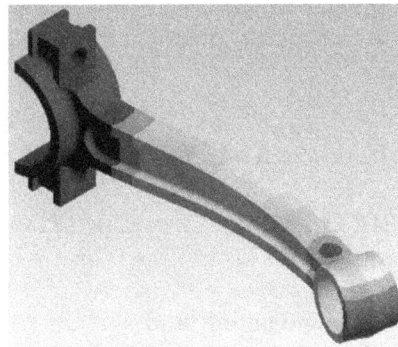

Figure 10-36 Second mode shape

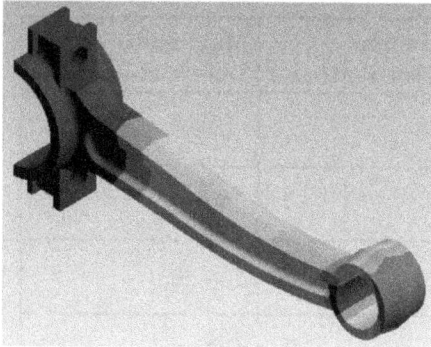

Figure 10-37 Fourth mode shape

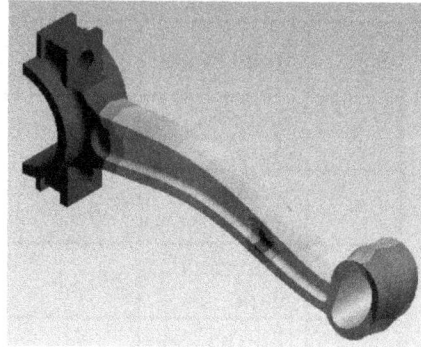

Figure 10-38 Sixth mode shape

Playing the Animation

Now you need to animate the behavior of the model.

1. Select **Total Deformation** under the **Solution** node from the **Outline** pane in the **Mechanical** window; first mode shape is displayed in the graphics screen, refer to Figure 10-32. Also, the **Graph** and **Tabular Data** windows corresponding to the first mode are displayed.

2. Choose the **Play** button available in the **Animation** area in the **Graph** window; the animation corresponding to the first mode shape is played in the graphics screen.

3. Choose the **Export Video File** button from the **Animation** area of the **Graph** window; the **Save As** dialog box is displayed.

4. Browse to the location *C:\ANSYS_WB\c10\Tut02*

5. Enter **Mode 1** in the **File name** edit box of the **Save As** dialog box and then choose the **Save** button to save the video of the animation in the specified location.

6. Next, select **Total Deformation 2** from the **Outline** pane and play animation for the second mode shape.

7. Save the animations for second to sixth mode shapes named as **Mode 2**, - - - **Mode 6** in the same folder.

8. Exit the **Mechanical** window; the **Workbench** window is displayed.

Setting the Boundary Condition for Case II Analysis System

After you retrieve the results of the case I analysis system, you now need to work for case II.

1. In the **Connecting Rod** analysis system of the **Project Schematic** window, right-click on **Modal**; a shortcut menu is displayed.

2.   Choose **Duplicate** from the shortcut menu, as shown in Figure 10-39; a copy of the **Modal** analysis system is added to the **Project Schematic** window, refer to Figure 10-40.

3.   Rename the newly created analysis system to **Connecting Rod 2**, refer to Figure 10-40.

Notice that **Setup**, **Solution**, and **Results** cells in the **Connecting Rod 2** analysis system have yellow thunderbolts attached to them indicating that they need to be updated.

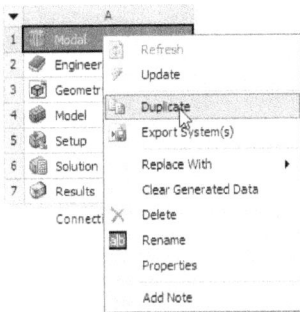

*Figure 10-39 Choosing the **Duplicate** option from the shortcut menu*

*Figure 10-40 A copy of the **Connecting Rod 2** analysis system added in the **Project Schematic** window*

4.   Double-click on the **Model** cell of the **Connecting Rod 2** analysis system in the **Project Schematic** window; the **Mechanical** window for **Connecting Rod 2** analysis system is displayed.

5.   Right-click on **Fixed Support** in the **Outline** pane; a shortcut menu is displayed.

6.   Choose the **Suppress** option from the shortcut menu displayed, as shown in Figure 10-41; a cross icon ( ) is placed on the left of **Fixed Support** in the **Outline** pane, indicating that this support is not available for analysis anymore.

7.   Select the **Modal** node in the **Outline** pane to display the **Environment** contextual tab.

8.   Choose the **Displacement** tool from the **Structural** group of the **Environment** contextual tab; **Displacement** is added under the **Modal** node in the **Outline** pane. Also, the **Details of "Displacement"** window is displayed.

9.   In the **Details of "Displacement"** window, select the **Geometry** selection box to display the **Apply** and the **Cancel** buttons, if they are not displayed already.

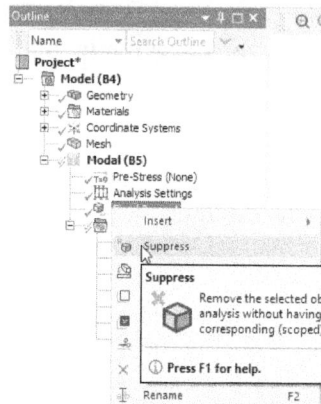

Figure 10-41 *Choosing the* **Suppress**
option from the shortcut menu

10. Choose the **Face** tool from the **Select** toolbar.

11. Select the circular face of cylindrical portion, refer to Figure 10-42.

12. Next, choose the **Apply** button from the **Geometry** selection box in the **Details of "Displacement"** window; **1 Face** is displayed in the **Geometry** selection box.

    In this tutorial, you will restrict the movement of the component along the Y and Z axes, whereas a displacement is set free along the X axis.

13. In the **Details of "Displacement"** window, expand the **Definition** node, if it is not already expanded.

14. Enter **0** in the **Y Component** edit box, indicating that the displacement is restricted along positive Y axis. Similarly, enter **0** in the **Z Component** edit box to restrict the displacement along Z direction, refer to Figure 10-43

    Notice that after entering 0 value for the **Y Component** and **Z Component** edit boxes, X component edit box is set to Free by default, indicating that displacement is allowed along X axis.

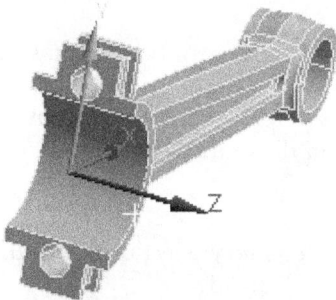

Figure 10-42 *Selecting the circular faces*

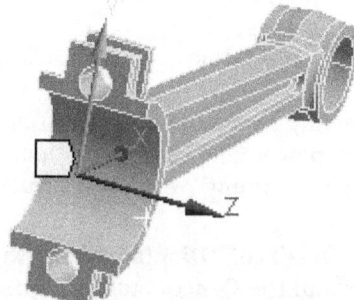

Figure 10-43 *Coordinate system placed in the model*

Solving the Analysis and Interpreting the Result

After you finish setting the boundary conditions, you now need to solve it in order to get results.

1. Next, choose the **Solve** tool from the **Solve** panel of the **Home** tab; the **ANSYS Workbench Solution Status** box is displayed. After some times, this box is closed and the analysis is solved and green tick marks are placed before the Total Deformation results available under the **Solution** node in the **Outline** pane.

2. Select **Total Deformation** from the **Solution** node to display the first mode shape in the graphics screen, as shown in Figure 10-44.

3. Similarly, select **Total Deformation 6** under the **Solution** node in the **Outline** pane; the corresponding mode shape is displayed in the graphics screen, refer to Figure 10-45.

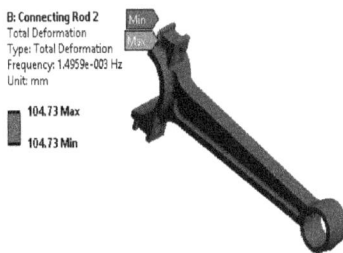

Figure 10-44 *First mode shape when **Total Deformation** is selected under the **Solution** node*

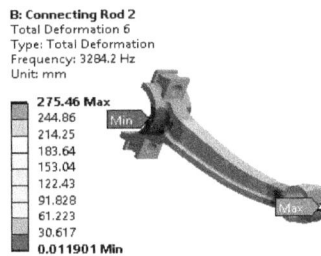

Figure 10-45 *Sixth mode shape when **Total Deformation 6** is selected under the **Solution** node*

Note that the maximum and minimum values of total deformation for the first mode shape is same which is equal to 104.73 mm. It is because of free displacement along X axis.

Figure 10-46 shows the **Tabular Data** window.

	Mode	✔ Frequency [Hz]
1	1.	1.4959e-003
2	2.	420.22
3	3.	546.06
4	4.	2461.6
5	5.	2800.5
6	6.	3284.2

Figure 10-46 *The **Tabular Data** window*

4. Select **Total Deformation** from the **Outline** pane to display the first mode shape in the graphics screen, as shown in Figure 10-47.

5. Similarly, select **Total Deformation 2**, **Total Deformation 4**, and **Total Deformation 6** to display the corresponding mode shapes.

    Figures 10-48, 10-49, and 10-50 display the second, fourth, and sixth mode shapes, respectively.

Figure 10-47 First mode shape   *Figure 10-48 Second mode shape*

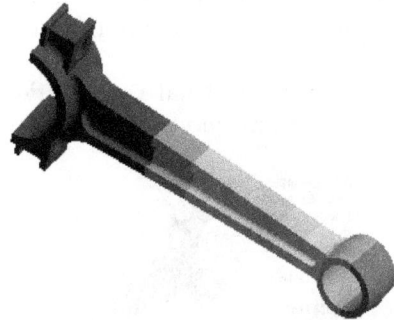

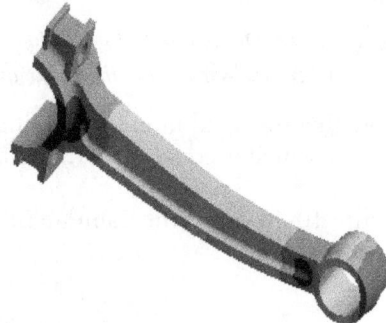

Figure 10-49 Fourth mode shape   *Figure 10-50 Sixth mode shape*

The results obtained from the three analysis systems are shown in the table given next.

S.N.	Mode Shape	Natural Frequency (Hz) & Corresponding Maximum Deformation (mm) for Case I	Natural Frequency (Hz) & Corresponding Maximum Deformation (mm) for Case II
1	First mode shape	450.76 Hz; 259.61 mm	1.4959e-003 Hz; 104.73 mm
2	Second mode shape	948.97 Hz; 269.19 mm	420.22 Hz; 254.67 mm
3	Third mode shape	2673.8 Hz; 258.26 mm	546.06 Hz; 235.98 mm
4	Fourth mode shape	2813.6 Hz; 390.25 mm	2461.6 Hz; 250.65 mm
5	Fifth mode shape	4265.6 Hz; 283.16	2800.5 Hz; 389.52 mm
6	Sixth mode shape	6961.7Hz; 298.19 mm	3284.2 Hz; 275.46 mm

The above table shows the natural frequency result for the case I ranges between 450.76 Hz and 6961.7 Hz. Whereas, natural frequency result for the case II ranges between 1.4959e-003 Hz and 3284.2 Hz. The first natural frequency for case I is 450.76 Hz, whereas the first natural frequency for case II is 1.4959e-003 Hz which is almost equals to zero.

Note

1. In case II, model has constrained partially hence the first natural frequency obtained is almost equal to zero and corresponding mode shape is a rigid body mode. The rigid body mode shape allows the model to move freely.

2. From the above table, you will notice that the fully constrained and partially constrained model has a different set of natural frequencies and corresponding mode shapes.

3. In general terms, vibration characteristics of a model are affected by the constraint conditions and therefore, you need to consider constraints carefully when setting up a model.

Tip

Modal analysis can be performed for any constrained conditions: Fully constrained, Unconstrained, and Partially constrained. Also, it is not recommended to use symmetry feature for further geometry modifications because symmetric model may have asymmetric mode shapes.

6. Select **Total Deformation** under the **Solution** node from the **Outline** pane in the **Mechanical** window; first mode shape is displayed in the graphics screen, refer to Figure 10-44. Also, the **Graph** and **Tabular Data** windows corresponding to the first mode are displayed.

7. Choose the **Play** button available in the **Animation** area in the **Graph** window; the animation corresponding to the first mode shape, rigid body motion along X direction, is played in the graphics screen.

8. Choose the **Export Video File** button from the **Animation** area of the **Graph** window; the **Save As** dialog box is displayed.

9. Browse to the location *C:\ANSYS_WB\c10\Tut02*

10. Enter **Case II_Mode 1** in the **File name** edit box of the **Save As** dialog box and then choose the **Save** button to save the video of the animation in the specified location.

11. Next, select **Total Deformation 2** from the **Outline** pane and play animation for the second mode shape.

12. Save the animations for the second to sixth mode shapes named as **Mode 2**, - - - **Mode 6** in the same folder.

13. Exit the **Mechanical** window; the **Workbench** window is displayed.

Save the Project and Closing the ANSYS Workbench Session

1. Choose the **Save Project** tool from the **Standard** toolbar; the model is saved.

2. Exit the **Workbench** window.

Tutorial 3 — Harmonic Response Analysis

In this tutorial, you will investigate the vibration characteristics of a motor cover component manufactured in Structural Steel. The cover is fastened at four bolt holes to a device operating at 1200 Hz. Download the model from *www.cadcim.com* and give it a thickness of 1.2 mm. Figure 10-51 shows the model of the motor cover component. Next, you will apply pressure load at the cylindrical face of the model, refer to Figure 10-51, and run the Harmonic Response analysis to evaluate total deformation, equivalent stress and frequency response with respect to deformation at different frequency values. **(Expected time: 60 min)**

The following steps are required to complete this tutorial:

a. Download and import the geometry into Workbench.
b. Generate a mesh and specify boundary conditions.
c. Solve the analysis and retrieve the results.
d. Analyze the data.
e. Play and save the animation.
f. Create and set the Harmonic Response analysis system.
g. Apply loading condition and solve the model.
h. Interpret and animate the results.
i. Save the project.

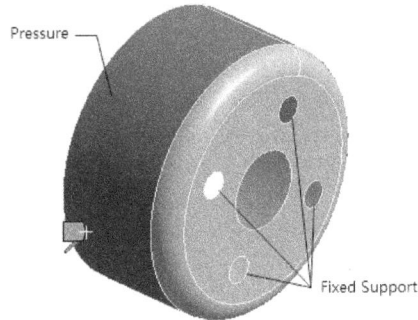

Figure 10-51 *Model with boundary and loading conditions for Tutorial 3*

Downloading the Part File and Importing It Into the Workbench

Before starting the tutorial, you have to download the file *c10_ansWB_tut03.zip* file from *www.cadcim.com*. Next, import the file into ANSYS Workbench.

1. Create a folder with the name **Tut03** at the location *C:\ANSYS_WB\c10*.

2. Download the file *c10_ansWB_tut03.zip* from *www.cadcim.com*. The complete path for the file is:
    Textbooks > CAE Simulation > ANSYS > ANSYS Workbench 2023 R2: A Tutorial Approach > Input Files

    Next, extract the zip file to save the *c10_ansWB_tut03.igs* file in the folder *C:\ANSYS_WB\c10*.

3. Open ANSYS Workbench.

4. Add the **Modal** analysis system to the **Project Schematic** window.

5. Right-click on the **Geometry** cell of the **Modal** analysis system and then choose **Import Geometry > Browse** from the shortcut menu displayed; the **Open** dialog box is displayed.

6. Browse to the location *C:\ANSYS_WB\c10\Tut03* and then select *c10_ansWB_tut03.igs*. Next, choose the **Open** button from the **Open** dialog box; the file is imported into the **Workbench** window. Also, a green tick mark is placed corresponding to the **Geometry** cell in the **Modal** analysis system.

7. Choose the **Save Project** button from the **Standard** toolbar; the **Save As** dialog box is displayed.

8. Save the project with the name **c10_ansWB_tut03** at the location *C:\ANSYS_WB\c10\Tut03*.

9. In the **Project Schematic** window, rename the **Modal** analysis system as **Motor_Cover**.

Generating the Mesh

After the geometry is imported, you need to apply thickness and then generate mesh for the model.

1. Double-click on the **Model** cell of the **Motor_Cover** analysis system; the **Mechanical** window is displayed.

Notice that the model is a surface and therefore, you need to add thickness to it.

2. In the **Outline** pane of the **Mechanical** window, expand the **Geometry** node and then select **Motor_Cover-Freeparts**; the **Details of "Motor_Cover-FreeParts"** window is displayed.

3. In the **Details of "Motor_Cover-FreeParts"** window, expand the **Definition** node if not already expanded.

4. Enter **1.2** in the **Thickness** edit box to provide a thickness of 1.2 mm to the model, refer to Figure 10-52.

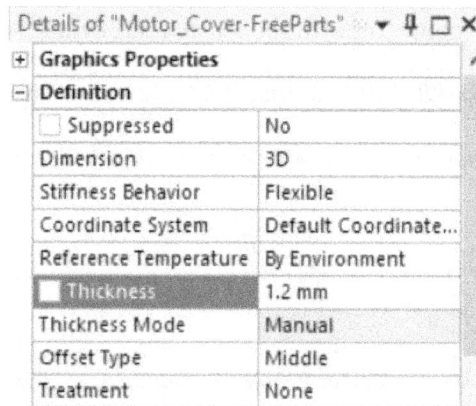

Details of "Motor_Cover-FreeParts"	
⊞ **Graphics Properties**	
⊟ **Definition**	
☐ Suppressed	No
Dimension	3D
Stiffness Behavior	Flexible
Coordinate System	Default Coordinate...
Reference Temperature	By Environment
☐ Thickness	1.2 mm
Thickness Mode	Manual
Offset Type	Middle
Treatment	None

Figure 10-52 The Details of "Motor_Cover-FreeParts" window

5. Select the **Mesh** node in the **Outline** pane; the **Details of "Mesh"** window is displayed.

6. Expand the **Sizing** node in the **Details of "Mesh"** window, if not already expanded.

7. Select the **Yes** option from the **Capture Curvature** option under the **Sizing** node if not selected.

8. Enter **1.50** in the **Curvature Min Size** edit box, refer to Figure 10-53.

9. Press **F5** button to generate the mesh; the mesh is generated, as shown in Figure 10-54.

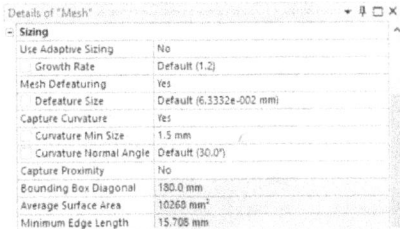

Figure 10-53 The *Details of "Mesh"* window

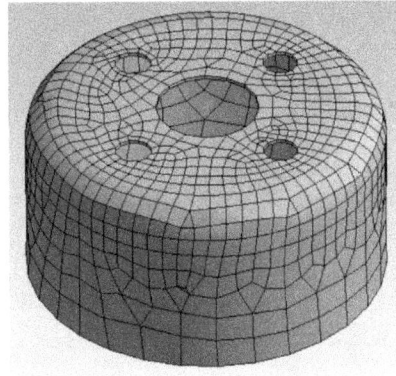

Figure 10-54 Mesh generated with the changed global mesh control settings

10. Expand the **Statistics** node in the **Details of "Mesh"** window to display the element count. Notice that the total number of elements after generating the mesh is 895.

To achieve a fine mesh along the curved faces, you need to provide local mesh controls to the geometry.

11. Right-click on **Mesh** in the **Outline** pane and then choose **Insert > Method** from the shortcut menu displayed; **Automatic Method** is added under the **Mesh** node with a question symbol attached to it. Also, the **Details of "Automatic Method"** window is displayed.

12. In the **Details of "Automatic Method"** window, select the **Geometry** selection box to display the **Apply** and **Cancel** buttons if not already displayed.

13. Select the model in the graphics screen, as shown in Figure 10-55.

14. Choose the **Apply** button in the **Details of "Automatic Method"** window; the model is selected.

15. In the **Details of "Automatic Method"** window, select the **MultiZone Quad/Tri** option from the **Method** drop-down list, refer to Figure 10-56; the **Details of "Automatic Method"** window is replaced by the **Details of "MultiZone Quad/Tri Method"** window.

Figure 10-55 The part selected in the Graphics screen

16. In this window, select the **Uniform** option from the **Surface Mesh Method** drop-down list and enter **1.50** in the **Element Size** edit box.

17. Choose the **Update** tool from the **Mesh** contextual tab; the **ANSYS Workbench Update Model Status** message box is displayed. After sometime, mesh is generated with local mesh control settings applied, as shown in Figure 10-57.

Details of "Automatic Method" - Method	▼ 🔎 ☐ ✕
Scope	
Scoping Method	Geometry Selection
Geometry	1 Body
Definition	
Suppressed	No
Method	Quadrilateral Dominant ▼
Element Order	Quadrilateral Dominant
Free Face Mesh Type	Triangles / MultiZone Quad/Tri

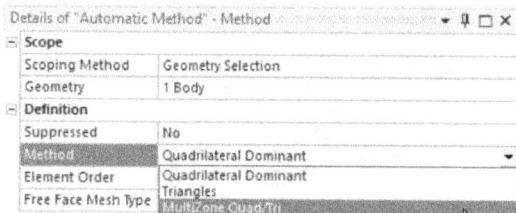

Figure 10-56 *Selecting the MultiZone Quad/Tri option from the* **Method** *drop-down list*

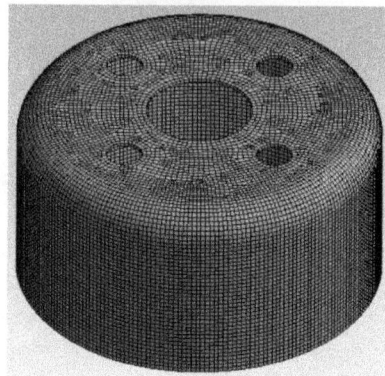

Figure 10-57 *Mesh generated with local mesh control settings applied*

18. Select the **Mesh** node in the **Outline** pane; the **Details of "Mesh"** window is displayed.

19. Expand the **Statistics** node in the **Details of "Mesh"** window to display the element count.

   Notice that the total number of elements after generating the mesh is approximately 14736.

Applying Boundary Conditions

After the surface model is meshed, you need to apply the boundary conditions for the model. The model is fixed to the motor with the help of bolts. Therefore, you need to provide fixed support at the bolt holes.

1. Right-click on the **Modal** node in the **Outline** pane and then choose **Insert > Fixed Support** from the shortcut menu displayed; **Fixed Support** with a question symbol is added under the **Modal** node in the **Outline** pane. Also, the **Details of "Fixed Support"** window is displayed.

2. In the **Details of "Fixed Support"** window, select the **Geometry** selection box to display the **Apply** and **Cancel** buttons if not already displayed.

3. Choose the **Edge** tool from the **Select** toolbar to select the edges for applying Fixed Support.

4. Select the edges of the holes provided for Fixed supports, as shown in Figure 10-58.

   Note
   To select all edges of the holes, you need to use the Ctrl key.

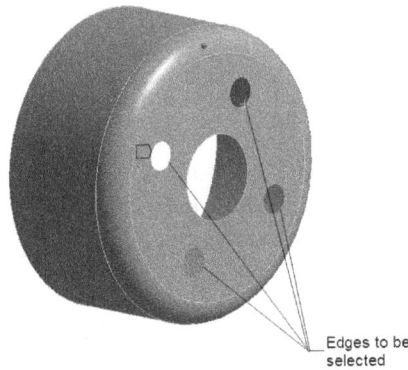

Figure 10-58 Edges of the bolt holes selected for applying Fixed support

5. Choose the **Apply** button in the **Geometry** selection box in the **Details of "Fixed Support"** window; the fixed support is applied to the four holes. Also a green tick mark is placed before **Fixed Support** in the **Outline** pane indicating that fixed support is applied on the model.

Solving the Analysis and Retrieving the Analysis Results

After the boundary conditions are defined in the **Mechanical** window, you now need to set the variables for the analysis and then retrieve the results.

1. Select **Analysis Settings** from the **Modal** node in the **Outline** pane; the **Details of "Analysis Settings"** window is displayed.

2. In the **Details of "Analysis Settings"** window, enter **8** in the **Max Modes to Find** edit box.

3. Next, choose the **Solve** tool from the **Environment** contextual tab; the analysis is solved. Also, the **Graph** and **Tabular Data** windows are displayed.

4. Select the **Solution** node in the **Outline** pane; the **Graph** and the **Tabular Data** windows with all their contents are displayed, refer to Figure 10-59.

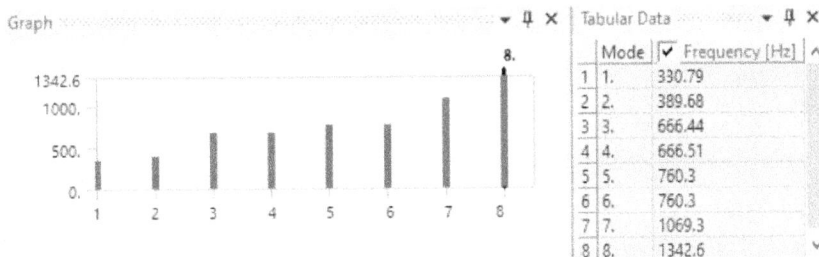

*Figure 10-59 The **Graph** and **Tabular Data** windows*

5. Right-click in the **Graph** window; a shortcut menu is displayed.

6. Choose **Select All** from this shortcut menu to select all the data available in the **Graph** window.

7. Right-click in the **Graph** window again and then choose the **Create Mode Shape Results** option from the shortcut menu displayed; Total Deformation results are added under the **Solution** node in the **Outline** pane with the names: **Total Deformation**, **Total Deformation 2, - - - Total Deformation 8**. Also, yellow thunderbolts are attached to each one of them indicating that they need to be evaluated.

8. Right-click on the **Solution** node and then choose the **Evaluate All Results** option from the shortcut menu displayed; the **ANSYS Workbench Solution Status** message box is displayed and the results are evaluated. Also, green tick marks are placed before the Total Deformation results added under the **Solution** node in the **Outline** pane.

9. Select **Total Deformation** under the **Solution** node in the **Outline** pane; first mode shape is displayed in the graphics screen, as shown in Figure 10-60.

10. Similarly, select **Total Deformation 2** from the **Outline** pane; second mode shape is displayed in the graphics screen, as shown in Figure 10-61.

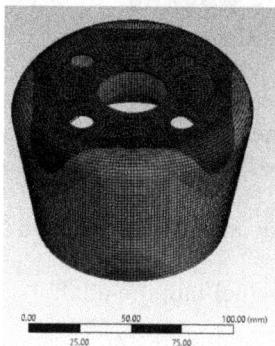

Figure 10-60 First mode shape displayed in the Graphics screen

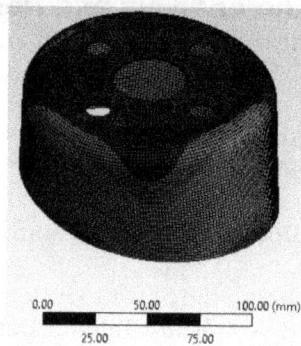

Figure 10-61 Second mode shape displayed in the Graphics screen

Each mode shape represents a frequency value. Figure 10-62 shows the **Tabular Data** window displaying the natural frequencies of different mode shapes.

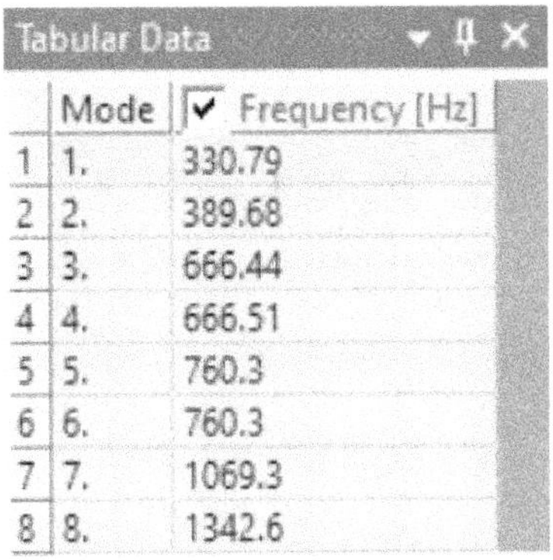

*Figure 10-62 The **Tabular Data** window displaying the natural frequencies of different mode shapes*

Table shown next displays the deformation in mode shapes and their corresponding natural frequencies.

S.N.	Mode Shape	Natural Frequency (Hz)	Deformation (mm)
1	First mode shape	330.79	145.87
2	Second mode shape	389.68	145.97
3	Third mode shape	666.44	107.83
4	Fourth mode shape	666.51	107.84
5	Fifth mode shape	760.3	165.66
6	Sixth mode shape	760.3	165.66
7	Seventh mode shape	1069.3	65.979
8	Eigth mode shape	1342.6	170.78

Minimum deformation in all the cases is 0 because the model is constrained with the help of bolts and there is no deformation near the constrained edges of the model.

Analyzing the Data

After the results are generated and mode shapes are displayed, it is advisable to compare the data with an analysis of a model that is more optimized.

1. In the **Project Schematic** window, right-click on the **Motor_Cover** analysis system; a shortcut menu is displayed.

2. Choose **Duplicate** from the shortcut menu, as shown in Figure 10-63; a copy of the **Motor_Cover** analysis system is added to the **Project Schematic** window, refer to Figure 10-64.

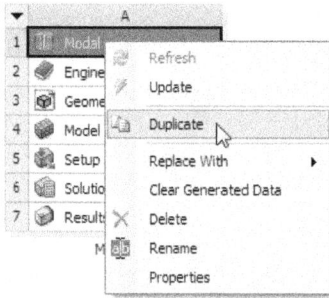

Figure 10-63 Choosing the **Duplicate** option from the shortcut menu

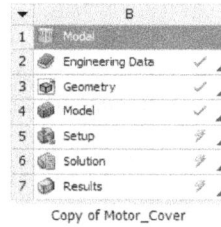

Figure 10-64 A copy of the **Motor_Cover** analysis system added in the **Project Schematic** window

Notice that **Setup, Solution**, and **Results** cells in the **Motor_Cover 2** analysis system have yellow thunderbolts attached to them indicating that they need to be updated.

3. Rename the newly created analysis system to **Motor_Cover 2**.

4. Double-click on the **Model** cell of the **Motor_Cover 2** analysis system in the **Project Schematic** window; the **Mechanical** window for **Motor_Cover 2** analysis system is displayed.

5. In the **Outline** pane, expand the **Geometry** node and then select **Motor_Cover-FreeParts** in it; the **Details of "Motor_Cover-FreeParts"** window is displayed.

> **Note**
> *Make sure that the unit system selected is **mm, Kg, N, s, mV, mA** in the **Units** menu.*

6. In the **Details of "Motor_Cover-FreeParts"** window, expand the **Definition** node and then enter **2** in the **Thickness** edit box; the thickness of the surface model changes to 2 mm.

7. Select the **Mesh** node in the **Outline** pane to display the **Details of "Mesh"** window.

8. In this window, expand the **Sizing** node and then enter **1.8** in the **Curvature Min Size** edit box.

9. Expand the **Mesh** node in the **Outline** pane to display the **MultiZone Quad/Tri Method**. Select **MultiZone Quad/Tri Method** to display the **Details of "MultiZone Quad/Tri Method"** window.

10. In the **Details of "MultiZone Quad/Tri Method"** window, expand the **Definition** node and then enter **1.2** in the **Element Size** edit box.

11. Choose the **Update** tool from the **Mesh** contextual tab; the mesh is generated with changed settings and a green tick mark is displayed before the **Mesh** node in the **Outline** pane.

Figure 10-65 shows the mesh generated with changed settings.

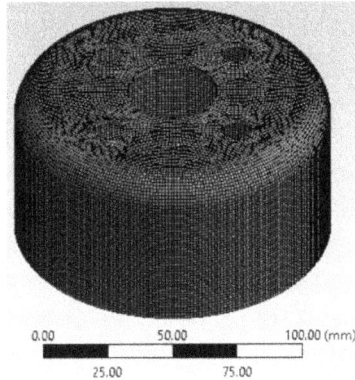

Figure 10-65 Mesh generated with changed settings

12. Next, choose the **Solve** tool from the **Solve** panel of the **Home** tab; the **ANSYS Workbench Solution Status** box is displayed. After sometime this box is closed and the analysis is solved and green tick marks are placed before the total deformation results available under the **Solution** node in the **Outline** pane.

13. Select **Total Deformation** from the **Solution** node to display the first mode shape in the graphics screen, as shown in Figure 10-66.

14. Similarly, select **Total Deformation 6** under the **Solution** node in the **Outline** pane; the corresponding mode shape is displayed in the graphics screen, refer to Figure 10-67.

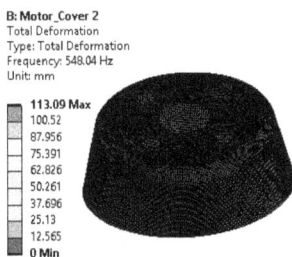

Figure 10-66 First mode shape when **Total Deformation** is selected under the **Solution** node

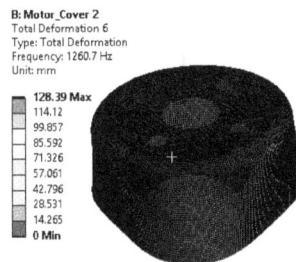

Figure 10-67 Sixth mode shape when **Total Deformation 6** is selected under the **Solution** node

The following table displays the results obtained from the **Motor_Cover 2** analysis system.

S.N.	Mode Shape	Natural Frequency (Hz)	Minimum Deformation (mm)	Maximum Deformation (mm)
1	First mode shape	548.04	0	113.09
2	Second mode shape	633.04	0	113.04
3	Third mode shape	999.03	0	78.232
4	Fourth mode shape	999.06	0	78.231
5	Fifth mode shape	1260.7	0	128.39
6	Sixth mode shape	1260.7	0	128.39
7	Seventh mode shape	1604.1	0	51.153
8	Eigth mode shape	2224.9	0	132.55

In this tutorial, the model is assumed to be attached to a motor which runs at a frequency of 1200 Hz. Therefore, it is important to analyze whether the model survives the induced frequency. If desired result is not obtained from the analysis, you can optimize the model. To do so, you can change the material, vary the thickness of the component, and so on. Next, run the analysis to evaluate results again and then compare the results obtained with the previously obtained data.

Playing the Animation

After the results are generated, you can play the animation to know how the model behaves.

1. Select **Total Deformation** in the **Solution** node in the **Outline** pane; the **Graph** window is displayed with the **Animation** area displayed in it.

2. In the **Animation** area of the **Graph** window, enter **20** in the **Frames** edit box and **4** in the **Time** edit box.

3. Choose the **Play** button from the **Animation** area in the **Graph** window; the animation with respect to the first mode shape is played in the graphics screen.

4. Choose the **Export Video File** button from the **Animation** area of the **Graph** window; the **Save As** dialog box is displayed.

5. Browse to the location *C:\ANSYS_WB\c10\Tut03* and enter **mode 1** in the **File name** input box in the **Save As** dialog box and then choose the **Save** button to save the video of the animation in the specified location.

6. Select **Total Deformation 2** under the **Solution** node in the **Outline** pane; second mode shape is displayed in the graphics screen.

7. Play the animation for the second mode shape.

8. Save the animation as discussed earlier.

9. Play all the animations from third mode shape to eigth mode shape and then save them.

10. Exit the **Mechanical** window; the **Workbench** window is displayed.

Creating and Setting a Harmonic Response Analysis System

After interpreting the results for modal analysis, you need to create and set a Harmonic Response analysis system.

1. In the **Project Schematic** window, drag the **Harmonic Response** analysis system and drop in the **Solution** cell of the **Motor_Cover** analysis system, as shown in Figure 10-68. As a result, the data is transferred   from the **Motor_Cover** analysis system to the **Harmonic Response** analysis system, as shown in Figure 10-69.

Note
*1. For this tutorial, the solution method is considered as **Mode Superposition** and therefore the **Modal** analysis is a prerequisite.*

2. The linked analysis setup allows to share the input and output data of one analysis with the next as a input.

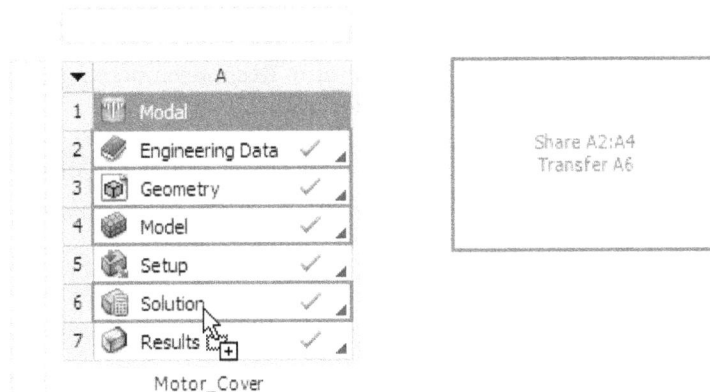

*Figure 10-68 Dragging the **Harmonic Response** analysis system into **Solution** cell of the **Motor_Cover** analysis system*

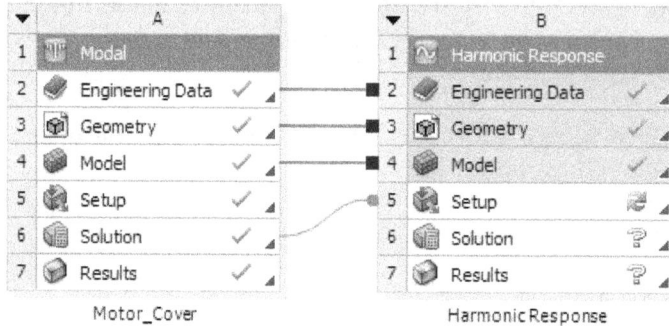

Figure 10-69 The **Motor_Cover** *analysis system showing the transfer of data with the* **Harmonic Response** *analysis system*

Notice that **Setup**, **Solution**, and **Results** cells in the **Eigenvalue Buckling** analysis system has refreshed and unfulfilled attached to them indicating that they need to be updated.

2. Next, right-click on the **Setup** cell of the **Harmonic Response** analysis system in the **Project Schematic** window; a shortcut menu is displayed. Choose **Update** from this shortcut menu to update the analysis system.

3. Double-click on the **Setup** cell of the **Harmonic Response** analysis system in the **Project Schematic** window; the **Mechanical** window for **Harmonic Response** analysis system is displayed.

    Notice that the **Modal (A5)** and **Solution (A6)** nodes in the outline pane has green tick marks indicating that they are updated and imported from initial analysis system, as shown in Figure 10-70.

    Notice that the **Modal (Modal)** is displayed under the **Harmonic Response** node in the **Outline** pane indicating that the **Modal** analysis is the pre-stressed analysis system.

4. Select **Analysis Settings** under the **Harmonic Response** node in tree of the **Outline** pane; the **Details of "Analysis Settings"** window is displayed.

5. In the **Details of "Analysis Settings"** window, expand the **Options** node if not already expanded.

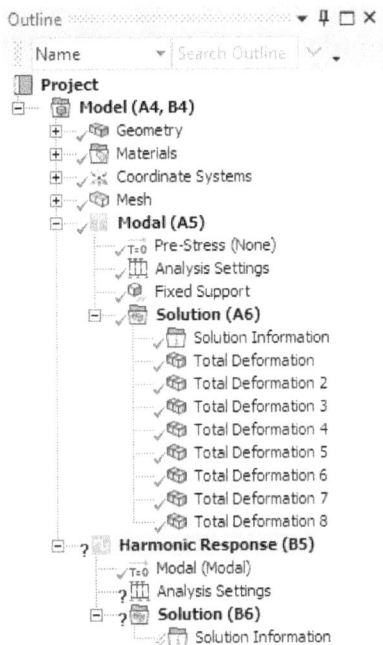

*Figure 10-70 The **Outline** pane displaying
imported data from initial analysis system*

6.   Enter **100** in the **Range Minimum** edit box and **2000** in the **Range Maximum** edit box.
Next, enter **100** in the **Solution Intervals** edit box, as shown in Figure 10-71.

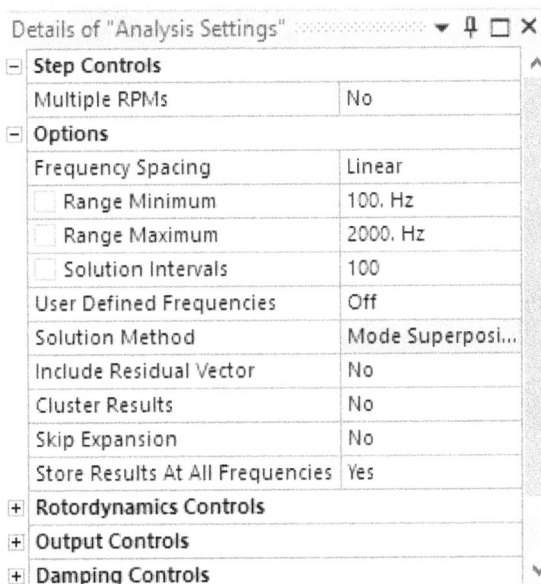

*Figure 10-71 The **Details of "Analysis Settings"**
window*

Applying Loading Condition and Solving the Model

After creating the Harmonic Response analysis, you will apply the loading condition and solve the model.

1. Select the **Harmonic Response** node in the **Outline** pane; the **Details of "Harmonic Response "** window is displayed. Also, the **Environment** contextual tab is displayed.

2. Right-click on the **Harmonic Response** node and then choose **Insert > Pressure** from the shortcut menu displayed; **Pressure** is added under the **Harmonic Response** node in the **Outline** pane and the **Details of "Pressure"** window is displayed.

3. In the **Details of Pressure"** window, select the **Geometry** selection box to display the **Apply** and **Cancel** buttons, if they are not already displayed.

4. Choose the **Face** tool from the **Select** toolbar to select a face from the graphics screen.

5. In the graphics screen, select the cylindrical face, refer to Figure 10-72.

6. Choose the **Apply** button from the **Geometry** selection box in the **Details of "Pressure"** window; the pressure is selected for applying the pressure load.

Figure 10-72 Selecting the face for applying a pressure load

7. Select the **Magnitude** edit box and then enter **1** as the magnitude of pressure load.

Note that the direction of pressure is always normal to surface.

8. Select the **Solution** node in the **Outline** pane; the **Solution** contextual tab is displayed. Also, the **Details of "Solution"** window is displayed.

9. Choose the **Total** tool from the **Deformation** drop-down in the **Solution** contextual tab; **Total Deformation** is attached under the **Solution** node. Also, the **Details of "Total Deformation"** window is displayed.

10. Choose the **Equivalent (von-Mises)** tool from the **Stress** drop-down; **Equivalent Stress** is attached to the Outline pane. Also, the **Details of "Equivalent Stress"** window is displayed.

11. Choose the **Deformation** tool from the **Frequency Response** drop-down; **Frequency Response** is attached to the **Outline** pane. Also, the **Details of "Frequency Response"** window is displayed.

12. In the **Details of "Frequency Response"** window, select the **Geometry** selection box to display the **Apply** and **Cancel** buttons, if they are not already displayed.

13. Choose the **Face** tool from the **Select** toolbar to select a face from the graphics screen.

14. In the graphics screen, select the cylindrical face, refer to Figure 10-73.

15. Choose the **Apply** button from the **Geometry** selection box in the **Details of "Frequency Response"** window, as shown in Figure 10-74. The face is selected for determining the deformation with respect to that surface.

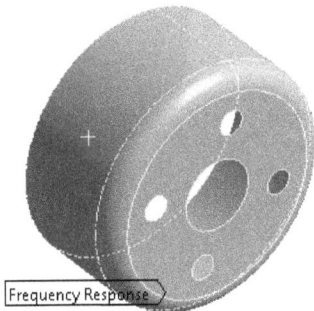

Details of "Frequency Response"	
− Scope	
Scoping Method	Geometry Selection
Geometry	1 Face
Spatial Resolution	Use Average
− Definition	
Type	Directional Deformation
Orientation	X Axis
Coordinate System	Global Coordinate System
Suppressed	No
− Options	
Frequency Range	Use Parent
Minimum Frequency	100. Hz
Maximum Frequency	2000. Hz
Display	Bode
Chart Viewing Style	Log Y
+ Results	

Figure 10-73 Selecting the face for Frequency Response

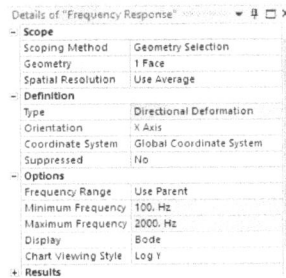

*Figure 10-74 The **Details of Frequency Response** window*

16. Choose the **Solve** tool from the **Solution** contextual tab; the parameters are evaluated.

Interpreting and Animating the Results

After solving the model, you will need to interpret the results.

1. Select the **Solution** node in the **Outline** pane; the **Graph** and **Tabular Data** windows with all their contents are displayed, refer to Figure 10-75.

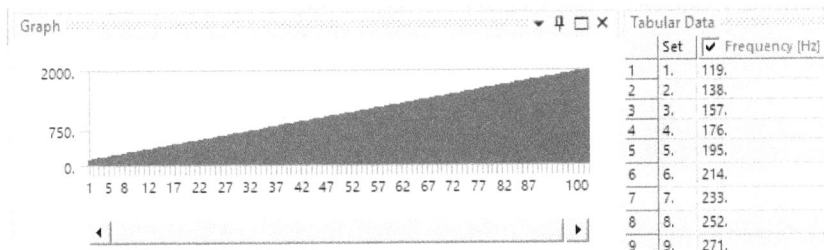

*Figure 10-75 The **Graph** and **Tabular Data** windows*

Figure 10-76 shows the model when the **Total Deformation** node is selected in the **Outline** pane and Figure 10-77 shows the model when the **Equivalent Stress** node is selected in the **Outline** pane.

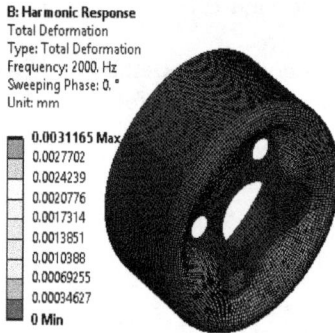

Figure 10-76 *Total deformation of the model*

Figure 10-77 *Equivalent stress of the model*

The above two figures show the results for last frequency value which is 2000 Hz. You can be able to retrieve the results at any other frequency value ranging from 100 Hz to 2000 Hz. To do this, select the **Total Deformation** node in the **Outline**; the **Details of "Total Deformation"** window is displayed. In this window, expand the **Definition** node if it is not already expanded. In the **Frequency** edit box, enter the frequency value as per your requirements. Next, right-click on the **Total Deformation** node and select the **Retrieve This Result** option from shortcut menu displayed. On doing so, you will retrieve the total deformation value at that specified frequency. Similar steps can be followed to retrieve the equivalent stress results. Alternatively, you can add multiple instances of total deformations or equivalent stress and set the frequency value as per your requirements in the **Frequency** edit box.

Figure 10-78 shows the **Worksheet** window when the **Frequency Response** node is selected in the **Outline** pane.

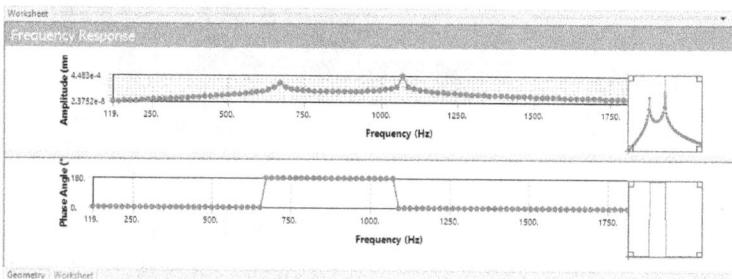

Figure 10-78 *The **Worksheet** windows*

Figure 10-78 shows two graphs, the first graph is plotted with displacement Amplitude versus Frequency and the second graph is plotted with Phase Angle versus Frequency parameters for specified face. Similarly, you can analyze the results for maximum amplitude, frequency, phase angle and so on from the **Details of "Frequency Response"** window, as shown in Figure 10-79.

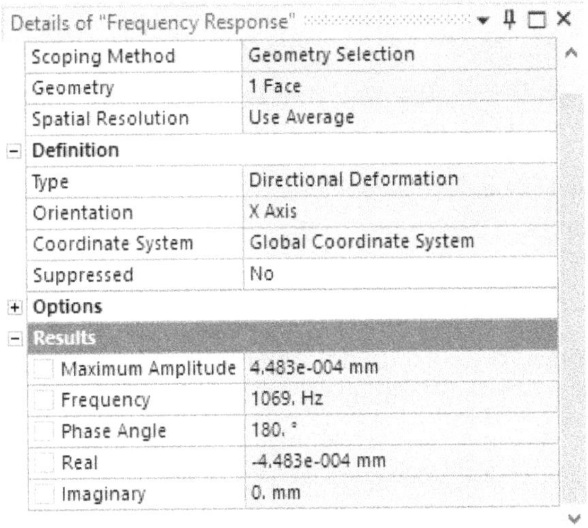

Figure 10-79 The Details of "Frequency Response" window

After the results are generated, you can play the animation to know how the model behaves.

2. Select **Total Deformation** in the **Solution** node in the **Outline** pane; the **Graph** window is displayed with the **Animation** area displayed in it.

3. In the **Animation** area of the **Graph** window, enter **40** in the **Frames** edit box and **4** in the **Time** edit box.

4. Choose the **Play** button from the **Animation** area in the **Graph** window; the animation with respect to the total deformation is played in the graphics screen.

5. Choose the **Export Video File** button from the **Animation** area of the **Graph** window; the **Save As** dialog box is displayed.

6. Browse to the location *C:\ANSYS_WB\c10\Tut03* and enter **Harmonic_Total Deformation** in the **File name** input box in the **Save As** dialog box and then choose the **Save** button to save the video of the animation in the specified location.

7. Select **Equivalent Stress** under the **Solution** node in the **Outline** pane; equivalent stresses contour is displayed in the graphics screen.

8. Play the animation and save the animation as **Harmonic_Equivalent Stress**.

9. Exit the **Mechanical** window; the **Workbench** window is displayed.

Saving the Project

Now, you need to save the project.

1. In the **Workbench** window, choose the **Save Project** button from the **Standard** toolbar to save the project *c10_ansWB_tut03*.

2. Exit the **Workbench** window.

Self-Evaluation Test

Answer the following questions and then compare them to those given at the end of this chapter:

1. When the intensity of the stress exceeds the elastic limit, the material looses its _____ property.

2. Modal analysis determines the vibration characteristics of a structural component. (T/F)

3. When a damper is added to a component, the damped modal analysis is carried out. (T/F)

4. Only natural frequencies are determined in a Modal analysis. (T/F)

5. Modal analysis determines the response of a structure to the vibrations that are time dependent. (T/F)

6. To start a Modal analysis in ANSYS Workbench, you first need to run a Static Structural analysis. (T/F)

7. In Dynamic analysis, the load or field conditions vary with time. (T/F)

8. When the natural frequency of a system is very close to the operating conditions or close to the excitation frequency, resonance may be induced in the component. (T/F)

Review Questions

Answer the following questions:

1. For every _____, there is a corresponding mode shape displayed in the graphics screen.

2. The number of modes to be extracted should be less than the number of DOFs. (T/F)

3. In Modal analysis, you can find the stresses induced in the model . (T/F)

4. In the stepped boundary condition, the defined force/moment is applied to the model at each frequency. (T/F)

5. If the loading is removed after exceeding the elastic limit, the geomtry will not retain its original shape. (T/F)

6. You cannot find more than 6 mode shapes in the Modal analysis. (T/F)

EXERCISES

Exercise 1

Open the project titled *c03_ansWB_tut03* from the location *C:\ANSYS_WB\c03\Tut03*. Save the project with the name *c10_ansWB_exr01* at the location *C:\ANSYS_WB\c10\Exr01*. Add the **Modal** analysis system to the project and then perform the Modal analysis. Apply Fixed support to the inner cylindrical faces of the holes. Number of modes to be found are 8. The model for Exercise 1 is shown in Figure 10-80. **(Expected time: 30 min)**

Figure 10-80 *Model for Exercise 1*

Exercise 2

Open the project titled *c04_ansWB_tut02* from the location *C:\ANSYS_WB\c04\Tut02*. Save the project with the name *c10_ansWB_exr02* at the location *C:\ANSYS_WB\c10\ Exr02*. Add the **Modal** analysis system to the project and then perform the Modal analysis. Apply Fixed support to the inner cylindrical faces of the holes. Number of modes to be found are 6. The model for Exercise 2 is shown in Figure 10-81. **(Expected time: 30 min)**

Figure 10-81 *Model for Exercise 2*

Chapter **11**

Thermal Analysis

Learning Objectives

After completing this chapter, you will be able to:

- *Understand the types of Thermal analysis*
- *Understand the Mechanical interface used for Thermal analysis*
- *Understand different terms used in Thermal analysis*
- *Perform Steady-State Thermal analysis*
- *Perform Transient Thermal analysis*
- *Perform Thermal Stress analysis*
- *Understand temperature distribution*
- *Run Probe*

INTRODUCTION TO THERMAL ANALYSIS

Before a model is set for production, it passes through several stages. Thermal analysis is one of them and plays an important role in product development. Various products such as engines, refrigerators, heat exchangers, and so on are designed based on the results of this analysis.

Thermal analysis is used to determine the temperature distribution and related thermal quantities in the model. In this analysis, all heat transfer modes, namely conduction, convection, and radiation are analyzed. The output from a thermal analysis can be the following:

1. Temperature distribution.
2. Amount of heat loss or gain.
3. Thermal gradients.
4. Thermal fluxes.

This analysis is used in many engineering industries such as automobile, piping, electronic, power generation, and so on. In ANSYS Workbench, two types of thermal analysis can be carried out, namely Steady-State and Transient Thermal analysis.

The following are the basic steps required to perform the thermal analysis:

1. Set the analysis preference.
2. Create or import solid model.
3. Define element attributes (element types, real constants, and material properties).
4. Mesh the model.
5. Specify the analysis type, analysis options, and the loads to be applied.
6. Solve the analysis problem.
7. Post-process results.

IMPORTANT TERMS USED IN THERMAL ANALYSIS

Before conducting thermal analysis, you should be familiar with the basic concepts and terminologies of thermal analysis. Following are some of the important terms used in thermal analysis:

Heat Transfer Modes

Whenever two bodies having different temperatures come in contact, then the heat transfer takes place from the body of higher temperature to the body of lower temperature. There are three modes of heat transfer: Conduction, Convection, and Radiation.

Conduction

Conduction is the process of heat transfer between bodies in contact. For example, hold an iron rod on a flame and then wait for sometime; the iron rod will be heated and you will feel the heat on your palm.

Convection
Convection is the process of heat transfer in which the medium of heat transfer is a fluid. For example, heating up water using an electric water heater is a good example of heat convection. In this case, water takes heat from the heater.

Radiation
Radiation is mode of heat transfer in which the heat is transferred through space. Heat from sun is a perfect example of this mode of heat transfer.

Thermal Gradient
The thermal gradient is the rate of change in temperature per unit depth in a material.

Thermal Flux
The thermal flux is defined as the rate of heat transfer per unit cross-sectional area. It is denoted by q.

Bulk Temperature
It is the temperature of a fluid flowing outside the material. It is denoted by Tb. The Bulk temperature is used in convective heat transfer.

Film Coefficient
It is a measure of the heat transfer through a fluid film.

Emissivity
The emissivity of a material is the ratio of energy radiated by the material to the energy radiated by a black body at the same temperature. Emissivity is the measure of a material's ability to absorb and radiate heat. It is denoted by e. Emissivity is a numerical value without any unit. For a perfect black body, $e = 1$. For any other material, $e < 1$.

Stefan–Boltzmann Constant
The Stefan-Boltzmann constant is a physical constant and defines the power per unit area emitted by a black body as a function of its thermodynamic temperature. It is denoted by s.

Thermal Conductivity
The thermal conductivity is the property of a material that indicates its ability to conduct heat. It is denoted by K.

Specific Heat
The specific heat is the amount of heat required per unit mass to raise the temperature of the body by one degree Celsius. It is denoted by c.

TYPES OF THERMAL ANALYSIS

In ANSYS Workbench, two types of thermal analysis can be carried out, namely Steady-State Thermal analysis and Transient Thermal analysis.

Steady-State Thermal Analysis

In the Steady-State Thermal analysis, thermal load does not vary with time and remains constant throughout the period of application. This analysis considers only steady loads and does not consider any thermal load that varies with time. In the Steady-State Thermal analysis, the system is studied under steady thermal loads with respect to time. These thermal loads include convection, radiation, heat flow rates, heat fluxes (heat flow per unit area), heat generation rates (heat flow per unit volume), and constant temperature boundaries.

The Steady-State Thermal analysis may be either linear or nonlinear, with respect to material properties that depend on temperature. The thermal properties of most of the materials do vary with temperature, therefore the analysis usually is nonlinear. Including radiation effects or temperature dependent convection in a model also makes the analysis nonlinear.

The steps to solve a problem related to the Thermal analysis are the same as that of the structural analysis except a few steps such as selecting the element type, applying the load, and postprocessing results.

Transient Thermal Analysis

Transient thermal analysis, the application of thermal loads is time dependent. Most of the engineering applications need Transient thermal analysis, such as engine blocks, pressure vessels, nozzles, piping systems, and so on. The process of solving the Transient thermal analysis problem is the same as that of the Steady-state thermal analysis. The only difference between these two analyses is that in Transient thermal analysis, the thermal load applied on a body is the function of time.

In the Transient Thermal analysis, the system is studied under varying thermal loads with respect to time. You can get the temperatures varying with time, thermal gradients, and thermal fluxes in a Transient thermal analysis.

The Transient Thermal analysis takes more time compared to other analyses types. It is necessary to understand the basic mechanism of the problem to reduce the time involved in getting its solution. For example, if the problems contain nonlinearity, then you first need to understand how they affect the response of structures by doing the Steady-State Thermal analysis.

THERMAL STRESS ANALYSIS

Heat Transfer analysis provides only the temperature distribution and related thermal quantities in the model as a result. Therefore, the thermo-mechanical coupled analysis must be carried out afterwards to determine the thermal stresses and strains subjected to a temperature change in the model. Similar to cyclic structural load, cyclic thermal load can also cause thermal fatigue of the model and lead to failures. For this analysis, create a structural analysis system that shares the data with thermal analysis system which acts as initial condition for subsequent analysis.

TUTORIALS

Tutorial 1   Steady-State Thermal

In this tutorial, you will open the existing project *c04_ansWB_tut02*. Next, you will save the project with a different name and run the Steady-State Thermal analysis on the model shown in Figure 11-1 to find out the effect of temperature on the whole body of the model. Also, you will evaluate the Total Heat Flux and the Directional Heat Flux with respect to the X Axis. The parameters required to run the analysis are given next.   **(Expected time: 45 min)**

Boundary Conditions:

Apply the Temperature load of 80 °C on the front flat face of the model, as shown in Figure 11-2.

Apply the Convection load on the inner faces of the model, as shown in Figure 11-3. The inner faces are exposed to air. The ambient temperature is 22 °C.

Material:

Structural Steel (Default)

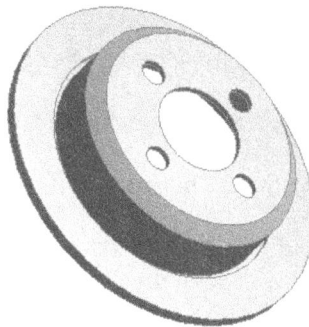

Figure 11-1 Model for Tutorial 1

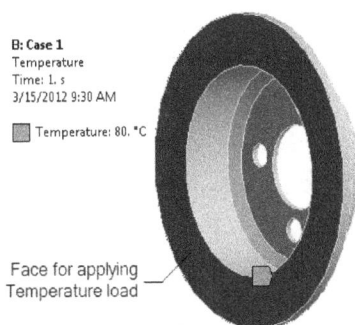

Figure 11-2 Face for applying the Temperature load

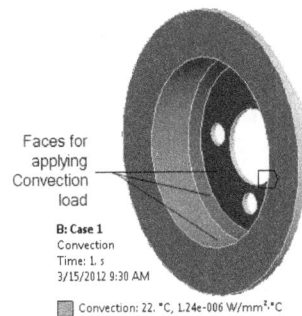

Figure 11-3 Faces for applying the Convection load

The following steps are required to complete this tutorial:

a. Open an existing project and save it with a different name.
b. Change the unit and generate the mesh.
c. Apply thermal boundary conditions.
d. Analyze the results.
e. Save the project and exit ANSYS Workbench.

Opening an Existing Project and Saving it with a New Name

First, you need to start ANSYS Workbench and then open an existing project.

1. Start ANSYS Workbench and then open the file *c04_ansWB_tut02* from the location *C:\ANSYS_WB\c04\Tut02*. Notice that the **Geometry** component system is already available in the **Project Schematic** window.

2. Choose the **Save Project As** button from the **Standard** toolbar; the **Save As** dialog box is displayed.

3. In this dialog box, browse to the location *C:\ANSYS_WB* and create a folder with the name **c11**.

4. Browse to the *c11* folder and then create a folder with the name **Tut01** in it.

5. In the *Tut01* folder, save the project with the name **c11_ansWB_tut01**.

   Now you need to add the **Steady-State Thermal** analysis system into the **Project Schematic** window.

6. In the **Workbench** window, double-click on the **Steady-State Thermal** analysis system under the **Analysis Systems** toolbox in the **Toolbox** window.

7. Rename the **Steady-State Thermal** analysis system as **Case 1**.

8. Drag the **Geometry** cell from the **Geometry** component system into the **Geometry** cell of the **Case 1** analysis system, refer to Figure 11-4; the geometry is shared.

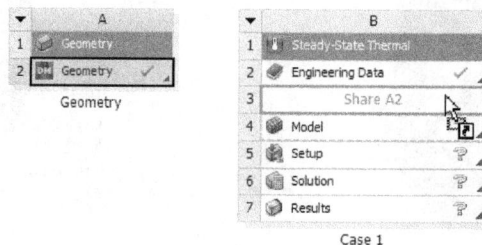

*Figure 11-4 The **Geometry** component system and the **Case 1** analysis system*

9. Double-click on the **Model** cell of the **Case 1** analysis system; the **Mechanical** window is displayed.

Changing the Units and Generating the Mesh

After the **Mechanical** window is displayed, you need to set units for the analysis and generate mesh on the model.

1. In the **Mechanical** window, choose the **Metric (mm, kg, N, s, mV, mA)** option from the **Units** drop-down list of the **Tools** group in the **Home** tab.

2. In the **Outline** pane, right-click on **Mesh** to display a shortcut menu.

3. Choose the **Generate Mesh** option from the shortcut menu; a mesh with default settings is generated, refer to Figure 11-5.

   Notice that the number of elements in the model after generating the mesh are 2734 approximately. For better results, you need to generate a mesh that is finer than the one already generated.

4. Select **Mesh** in the **Outline** pane; the **Details of "Mesh"** window is displayed.

5. In the **Details of "Mesh"** window, expand the **Sizing** node.

6. Select the **Yes** option from the **Use Adaptive Sizing** drop-down list if not selected.

7. Enter **10** in the **Element Size** edit box in the **Details of "Mesh"** window.

8. Right-click on **Mesh** in the **Outline** pane and choose the **Update** option from the shortcut menu displayed, refer to Figure 11-6.   ⌁ Update

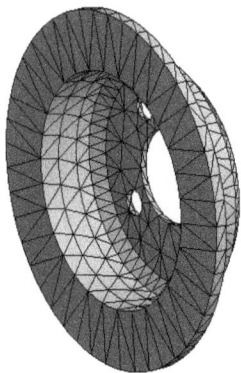

Figure 11-5 Mesh generated with the default settings

Figure 11-6 Mesh generated with the modified settings

Note that the number of elements in this mesh is 4069 approximately. This number is very large as compared to the element count that was achieved in the mesh generated with default mesh control settings.

Note

As the Structural Steel material is applied by default, you can skip the steps used for assigning material to the geometry.

Applying Thermal Boundary Conditions

After the model is meshed, you now need to apply the boundary condition for the analysis.

1. Select the **Steady-State Thermal** node in the **Outline** pane to display the **Environment** contextual tab.

2. From the **Environment** contextual tab, choose the **Temperature** tool; **Temperature** is added under the **Steady-State Thermal** node. Also, the **Details of "Temperature"** window is displayed.

3. In the **Details of "Temperature"** window, select the **Geometry** selection box to display the **Apply** and **Cancel** buttons if not already displayed.

4. Select the front face of the model, as shown in Figure 11-7.

5. Choose the **Apply** button from the **Geometry** selection box in the **Details of "Temperature"** window; the face is selected for applying the Temperature boundary condition.

6. In the **Details of "Temperature"** window, specify **80** in the **Magnitude** edit box.

Figure 11-7 Selecting the front face for applying Temperature load

7. Select **Temperature** from the **Steady-State Thermal** node in the **Outline** pane; the **Graph** and **Tabular Data** windows are displayed, as shown in Figures 11-8 and 11-9.

Notice that in Figures 11-8 and 11-9, the total time taken for applying the Thermal boundary condition is 1 second which means the temperature value is constant over a period of 1 second.

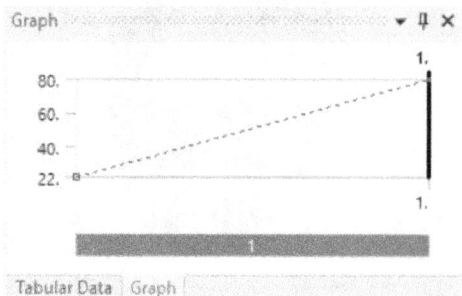

Steps	Time [s]	☑ Temperature [°C]
1	0.	= 22.
1	1.	80.

*Figure 11-8 The **Graph** window displayed on selecting **Temperature** from the **Steady-State Thermal** node in the **Outline** Pane*

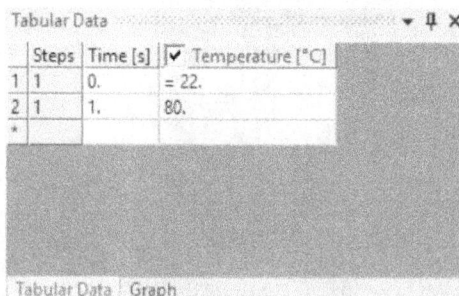

*Figure 11-9 The **Tabular Data** window displayed on selecting **Temperature** from the **Steady-State Thermal** node in the **Outline** Pane*

8. Choose the **Convection** tool from the **Environment** contextual tab; **Convection** is added under the **Steady-State Thermal** node. Also, the **Details of "Convection"** window is displayed.

9. In the **Details of "Convection"** window, select the **Geometry** selection box if not already selected; the **Apply** and **Cancel** buttons are displayed.

10. Select the inner faces of the model, refer to Figure 11-10.

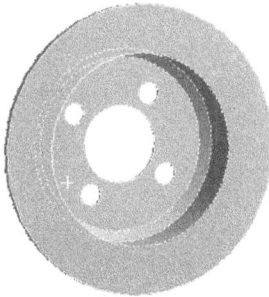

Figure 11-10 Selecting inner faces for applying the Convection load

11. In the **Details of "Convection"** window, choose the **Apply** button; the faces are specified for applying the Convection load.

12. Now, choose the **Import Temperature Dependent** option from the **Film Coefficient** drop-down list, refer to Figure 11-11; the **Import Convection Data** dialog box is displayed, as shown in Figure 11-12.

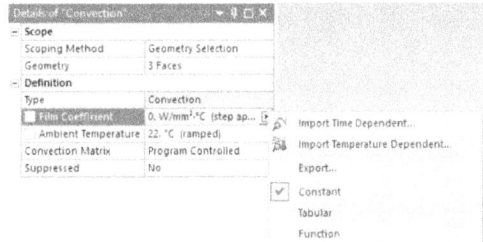

*Figure 11-11 The **Import Temperature Dependent** option chosen from the **Film Coefficient** drop-down list*

13. In the **Import Convection Data** dialog box, select the **Stagnant Air - Horizontal Cyl** radio button, as shown in Figure 11-12.

The **Stagnant Air - Horizontal Cyl** radio button is selected when the surrounding air is considered to be stagnant.

14. Next, choose the **OK** button from the **Import Convection Data** dialog box to exit the window and apply the changes.

Notice that the **Details of "Convection"** window is also modified, as shown in Figure 11-13. Also, a green tick mark is placed before **Convection** added under the **Steady-State Thermal** node in the **Outline** pane indicating that the Convection boundary condition has been applied.

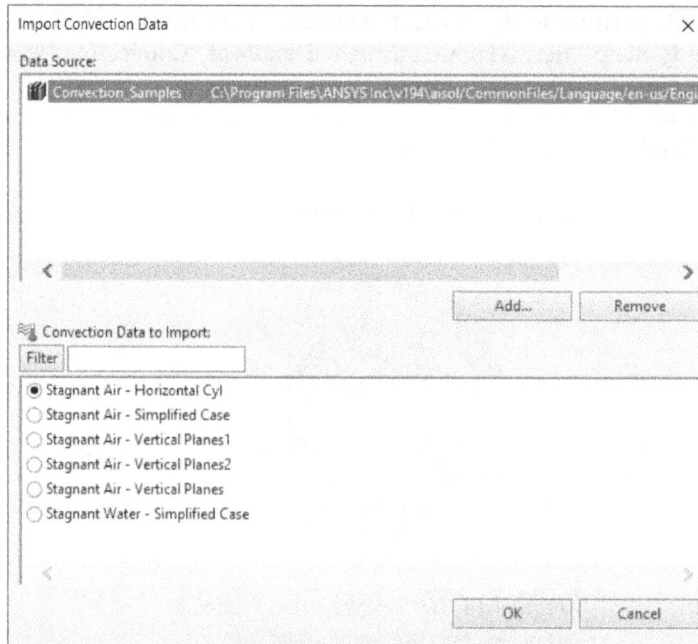

Figure 11-12 The ***Stagnant Air - Horizontal Cyl*** *radio button selected from the* ***Import Convection Data*** *dialog box*

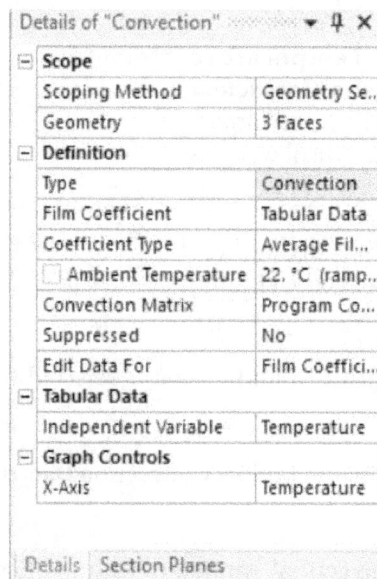

Figure 11-13 The ***Details of "Convection"*** *window after the changes are made*

15. Select **Convection** from the **Steady-State Thermal** node in the **Outline** pane; the **Graph** and **Tabular Data** windows are displayed, as shown in Figures 11-14 and 11-15.

Figure 11-14 shows the graph for distribution of Convection Coefficient with respect to Temperature with **Stagnant Air - Horizontal Cyl** considered as the film of the fluid. This graph shows that as the Temperature increases, the Coefficient of Convection also changes.

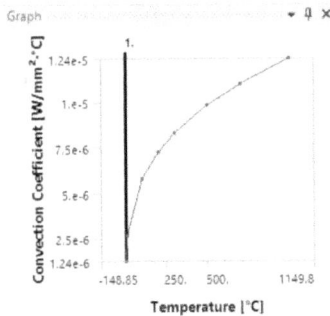

Figure 11-14 *The* ***Graph*** *window displayed on selecting* ***Convection*** *from the* ***Steady-State Thermal*** *node in the* ***Outline*** *pane*

Figure 11-15 *The* ***Tabular Data*** *window displayed on selecting* ***Convection*** *under the* ***Steady-State Thermal*** *node in the* ***Outline*** *pane*

Analyzing the Results

After the thermal boundary conditions are applied, it is now important to analyze the behavior of the model with respect to the boundary conditions applied.

1. Right-click on the **Solution** node in the **Outline** pane to display a shortcut menu.

2. From this shortcut menu, choose **Insert > Thermal > Temperature**; **Temperature** is added under the **Solution** node in the **Outline** pane. Also, the **Details of "Temperature"** window is displayed.

3. In this window, select the **Geometry** selection box to display the **Apply** and **Cancel** buttons.

4. Choose the **Body** tool from the **Select** toolbar and then select the model, as shown in Figure 11-16.

5. Choose the **Apply** button from the **Geometry** selection box in the **Details of "Temperature"** window; **1 Body** is displayed in the **Geometry** selection box.

Note
To understand the temperature distribution better in complex models, you need to insert more Temperature parameters at different regions of the model and then evaluate them.

6. Choose **Thermal > Temperature** from the **Solution** contextual tab; **Temperature 2** is added under the **Solution** node in the **Outline** pane. Also, the **Details of "Temperature 2"** window is displayed.

7. In the **Details of "Temperature 2"** window, select the **Geometry** selection box to display the **Apply** and **Cancel** buttons.

8. Choose the **Face** tool from the **Select** toolbar and then select the cylindrical face of the model, as shown in Figure 11-17.

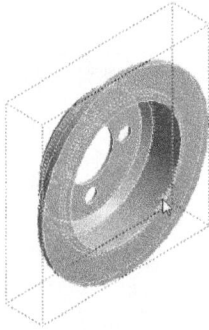

Figure 11-16 Body selected

Figure 11-17 Face selected

9. Choose the **Apply** button from the **Details of "Temperature 2"** window; **1 Face** is displayed in the **Geometry** selection box in the **Details of "Temperature 2"** window.

10. Choose the **Solve** tool from the **Solve** group of the **Home** tab; the **ANSYS Workbench Solution Status** message box is displayed and the temperature distribution for the Temperature 2 boundary condition is displayed in the graphics screen.

11. Select **Temperature** under the **Solution** node in the **Outline** pane; temperature distribution in the model is displayed in the graphics screen, as shown in Figure 11-18.

12. Similarly, select **Temperature 2** under the **Solution** node in the **Outline** pane; the temperature distribution in the model is displayed in the graphics screen, as shown in Figure 11-19.

Figure 11-18 Temperature distribution across the whole body

Figure 11-19 Temperature distribution across the selected face

The table given next shows the data generated from the analysis so far.

S. No.	Parameter	Max Value	Min Value
1	Temperature	80 °C	73.23 °C
2	Temperature 2	79.89 °C	77.49 °C

Note that the temperature in the region where the Temperature boundary condition is applied is higher as compared to the other regions. The region marked red shows the maximum temperature and the region marked blue shows the minimum temperature in the model. Similarly, the other colors display different values of temperature distribution in the model.

13. Select the **Solution** node in the **Outline** pane to display the **Details of "Convection"** window.

14. Choose **Total Heat Flux** from the **Thermal** drop-down list in the **Solution** contextual **tab**, as shown in Figure 11-20; **Total Heat Flux** with a yellow thunderbolt is added under the **Solution** node in the **Outline** pane. Also, the **Details of "Total Heat Flux"** window is displayed.

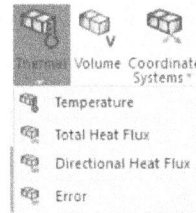

*Figure 11-20 Choosing **Total Heat Flux** from the **Thermal** drop-down list*

15. In the **Details of "Total Heat Flux"** window, select the **Geometry** selection box to display the **Apply** and **Cancel** buttons.

16. Choose the **Body** tool from the **Select** toolbar and then select the model, refer to Figure 11-16.

17. Choose the **Apply** button from the **Geometry** selection box in the **Details of "Total Heat Flux"** window; **1 Body** is displayed in the **Geometry** selection box.

18. Similarly, choose **Directional Heat Flux** from the **Thermal** drop-down in the **Solution** contextual tab; the **Directional Heat Flux** with a yellow thunderbolt is added under the **Solution** node in the **Outline** pane. Also, the **Details of "Directional Heat Flux"** window is displayed.

Note
*In this tutorial, the Directional Heat Flux is calculated along the X axis. However, you can determine the Directional Heat Flux along the Y or Z axis by selecting required option from the **Orientation** drop-down list in the **Details of "Directional Heat Flux"** window.*

19. Choose the **Solve** tool from the **Solve** group of the **Solution** contextual tab; the **ANSYS Workbench Solution Status** window is displayed. Notice that a tick mark is placed before **Total Heat Flux** and **Directional Heat Flux** in the tree of the **Outline** pane.

20. Select **Total Heat Flux** from the **Solution** node in the **Outline** pane; corresponding contours displaying the distribution of heat flux are displayed in the graphics screen, as shown in Figure 11-21.

21. Similarly, select **Directional Heat Flux** from the **Solution** node in the **Outline** pane; corresponding contours displaying the distribution of heat flux along the X axis are displayed in the graphics screen, as shown in Figure 11-22.

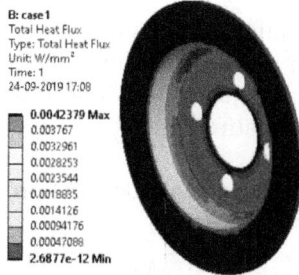

Figure 11-21 *Total Heat Flux for the whole body*

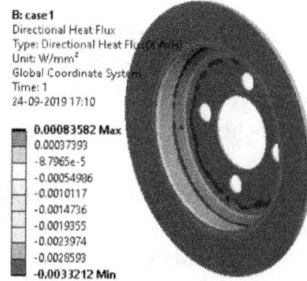

Figure 11-22 *Directional Heat Flux along the X Axis for the whole body*

The table given next displays the results obtained for the Total Heat Flux and Directional Heat Flux.

S. No.	Parameter	Max Value (W/mm^2)	Min Value (W/mm^2)
1	Total Heat Flux	0.0042397	2.6877E-12
2	Directional Heat Flux (X Axis)	0.00083582	-0.0033212

It is obvious from the above results that the Total Heat Flux is maximum in the region where the model is marked red and minimum where it is marked blue. Each color contour depicts a value and can be seen in the legend in the graphics screen.

Tip
*The material used so far in this tutorial is Structural Steel. However, you can also analyze a component by assigning a different material to it. You can do so by duplicating the **Case 1** analysis system in the **Project Schematic** window and then opening the **Mechanical** Workspace to apply the material and then solve the model.*

22. Exit the **Mechanical** window to display the **Workbench** window.

Saving the Project and Exiting ANSYS Workbench

After analyzing the results, save the entire project and exit ANSYS Workbench.

1. Choose the **Save Project** button from the **Standard** toolbar to save the project with the name already specified.

2. Close the **Workbench** window to exit the ANSYS Workbench session.

Tutorial 2 | Transient Thermal

In this tutorial, you will open the existing project *c04_ansWB_tut03* which is a piston model shown in Figure 11-23. Next, you will perform the Transient Thermal analysis on the piston model. You will assume the initial temperature of the system before combustion to be 22 °C. The temperature of the model at the time of combustion is 2000 °C. Consider the system to be water cooled. You will evaluate the temperature distribution on the body and also the Total Heat Flux for the complete model. **(Expected time: 45 min)**

Figure 11-23 The model for Tutorial 2

The following steps are required to complete this tutorial:

a. Open an existing project and save it with a new name.
b. Generate mesh on the model.
c. Set boundary conditions.
d. Set analysis results.
e. Analyze the results.
f. Save the project and exit ANSYS Workbench.

Opening an Existing Project and Saving it with a New Name

First, you need to start ANSYS Workbench, open the existing project, and then save it with a new name.

1. Start ANSYS Workbench to display the **Workbench** window.

2. Choose the **Open** button from the **Standard** toolbar; the **Open** dialog box is displayed.

3. In this dialog box, browse to the location *C:\ANSYS_WB\c04\Tut03* and then open the project *c04_ansWB_tut03*. Notice that the **Piston** component system is displayed in the **Project Schematic** window.

4. Choose the **Save Project As** button from the **Standard** toolbar; the **Save As** dialog box is displayed.

5. In this dialog box, browse to the location *C:\ANSYS_WB\c11* and then create a folder with the name **Tut02**.

6. Browse to the *Tut02* folder and save the project with the name **c11_ansWB_tut02** in it.

    After the project is saved with a new name, you need to add the Transient Thermal analysis system into the project.

7. In the **Workbench** window, double-click on the **Transient Thermal** analysis system in the **Analysis Systems** toolbox in the **Toolbox** window; the **Transient Thermal** analysis system is added to the **Project Schematic** window.

8. Drag the **Geometry** cell from the **Piston** component system into the **Geometry** cell of the **Transient Thermal** analysis system, refer to Figure 11-24. The geometry is now shared for the Transient Thermal analysis.

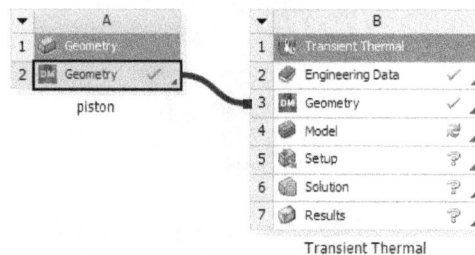

*Figure 11-24 The **Piston** component system and the **Transient Thermal** analysis system*

9. Next, double-click on the **Model** cell of the **Transient Thermal** analysis system; the **Mechanical** window is displayed.

Generating Mesh for the Model

After the **Mechanical** window is displayed, you need to generate mesh for the model.

1. Make sure **Metric (mm, kg, N, s, mV, mA)** is selected from the **Units** menu of the **Tools** group in the Menu bar.

2. In the **Outline** pane, right-click on **Mesh** to display a shortcut menu. Figure 11-25 shows the default **Outline** pane in the **Mechanical** window.

3. Choose **Generate Mesh** from the shortcut menu; the **ANSYS Workbench Mesh Status** box is displayed. After sometime, this box is closed and the mesh with default settings is generated, as shown in Figure 11-26.

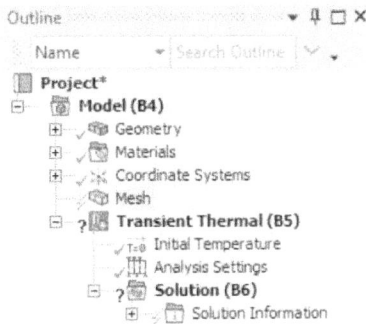

Figure 11-25 The default **Outline** pane in the **Mechanical** window

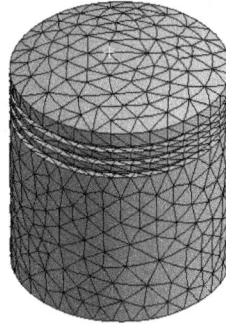

Figure 11-26 Mesh generated on the model with default settings

Note that, the number of elements in this case is 3,493 approximately. Now, you need to refine the mesh. You can do so by inserting localized mesh controls in various regions of the model or by changing the global variables in the **Details of "Mesh"** window.

4. Select **Mesh** from the **Outline** pane to display the **Details of "Mesh"** window.

5. In the **Details of "Mesh"** window, expand the **Sizing** node if not already expanded.

6. Select the **No** option from the **Use Adaptive Sizing** drop-down edit box; the **Details of "Mesh"** window is modified. Then, select the **Yes** option from the **Capture Proximity** drop-down list.

7. Enter **10** in the **Element Size** edit box of the **Defaults** node.

8. Enter **16** in the **Max Size** edit box.

9. Enter **6** in the **Proximity Min Size** edit box and enter **1** in the **Num Cells Across Gap** edit box.

10. Choose **Update** from the **Mesh** contextual tab; the mesh is updated, as shown in Figure 11-27.

Notice that the total number of elements created after generating the mesh is approximately **17,979** which is more than the previous element count.

Applying Boundary Conditions

After the piston model is meshed, you need to apply the boundary condition under which the thermal analysis will be performed. Note that you need to add Temperature load to the cylindrical face of the model.

1. Right-click on the **Transient Thermal** node in the **Outline** pane and then choose **Insert > Temperature** from the shortcut menu displayed; **Temperature** is added with a question symbol attached to it under the **Transient Thermal** node.

2. Select **Temperature** under the **Transient Thermal** node; the **Details of "Temperature"** window is displayed.

3. In the **Details of "Temperature"** window, click on the **Geometry** selection box to display the **Apply** and **Cancel** buttons if not already displayed.

4. Select the head of the piston, as shown in Figure 11-28, and then choose the **Apply** button from the **Geometry** selection box in the **Details of "Temperature"** window to specify the dome face of the model for applying the Temperature load.

Figure 11-27 Mesh generated with changed global variables

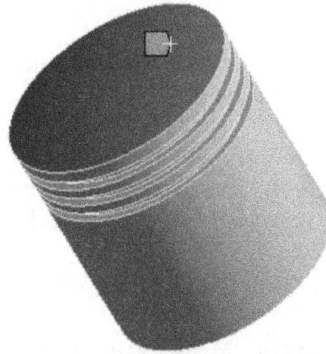

Figure 11-28 Face selected for applying Temperature load

5. In the **Tabular Data** window displayed at the bottom of the graphics screen, enter **2000** under the **Temperature [°C]** column corresponding to the row in the **Time** column where **1** is displayed, refer to Figure 11-29.

	Steps	Time [s]	✔ Temperature [°C]
1	1	0.	22.
2	1	1.	2000.
*			

Tabular Data

*Figure 11-29 The **Tabular Data** window*

6. Also, enter **22** under the **Temperature [°C]** column in the **0 Time [s]** row, refer to Figure 11-29.

The Temperature load is now specified. Next, you need to define the Convection load in the system.

7. Select the **Transient Thermal** node in the **Outline** pane to display the **Environment** contextual tab.

8. Choose the **Convection** tool from this tab; **Convection** with a question symbol is added under the **Transient Thermal** node in the **Outline** pane.

9. Choose the **Face** tool from the **Select** toolbar and then select all the outer faces of the piston, refer to Figure 11-30.

10. Choose the **Apply** button from the **Geometry** selection box in the **Details of "Convection"** window; **13 Faces** is displayed in the **Geometry** selection box, as shown in Figure 11-31.

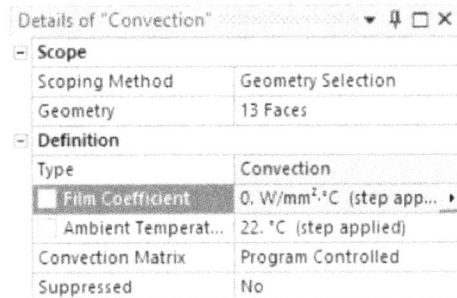

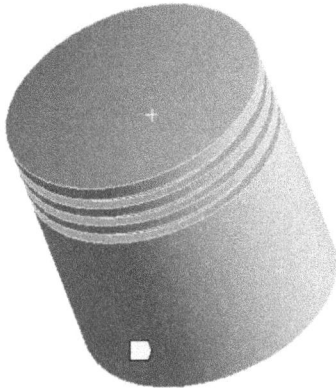

Figure 11-30 *Selecting all the outer faces for applying Convection load*

Figure 11-31 *13 Faces displayed in the Geometry selection box*

11. Next, choose the right arrow displayed next to the **Film Coefficient** edit box in the **Details of "Convection"** window; a flyout is displayed.

12. In this flyout, choose the **Import Temperature Dependent** option, as shown in Figure 11-32; the **Import Convection Data** dialog box is displayed.

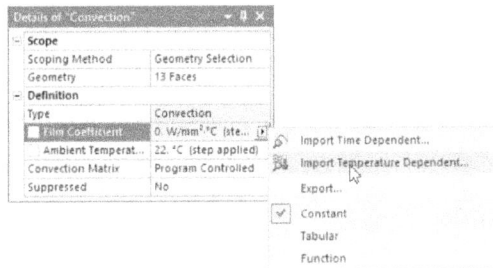

Figure 11-32 *Choosing the **Import Temperature Dependent** option from the **Film Coefficient** flyout*

13. In this dialog box, select the **Stagnant Water - Simplified Case** radio button, refer to Figure 11-33, and then choose the **OK** button to close the dialog box and apply the film coefficient; **1.2e-003 W/mm² °C (step applied)** is displayed in the **Film Coefficient** edit box in the **Details of "Convection"** window.

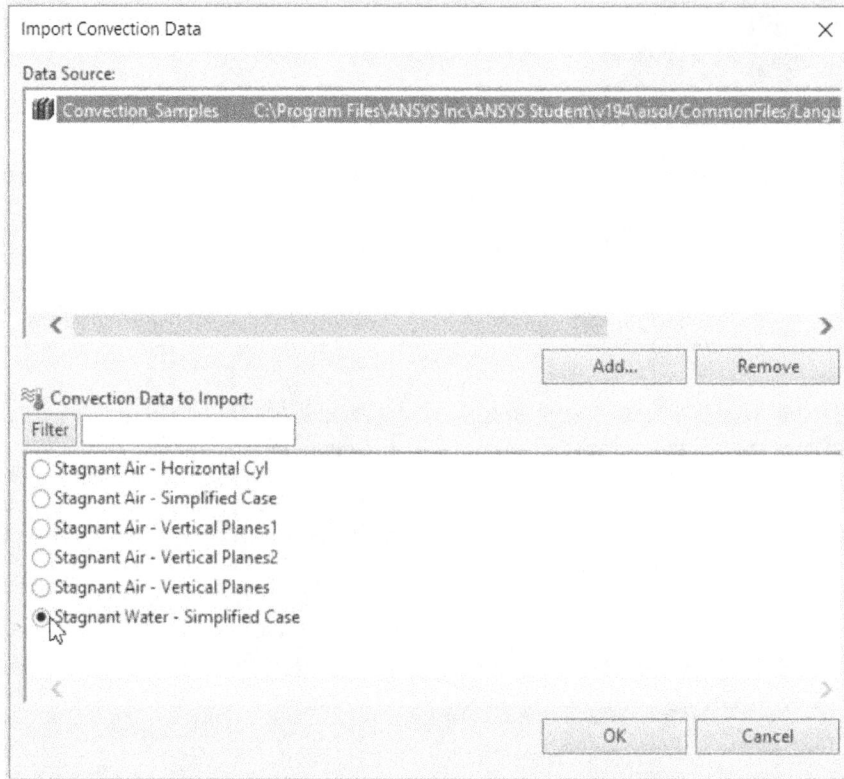

Figure 11-33 *The* **Import Convection Data** *dialog box with the* **Stagnant Water - Simplified Case** *radio button selected*

The **Stagnant Water - Simplified Case** radio button is used when the convection is assumed to take place through stagnant water.

Setting Analysis Results

After the Temperature and Convection loads are applied, you need to specify the parameters which you want to evaluate.

1. Drag **Temperature** from the **Transient Thermal** node in the Outline pane and drop it in the **Solution** node, refer to Figure 11-34, **Reaction Probe** is added to the **Solution** node.

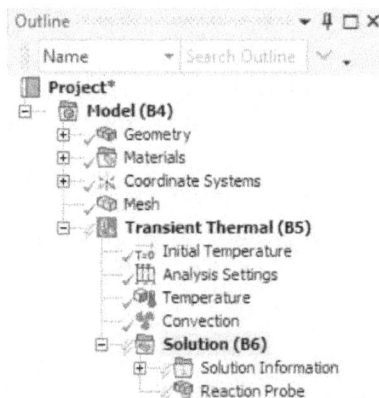

Figure 11-34 *Dragging and dropping* **Temperature** *from the* **Transient Thermal** *node into the* **Solution** *node*

Reaction Probe allows the user to find the reaction of a boundary condition at a certain point and at a certain instance of time. In thermal analysis, you can use probes to display reaction for the boundary conditions such as Convection, Temperature, and Radiation. Figure 11-35 shows the **Details of "Reaction Probe"** window with the **Boundary Condition** drop-down list displayed. In this drop-down list, the **Temperature** option is selected by default. You can also select **Convection** in this tutorial. The options in the **Boundary Condition** drop-down list are displayed based on the boundary conditions present under the **Transient Thermal** node in the **Outline** pane. Note that if there was a boundary condition named **Radiation** under the **Transient Thermal** node, the **Radiation** option would be displayed in the **Boundary Condition** drop-down list.

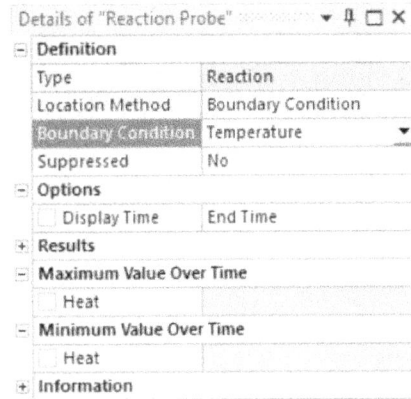

Figure 11-35   *The* **Details of "Reaction Probe"** *window*

2. Right-click on the **Solution** node in the **Outline** pane; a shortcut menu is displayed.

3. In this shortcut menu, choose **Insert > Thermal > Temperature**; **Temperature** is added under the **Solution** node.

4. Similarly, right-click on the **Solution** node again to display the shortcut menu and then choose **Insert > Thermal > Total Heat Flux**; **Total Heat Flux** is added under the **Solution** node.

5. Choose the **Solve** tool from the **Solution** contextual tab; the **ANSYS Workbench Solution Status** window is displayed. After sometime, the **ANSYS Workbench Solution Status** window is closed and a green tick mark is displayed before the results in the **Solution** node in the **Outline** pane indicating that the analysis has been solved.

6. Select **Temperature** under the **Solution** node in the **Outline** pane; the **Details of "Temperature"** window is displayed, as shown in Figure 11-36. Also, the temperature contours are displayed in the graphics screen, as shown in Figure 11-37.

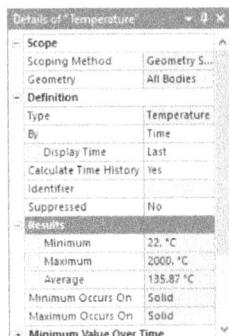

Figure 11-36   *The* **Details of "Temperature"** *window*

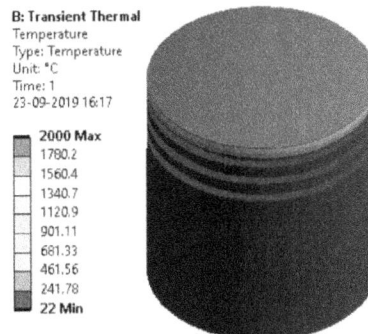

Figure 11-37   *The color contours displayed in the Graphics screen*

Notice that the **Graph** window shows the graphical representation of the data available in the **Tabular Data** window. Figures 11-38 and 11-39 show the **Graph** and **Tabular Data** windows, respectively.

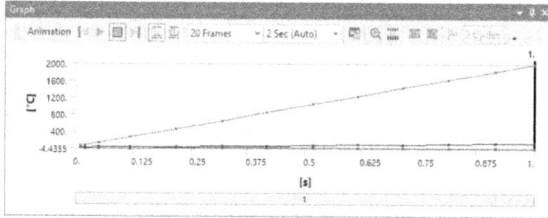

*Figure 11-38 The **Graph** window*

*Figure 11-39 The **Tabular Data** window with the distribution of temperature over time*

7. Right-click in the graphics screen; a shortcut menu is displayed.

8. Choose **View > Top** from this shortcut menu; the view is changed, as shown in Figure 11-40.

9. Choose the **Section Plane** tool from the **Home** tab of **Menu** Bar; the **Section Planes** window is displayed.

10. Create a section of the model, refer to Figure 11-41 and then choose the ISO ball at the bottom right corner of the graphics screen; the model is positioned, as shown in Figure 11-41.

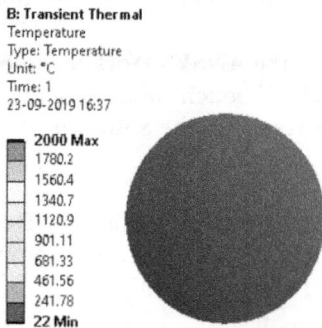

Figure 11-40 Top view of the model

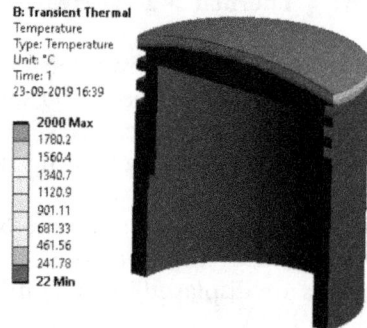

Figure 11-41 Isometric view of the sectioned model

Figure 11-41 shows the color contours that indicate the distribution of temperature along various regions of the model. Red color contour displays the region with the maximum temperature, whereas blue color contour shows the region with minimum temperature. The other colors represent various temperatures that lie within the maximum (red contour) and minimum (blue contour) temperatures. The color contours and their respective values can be seen in the legend, refer to Figure 11-42. The temperature on the dome of the piston is maximum because the Temperature boundary condition is applied on it and the component has attained maximum temperature in this region. The effect of this temperature on the farther regions is minimum, refer to Figure 11-41. In this model, the maximum temperature is 2000 °C and the minimum temperature is 22 °C.

B: Transient Thermal
Temperature
Type: Temperature
Unit: °C
Time: 1
23-09-2019 16:39

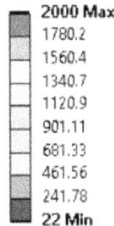

2000 Max
1780.2
1560.4
1340.7
1120.9
901.11
681.33
461.56
241.78
22 Min

*Figure 11-42 The legend displaying the
temperature distribution across the model*

11. Click on **Total Heat Flux** under the **Solution** node in the **Outline** pane; the **Details of "Total Heat Flux"** window is displayed, as shown in Figure 11-43. Also, the contours of Total Heat Flux are displayed in the graphics screen, refer to Figure 11-44.

Notice that when the contours are displayed in the graphics screen, the **Graph** and **Tabular Data** windows are also displayed below this screen, as shown in Figures 11-45 and 11-46.

Details of "Total Heat Flux"	▾ 및 ×
− Scope	
Scoping Method	Geometry Selection
Geometry	All Bodies
− Definition	
Type	Total Heat Flux
By	Time
Display Time	Last
Calculate Time History	Yes
Identifier	
Suppressed	No
+ Integration Point Results	
− Results	
Minimum	2.5391e-016 W/mm²
Maximum	31.43 W/mm²
Average	1.8486 W/mm²
Minimum Occurs On	Solid
Maximum Occurs On	Solid

*Figure 11-43 The **Details of "Total Heat Flux"** window*

B: Transient Thermal
Total Heat Flux
Type: Total Heat Flux
Unit: W/mm²
Time: 1
23-09-2019 16:45

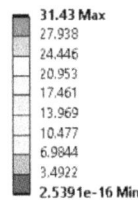

31.43 Max
27.938
24.446
20.953
17.461
13.969
10.477
6.9844
3.4922
2.5391e-16 Min

Figure 11-44 Contours of the Total Heat Flux displayed on the model

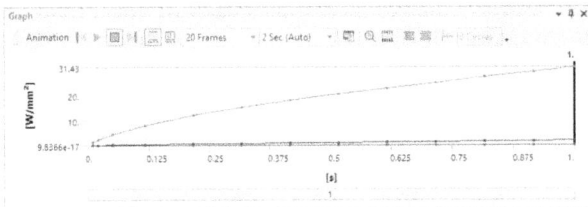

*Figure 11-45 The **Graph** window*

	Time [s]	✔ Minimum [W/mm²]	✔ Maximum [W/mm²]	✔ Average [W/m ▲
1	1.e-002	2.3085e-016	1.2251	4.8944e-002
2	2.e-002	3.0239e-016	2.165	8.8467e-002
3	5.e-002	4.5536e-016	4.4994	0.19169
4	0.11715	1.4742e-016	8.0597	0.37749
5	0.21715	1.3071e-016	12.097	0.59529
6	0.31715	9.8966e-017	15.374	0.77814
7	0.41715	1.567e-016	18.175	0.94114
8	0.51715	1.8793e-016	20.65	1.0937
9	0.61715	2.7766e-016	22.884	1.2477
10	0.71715	2.7042e-016	25.159	1.4035
11	0.81715	2.3973e-016	27.471	1.561
12	0.91715	1.9257e-016	29.673	1.7187
13	1.	2.5391e-016	31.43	1.8486

*Figure 11-46 The **Tabular Data** window*

Note that the Thermal analysis of a piston plays an important role in understanding the behavior of the component under the thermal loading conditions.

12. Close the **Mechanical** window; the **Workbench** window is displayed.

Saving the Project and Exiting ANSYS Workbench
Now you need to save the project.

1. In the **Workbench** window, choose the **Save Project** button to save the project as *c11_ansWB_tut02*.

2. Close the **Workbench** window to exit the ANSYS Workbench session.

Tutorial 3   Steady-State Thermal

In this tutorial, you will create the 3D model of the Heat Sink shown in Figure 11-47. The dimensions and views of the model are given in Figure 11-48. After creating the model, you will run the analysis under two conditions as stated below.

Case I
Run a Steady-State Thermal analysis and evaluate the Temperature Distribution and Total Heat Flux for the component on the basis of the boundary conditions given next.

Boundary Conditions:

Heat Flow: 13 W
Convection on Heat Sink walls with film coefficient 12e-6 W/mm²°C
Convection on Heat Sink floor with film coefficient 6e-6 W/mm²°C

Case II
Run Thermo-Mechanical Coupled analysis to evaluate the Thermal Stress, Deformation, and Fatigue life of the heat sink model by using the steady-state analysis solution as the thermal load. Apply fixed support boundary condition at the base of heat sink.

The following steps are required to complete this tutorial:

a. Start ANSYS Workbench and create the model.
b. Generate mesh and apply boundary conditions to the model.
c. Solve the FEA model and Analyze the results.
d. Add Static Structural analsys system and set the boundary condition for Case II.
e. Solve the model and analyze the results.
f. Save the project and exit ANSYS Workbench.

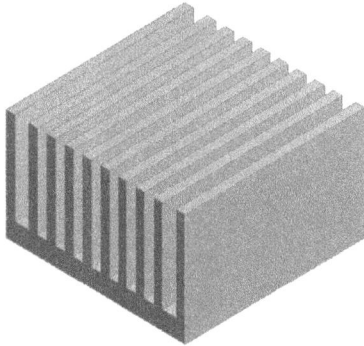

Figure 11-47 *The Heat Sink model*

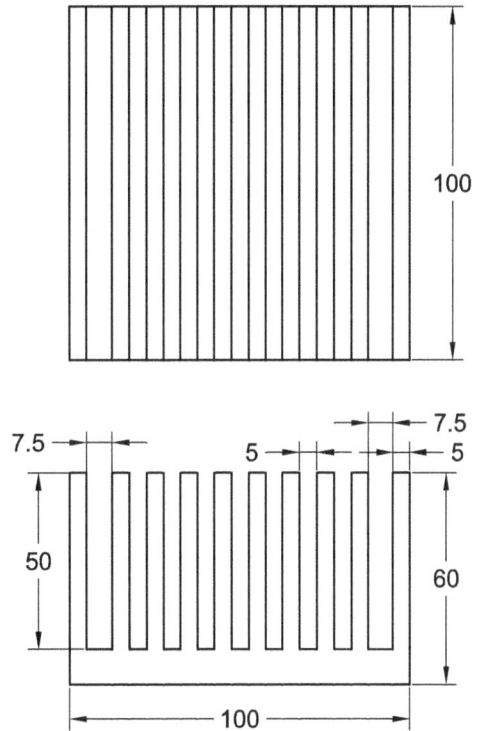

Figure 11-48 *The Top and Front views of the Heat Sink with dimensions*

Starting ANSYS Workbench and Creating the Model

First, you need to start ANSYS Workbench and then create the model.

1. Start ANSYS Workbench to display the **Workbench** window.

2. Add the **Steady-State Thermal** analysis system to the **Project Schematic** window.

3. Choose the **Save Project** button from the **Standard** toolbar; the **Save As** dialog box is displayed.

4. Browse to the location *C:\ANSYS_WB\c11* and then create a sub folder with the name **Tut03**.

5. Browse to the *Tut03* folder and save the project with the name **c11_ansWB_tut03** in this folder.

6 In the **Project Schematic** window, right-click on the **Geometry** cell of the **Steady-State Thermal** analysis system and choose the **New DesignModeler Geometry** option; the **DesignModeler** window is displayed.

7. In the **DesignModeler** window, draw a sketch on the XY plane, refer to Figure 11-49.

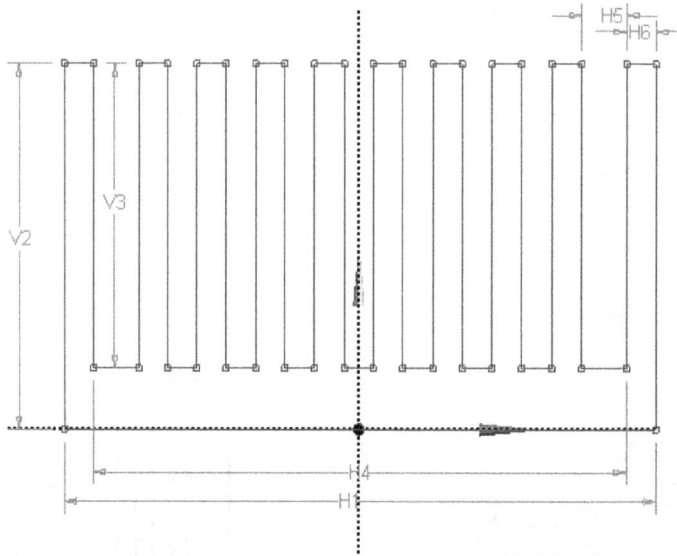

Figure 11-49 Sketch for creating the model

8. Extrude the sketch upto a distance of 100 mm to create the model, refer to Figure 11-47.

9. Exit the **DesignModeler** window to display the **Workbench** window.

Generating the Mesh and Specifying the Boundary Conditions

After the model is created, you need to mesh the model and also specify the thermal boundary conditions. The Heat Sink extracts the heat from the device it is attached to by conduction, and then dissipates the heat through the fins by convection. So you need to apply Heat Flow at the bottom face of the model. Also, you will apply convection on the floor and walls of the Heat Sink.

1. In the **Project Schematic** window of the **Workbench** window, double-click on the **Model** cell of the **Steady-State Thermal** analysis system; the **Mechanical** window is displayed.

2. Select **Mesh** in the **Outline** pane, as shown in Figure 11-50, to display the **Details of "Mesh"** window.

3. In the **Details of "Mesh"** window, expand the **Defaults** node to display its contents.

4. Under this node, enter **5** in the **Element Size** edit box.

5. Choose the **Generate** tool from the **Mesh** group of the **Mesh** contextual tab; the mesh is generated, as shown in Figure 11-51.

Generate

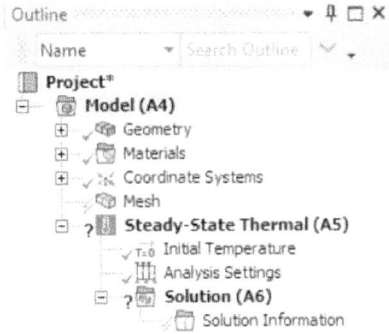

Figure 11-50 *Selecting* **Mesh** *in the* **Outline** *pane*

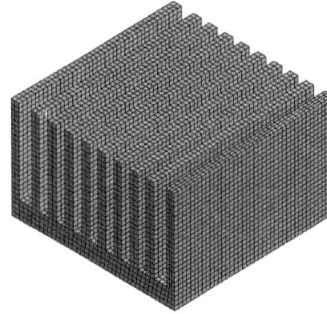

Figure 11-51 *Mesh generated on the model*

6. Expand the **Statistics** node in the **Details of "Mesh"** window to display the total number of elements created.

   Notice that the total number of elements created after generating the mesh are approximately 2840. After the model is meshed, you now need to specify the boundary conditions.

7. Right-click on the **Steady-State Thermal** node in the **Outline** pane and then choose **Insert > Heat Flow** from the shortcut menu displayed, refer to Figure 11-52; **Heat Flow** is added under the **Steady-State Thermal** node in the **Outline** pane. Also, the **Details of "Heat Flow"** window is displayed, as shown in Figure 11-53.

Figure 11-52 *The shortcut menu displayed on right-clicking on the* **Steady-State Thermal** *node in the* **Outline** *pane*

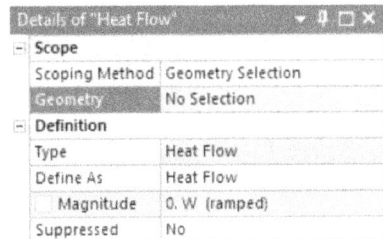

Figure 11-53 *The* **Details of "Heat Flow"** *window*

8. In the **Details of "Heat Flow"** window, select the **Geometry** selection box to display the **Apply** and **Cancel** buttons.

9. Next, select the bottom flat face of the Heat Sink in the graphics screen, as shown in Figure 11-54.

10. Choose the **Apply** button in the **Geometry** selection box, refer to Figure 11-55; the bottom face turns purple indicating that Heat Flow has been applied to this face.

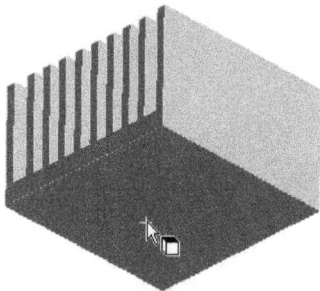

Figure 11-54 Selecting the bottom face of the Heat Sink to apply Heat Flow

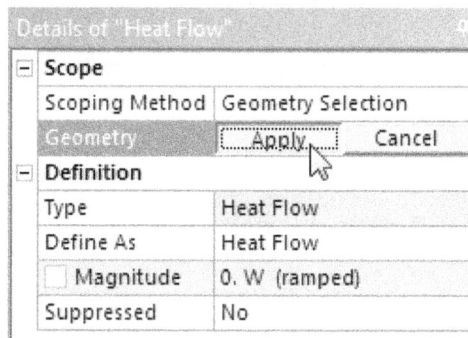

*Figure 11-55 Choosing the **Apply** button from the **Details of "Heat Flow"** window*

Notice that an arrow pointed upward is displayed at the bottom face of the model, as shown in Figure 11-56. This indicates that the direction of heat flow is upward.

11. In the **Details of "Heat Flow"** window, enter **13** in the **Magnitude** edit box, refer to Figure 11-57.

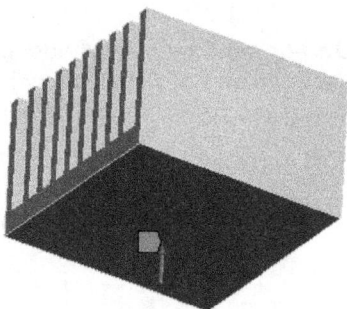

Figure 11-56 The arrow indicating the direction of Heat Flow

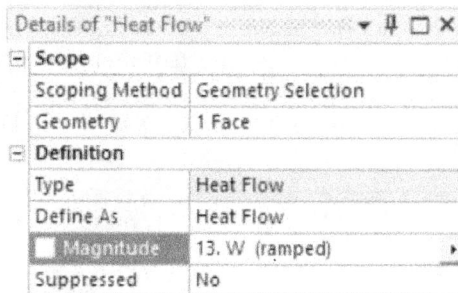

*Figure 11-57 The **Details of "Heat Flow"** window displayed with all the parameters specified*

After applying Heat Flow on the model, you need to specify convection where the heat will be dissipated.

12. Right-click on the **Steady-State Thermal** node in the **Outline** pane and then choose **Insert > Convection** from the shortcut menu displayed; **Convection** is added under the **Steady-State Thermal** node. Also, the **Details of "Convection"** window is displayed.

13. In the **Details of "Convection"** window, click on the **Geometry** selection box to display the **Apply** and **Cancel** buttons if not already displayed.

14. Select the top face (excluding the fin surface) of the Heat Sink, as shown in Figure 11-58 and then choose the **Apply** button in the **Geometry** selection box; **9 Faces** is displayed in the **Geometry** selection box indicating that 9 faces are selected on which convection is to be applied.

15. In the **Film Coefficient** edit box of the **Details of "Convection"** window, enter **6E-6**, refer to Figure 11-59.

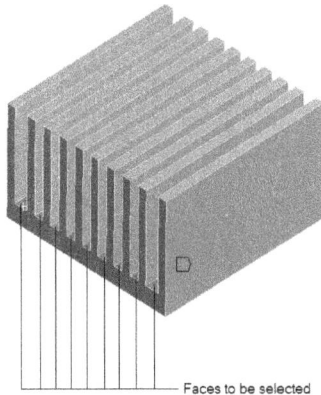

Figure 11-58 Faces selected for applying convection

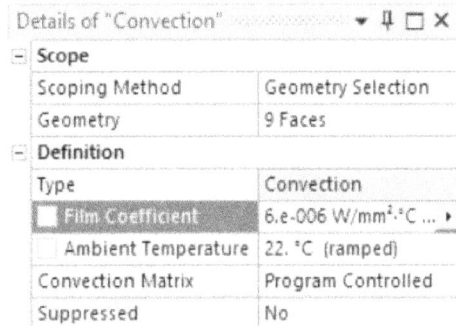

*Figure 11-59 The **Details of "Convection"** window displaying the parameters specified in it*

Leave the other options in the **Details of "Convection"** window as set by default. The Convection is now specified for the floor of the Heat Sink. In this tutorial, it is considered that convection at the floor is less as compared to convection on the walls of the fins. Therefore you need to specify a different film coefficient for the walls of the fins, to apply Convection load.

16. Right-click on the **Steady-State Thermal** node in the **Outline** pane and then choose **Insert > Convection** from the shortcut menu displayed; **Convection 2** is added under the **Steady-State Thermal** node. Also, the **Details of "Convection 2"** window is displayed.

17. Click on the **Geometry** selection box to display the **Apply** and **Cancel** buttons if not already displayed.

18. Select all the surfaces of the fins, refer to Figures 11-60.

Note
*1. Use the tools available in the **Graphics** toolbar to rotate the model.*

*2. Make sure that the **Face** tool is selected from the **Select** toolbar to select faces of the fins.*

19. Choose the **Apply** button in the **Geometry** selection box; the faces are specified for applying convection.

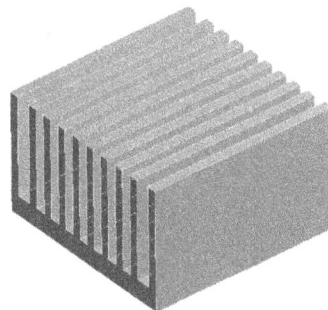

20. In the **Details of "Convection 2"** window, enter **12E-6** in the **Film Coefficient** edit box to specify the film coefficient.

Figure 11-60 Selecting the faces of the fins

Solving the FEA Model and Analyzing the Results

After all the boundary conditions are specified, you need to specify the outcomes of the analysis.

1. Right-click on the **Solution** node in the **Outline** pane and then choose **Insert > Thermal > Temperature** from the shortcut menu displayed, refer to Figure 11-61; **Temperature** with a yellow thunderbolt symbol is added under the **Solution** node in the **Outline** pane. Also, the **Details of "Temperature"** window is displayed.

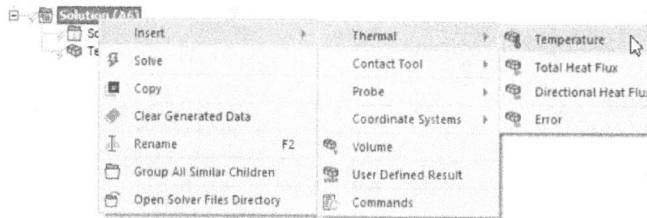

*Figure 11-61 Shortcut menu displayed by right-clicking on the **Solution** node*

2. Right-click again on the **Solution** node in the **Outline** pane and then choose **Insert > Thermal > Total Heat Flux** from the shortcut menu displayed, refer to Figure 11-61; **Total Heat Flux** is added under the **Solution** node. Also, the **Details of "Total Heat Flux"** window is displayed.

   Now, all the parameters are specified and the next step is to solve the FEA model.

3. Choose the **Solve** tool from the **Home** tab of the **Menu** bar; the **ANSYS Workbench Solution Status** box is displayed.

   Notice that after sometime, the **ANSYS Workbench Solution Status** box gets closed and the analysis is solved.

4. Right-click again on the **Solution** node in the **Outline** pane and then choose **Insert > Probe > Temperature** from the shortcut menu displayed, refer to Figure 11-62; **Temperature Probe** with a question symbol is added under the **Solution** node. Also, the **Details of "Temperature Probe"** window is displayed.

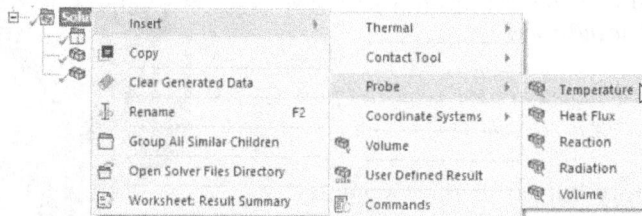

*Figure 11-62 Selecting the **Probe** option from the shortcut menu*

To find out the result at a particular point, edge, face, or complete body, use the probe tools. By using these tools you can find the maximum or minimum values of a result at a particular location.

5. In the **Details of "Temperature Probe"** window, select the **Geometry** selection box to display the **Apply** and **Cancel** buttons, as shown in Figure 11-63 if not already displayed.

6. Choose the **Face** tool from the **Select** toolbar and then select the top-face of the fins in the graphics screen, as shown in Figure 11-64.

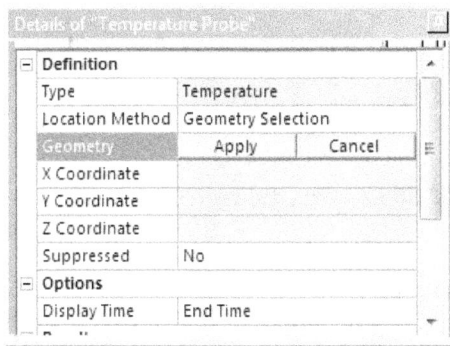

Figure 11-63 The **Details of "Temperature Probe"** *window*

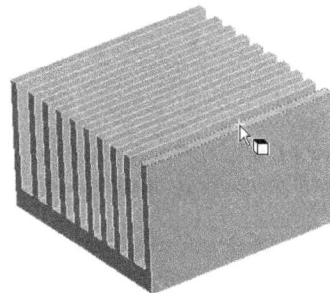

Figure 11-64 *Selecting a face of the fin in the Graphics screen*

7. Choose the **Apply** button from the **Geometry** selection box; **1 Face** is displayed in the **Geometry** selection box indicating that the 1 face is selected for a temperature probe now. Leave all other parameters as set by default in the **Details of "Temperature Probe"** window.

8. Right-click on the **Solution** node again and then choose **Evaluate All Results** from the shortcut menu displayed to complete the temperature probe.

9. In the **Details of "Temperature Probe"** window, expand the **Maximum Value Over Time** node; the **Temperature** edit box displays **31.458 ° C** as the maximum temperature at the selected face.

    After the model is solved, you need to check the results specified in the last section.

10. Select **Temperature** under the **Solution** node in the **Outline** pane; corresponding legend and the temperature contours are displayed in the graphics screen, as shown in Figures 11-65 and 11-66.

A: Steady-State Thermal
Temperature
Type: Temperature
Unit: °C
Time: 1
24-09-2019 11:44

32.826 Max
32.674
32.522
32.37
32.217
32.065
31.913
31.761
31.608
31.456 Min

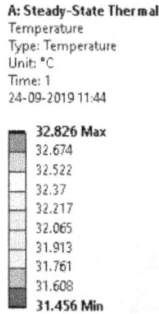

Figure 11-65 Legend showing the
values of temperature distribution

Figure 11-66 Contours displayed on the model

In the legend, red color shows maximum value of temperature attained in the model, which is 32.82 °C. Similarly, blue color displays minimum value of temperature in the model, which is 31.456 °C. Other colors display various temperatures attained in the model for the given boundary conditions.

11. Similarly, select **Total Heat Flux** displayed under the **Solution** node; corresponding legend and contours on the model are displayed, as shown in Figures 11-67 and 11-68, respectively.

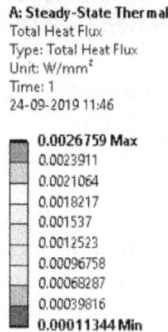

A: Steady-State Thermal
Total Heat Flux
Type: Total Heat Flux
Unit: W/mm²
Time: 1
24-09-2019 11:46

0.0026759 Max
0.0023911
0.0021064
0.0018217
0.001537
0.0012523
0.00096758
0.00068287
0.00039816
0.00011344 Min

Figure 11-67 Legend showing the
values of heat flux distribution

Figure 11-68 Contours displayed on the model

12. Close the **Mechanical** window to display the **Workbench** window.

Adding the Static Structural Analysis and Setting the Boundary Condition for Case II

After you retrieve the results of the Case I analysis system, you now need to work for Case II.

1. In the **Project Schematic** window, drag the **Static Structural** analysis system and drop in the **Solution** cell of the **Steady-State Thermal** analysis system, as shown in Figure 11-69. On doing so, the **Project Schematic** window will show that the **Steady-State Thermal** analysis system has transferred the data to the **Static Structural** analysis system, as shown in Figure 11-70.

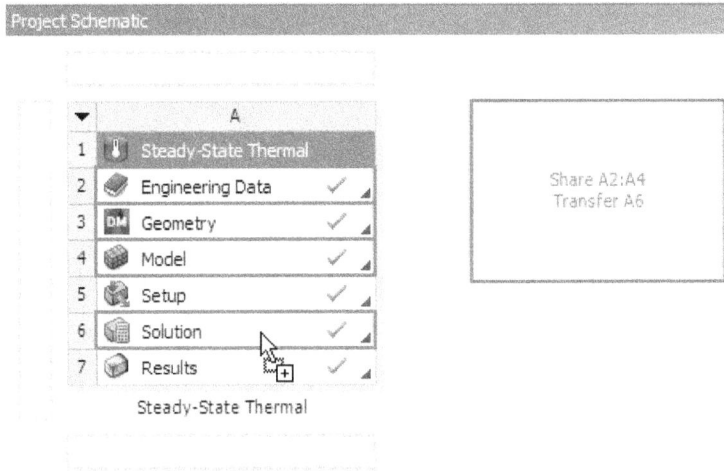

Figure 11-69 *Dragging the* **Static Structural** *analysis system into* **Solution** *cell of the* **Steady-State Thermal** *analysis system*

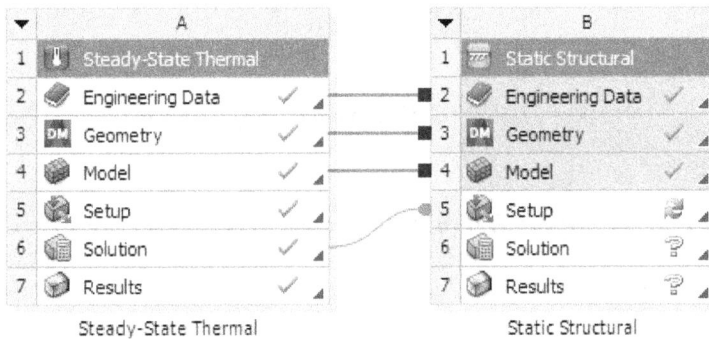

Figure 11-70 *The* **Steady-State Thermal** *analysis system showing transfer of data with the* **Static Structural** *analysis system*

Notice that **Setup, Solution**, and **Results** cells in the **Static Structural** analysis system have refresh and unfulfilled attached to them indicating that they need to be updated.

2. Double-click on the **Solution** cell of the **Static Structural** analysis system in the **Project Schematic** window; the **Mechanical** window for the **Static Structural** analysis system is displayed.

   Notice that the **Steady-State Thermal (A5)** and **Solution (A6)** nodes in the outline pane has green tick mark indicating that they are updated and imported from initial analysis system, as shown in Figure 11-71.

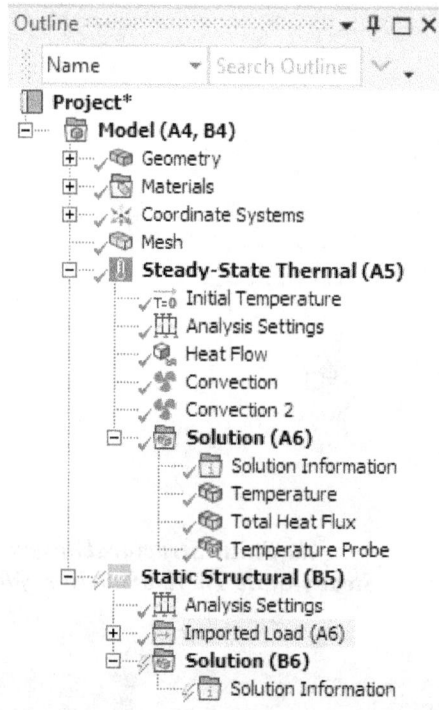

Figure 11-71 The **Outline** pane displaying
imported data from initial analysis system

3. Right-click on the **Static Structural (B5)** node in the **Outline** pane and then choose **Insert >
Fixed Support** from the shortcut menu displayed; **Fixed Support** with a question symbol is
added under the **Static Structural** node in the **Outline** pane. Also, the **Details of "Fixed
Support"** window is displayed.

4. In the **Details of "Fixed Support"** window, click on the **Geometry** cell to display the **Apply**
and **Cancel** buttons if not already displayed, as shown in Figure 11-72.

5. Select the base face of the model, as shown in Figure 11-73.

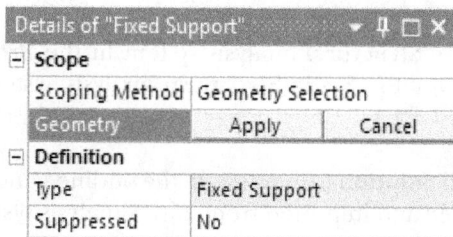

Figure 11-72 The **Details of "Fixed Support"**
window

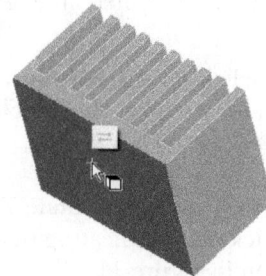

Figure 11-73 Selecting the base face of
the heat sink in the graphics screen

6. Next, choose the **Apply** button from the **Geometry** selection box in the **Details of "Fixed Support"** window; Fixed Support is applied to the selected face.

Solving the FE Model and Analyzing the Results

After the boundary conditions are specified for the model, you need to solve the analysis. After solving, you will get Equivalent Stress, Total Deformation and Fatigue Life, Damage, and Safety Factor of the model for thermal load input condition.

1. Select the **Solution** node in the **Outline** pane; the **Solution** contextual tab is displayed. Also, the **Details of "Solution (B6)"** window is displayed.

2. Choose the **Equivalent (von-Mises)** tool from the **Stress** drop-down in the **Solution** contextual tab.

3. Choose the **Total** tool from the **Deformation** drop-down of the **Solution** contextual tab; **Total Deformation** is added under the **Solution** node.

4. Choose the **Fatigue Tool** from the **Toolbox** drop-down in the **Solution** contextual tab; the **Fatigue Tool** node is attached to the **Solution** node in the **Outline** pane. Also, the **Details of "Fatigue Tool"** window is displayed.

> **Note**
> *The **Details of "Fatigue Tool"** window is displaying **Fully Reversed** from the **Type** drop-down under the **Loading** node by default. Similarly, **Stress Life** is chosen in the **Analysis Type** drop-down under **Options**, and the **None** option is chosen in the **Mean Stress Theory** drop-down list under the **Options** node by default.*

5. In the **Fatigue Tool** contextual tab, choose the **Life** tool from the **Contour** group, the **Life** is attached to the **Outline** pane. Also, the **Details of "Life"** window is displayed.

6. Choose the **Damage** tool from the **Contour** group, the **Damage** is attached to the **Outline** pane. Also, the **Details of "Damage"** window is displayed.

7. In this window, expand the **Definition** node if it is not already expanded. Next, select the **Design Life** selection box, and enter 10^6, as shown in Figure 11-74.

8. Choose the **Safety Factor** tool from the **Contour** group; the **Safety Factor** is attached to the **Outline** pane. Also, the **Details of "Safety Factor"** window is displayed.

9. In this window, expand the **Definition** node if it is not already expanded. Next, select the **Design Life** selection box, and enter 10^6, as shown in Figure 11-75.

Details of "Damage"	▾ ⊕ □ ×
Scope	
Scoping Method	Geometry Selection
Geometry	All Bodies
Definition	
Design Life	1.e+006 cycles
Type	Damage
Identifier	
Suppressed	No
Results	
☐ Maximum	
Maximum Occurs On	

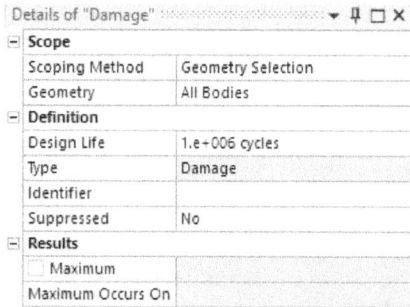

Figure 11-74 Displaying **Design Life** for the **Damage** tool

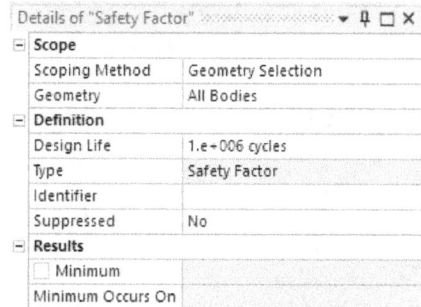

Details of "Safety Factor"	▾ ⊕ □ ×
Scope	
Scoping Method	Geometry Selection
Geometry	All Bodies
Definition	
Design Life	1.e+006 cycles
Type	Safety Factor
Identifier	
Suppressed	No
Results	
☐ Minimum	
Minimum Occurs On	

Figure 11-75 Displaying **Design Life** for the **Safety Factor** tool

10. Choose the **Solve** tool from the **Solution** contextual tab; the parameters are evaluated. Wait for sometime, the results are evaluated against the given boundary and loading conditions.

Figure 11-76 shows the model when the **Equivalent Stress** node is selected in the **Outline** pane and Figure 11-77 shows the model when the **Total Deformation** node is selected in the **Outline** pane.

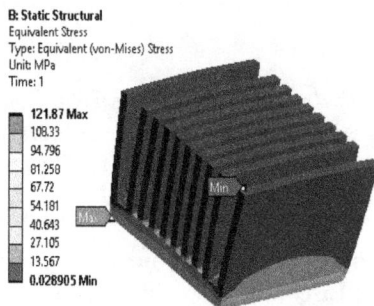

Figure 11-76 Equivalent stress of the model

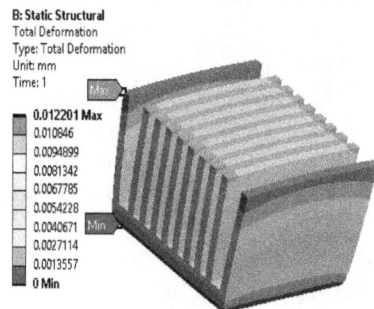

Figure 11-77 Total deformation of the model

Figure 11-78 shows the model when the **Life** node is selected in the **Outline** pane. Figure 11-79 shows the model when the **Damage** node is selected in the **Outline** pane and Figure 11-80 shows the model when the **Safety Factor** node is selected in the **Outline** pane.

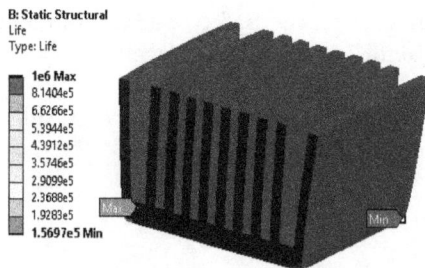

Figure 11-78 Life of the model

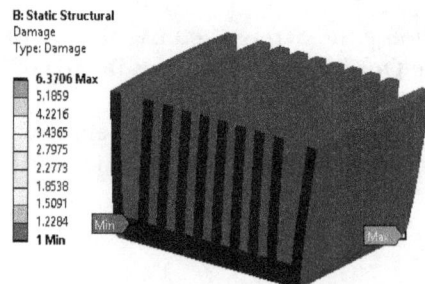

Figure 11-79 Damage of the model

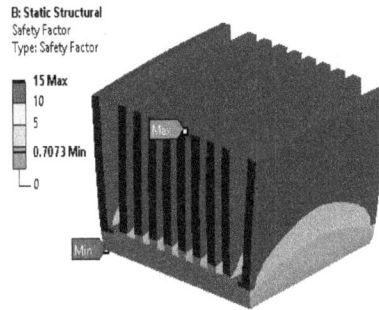

Figure 11-80 Safety factor of the model

The table displayed next shows all the values obtained from the analysis.

Parameter	Max. Value	Min. Value
Equivalent Stress	121.87 MPa	0.028905 MPa
Total Deformation	0.012201 mm	0 mm
Life	1e6 cycles	1.5697e5 cycles
Damage	6.3706	1
Safety Factor	15	0.7073

The above table shows that the maximum equivalent (von-mises) stress developed is 121.87 MPa indicating the maximum thermal stress in the heat sink for the given thermal conditions. Whereas, maximum total deformation of the model when it is provided Fixed support is 0.012201 mm.

Similarly, The minimum life of the model is 1.5697e5 cycles, maximum damage is 6.3706, and minimum safety factor is 0.7073. It indicates that the model will fail for 1e6 cycles design life.

Note
For fatigue analysis criteria, the minimum life, maximum damage, and minimum safety factor is of interest. As the design life is set to 1e6 cycles, all the fatigue tools such as Life, Damage, and Safety Factor are evaluated corresponds to 1e6 cycles.

11. Exit the **Mechanical** window to display the **Workbench** window.

Saving the Project and Exiting ANSYS Workbench

After evaluating all the results, you need to save the project.

1. Choose the **Save Project** button from the **Standard** toolbar; the model is saved with the name *c11_ansWB_tut03*.

2. Exit the **Workbench** window to close the ANSYS session.

Self-Evaluation Test

Answer the following questions and then compare them to those given at the end of this chapter:

1. To apply convection on any surface, you need to define the _____ in the **Details of "Convection"** window.

2. You can insert Temperature load by choosing the _____ tool in the **Environment** contextual toolbar.

3. The unit of heat flow is _____.

4. In ANSYS Workbench, only the Steady-State Thermal analysis system is used to carry out all types of thermal analyses. (T/F)

5. The Steady-State Thermal analysis system is used to carry out a thermal analysis where conditions do not change over time. (T/F)

6. In Thermal analysis, only thermal degrees of freedom of the elements are considered. (T/F)

7. You can insert the boundary conditions for a thermal analysis by using the tools available in the **Environment** contextual tab. (T/F)

8. A probe is used to evaluate results only along the edges of the geometry. (T/F)

9. You can apply different convection loads at different regions of the model. (T/F)

10. You can import various film coefficients in ANSYS Workbench. (T/F)

Review Questions

Answer the following questions:
1. By default, the _____ color in a legend represents the maximum value.

2. To insert a Temperature load, right-click on the analysis system node in the **Outline** pane and choose the **Temperature** option from the shortcut menu displayed.(T/F)

3. To view temperature distribution over a body or surface, right-click on the **Solution** node in the **Outline** pane and choose the **Temperature** option from the shortcut menu displayed. (T/F)

4. The minimum and maximum values obtained from an analysis are displayed in the corresponding legend and also in the corresponding **Details View** window. (T/F)

5. In ANSYS Workbench, you can carry thermal analysis for time-dependent loads as well. (T/F)

6. You can add the **Transient Thermal** analysis to the **Project Schematic** window from the **Analysis Systems** toolbox in the **Workbench** window. (T/F)

7. In Transient Thermal analysis, the loads are time-dependent. (T/F)

EXERCISE

Exercise 1

Download the *c11_ansWB_exr01.zip* file from *www.cadcim.com* and then extract it to save the *c11_ansWB_exr01.stp* file in the project folder. Next, run the Steady-State Thermal analysis on the model. Consider a hot fluid of film coefficient 1200 W/m² K at 450 °C is running inside the pipe. Find out the effect of thermal loading on the bracket attached to it. Figure 11-81 shows the model for Exercise 1.   **(Expected time: 45 min)**

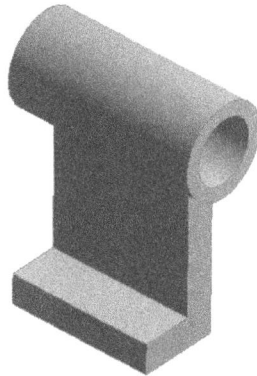

Figure 11-81 Model for Exercise 1

Answers to Self-Evaluation Test

1. Film Coefficient, 2. Temperature, 3. Watt, 4. F, 5. T, 6. T, 7. T, 8. F, 9. T, 10. T

Other Publications by CADCIM Technologies

The following is the list of some of the publications by CADCIM Technologies. Please visit *www.cadcim.com* for the complete listing.

AutoCAD Textbooks
- AutoCAD 2024: A Problem-Solving Approach, Basic and Intermediate, 30th Edition
- AutoCAD 2023: A Problem-Solving Approach, Basic and Intermediate, 29th Edition
- AutoCAD 2022: A Problem-Solving Approach, Basic and Intermediate, 28th Edition
- AutoCAD 2021: A Problem-Solving Approach, Basic and Intermediate, 27th Edition
- Advanced AutoCAD 2021: A Problem-Solving Approach (3D and Advanced), 25th Edition

Autodesk Inventor Textbooks
- Autodesk Inventor Professional 2024 for Designers, 24th Edition
- Autodesk Inventor Professional 2023 for Designers, 23rd Edition
- Autodesk Inventor Professional 2022 for Designers, 22nd Edition
- Autodesk Inventor Professional 2021 for Designers, 21st Edition

AutoCAD MEP Textbooks
- AutoCAD MEP 2023 for Designers, 7th Edition
- AutoCAD MEP 2022 for Designers, 6th Edition
- AutoCAD MEP 2020 for Designers, 5th Edition

AutoCAD Plant 3D Textbooks
- AutoCAD Plant 3D 2023 for Designers, 7th Edition
- AutoCAD Plant 3D 2021 for Designers, 6th Edition
- AutoCAD Plant 3D 2020 for Designers, 5th Edition

Autodesk Fusion 360 Textbook
- Autodesk Fusion 360: A Tutorial Approach, 4th Edition
- Autodesk Fusion 360: A Tutorial Approach, 3rd Edition

Solid Edge Textbooks
- Solid Edge 2023 for Designers, 20th Edition
- Solid Edge 2022 for Designers, 19th Edition
- Solid Edge 2021 for Designers, 18th Edition

NX Textbooks
- Siemens NX 2021 for Designers, 14th Edition
- Siemens NX 2020 for Designers, 13th Edition
- Siemens NX 2019 for Designers, 17th Edition

NX Mold Textbook
- Mold Design Using NX 11.0: A Tutorial Approach

NX Nastran Textbook
• NX Nastran 9.0 for Designers

SOLIDWORKS Textbooks
• SOLIDWORKS 2023 for Designers, 21st Edition
• Advanced SOLIDWORKS 2022 for Designers, 20th Edition
• SOLIDWORKS 2022: A Tutorial Approach, 6th Edition
• Learning SOLIDWORKS 2022: A Project Based Approach
• Advance SOLIDWORKS 2022 for Designers, 20th Edition

SOLIDWORKS Simulation Textbooks
• SOLIDWORKS Simulation 2022: A Tutorial Approach
• SOLIDWORKS Simulation 2018: A Tutorial Approach

CATIA Textbooks
• CATIA V5-6R2022 for Designers, 20th Edition
• CATIA V5-6R2021 for Designers, 19th Edition
• CATIA V5-6R2020 for Designers, 18th Edition

Creo Parametric Textbooks
• Creo Parametric 9.0 for Designers, 9th Edition
• Creo Parametric 8.0 for Designers, 8th Edition

ANSYS Textbooks
• ANSYS Workbench 2022 R1: A Tutorial Approach
• ANSYS Workbench 2021 R1: A Tutorial Approach

Creo Direct Textbook
• Creo Direct 2.0 and Beyond for Designers

Autodesk Alias Textbooks
• Learning Autodesk Alias Design 2016, 5th Edition
• Learning Autodesk Alias Design 2015, 4th Edition

AutoCAD LT Textbooks
• AutoCAD LT 2023 for Designers, 15th Edition
• AutoCAD LT 2022 for Designers, 14th Edition

EdgeCAM Textbooks
• EdgeCAM 11.0 for Manufacturers
• EdgeCAM 10.0 for Manufacturers

Autodesk Revit MEP Textbooks
• Exploring Autodesk Revit 2022 for MEP, 8th Edition
• Exploring Autodesk Revit 2021 for MEP, 7th Edition

AutoCAD Civil 3D Textbooks
- Exploring AutoCAD Civil 3D 2023, 12th Edition
- Exploring AutoCAD Civil 3D 2022, 11th Edition
- Exploring AutoCAD Civil 3D 2020, 10th Edition

AutoCAD Map 3D Textbooks
- Exploring AutoCAD Map 3D 2022, 9th Edition
- Exploring AutoCAD Map 3D 2018, 8th Edition
- Exploring AutoCAD Map 3D 2017, 7th Edition

RISA-3D Textbook
- Exploring RISA-3D 14.0

Autodesk Navisworks Textbooks
- Exploring Autodesk Navisworks 2022, 9th Edition
- Exploring Autodesk Navisworks 2021, 8th Edition
- Exploring Autodesk Navisworks 2020, 7th Edition

AutoCAD Raster Design Textbooks
- Exploring AutoCAD Raster Design 2017
- Exploring AutoCAD Raster Design 2016

Bentley STAAD.Pro Textbooks
- Exploring Bentley STAAD.Pro CONNECT Edition, 5th Edition
- Exploring Bentley STAAD.Pro CONNECT Edition, 4th Edition
- Exploring Bentley STAAD.Pro (CONNECT) Edition

Autodesk 3ds Max Design Textbooks
- Autodesk 3ds Max Design 2015: A Tutorial Approach, 15th Edition
- Autodesk 3ds Max Design 2014: A Tutorial Approach

Autodesk 3ds Max Textbooks
- Autodesk 3ds Max 2022 for Beginners: A Tutorial Approach, 22nd Edition
- Autodesk 3ds Max 2021: A Comprehensive Guide, 21st Edition
- Autodesk 3ds Max 2020: A Comprehensive Guide, 20th Edition

Autodesk Maya Textbooks
- Autodesk Maya 2022: A Comprehensive Guide, 13th Edition
- Autodesk Maya 2020: A Comprehensive Guide, 12th Edition

Pixologic ZBrush Textbooks
- Pixologic ZBrush 2022: A Comprehensive Guide, 8th Edition
- Pixologic ZBrush 2021: A Comprehensive Guide, 7th Edition
- Pixologic ZBrush 2020: A Comprehensive Guide, 6th Edition

Fusion Textbooks
- Blackmagic Design Fusion 7 Studio: A Tutorial Approach, 3rd Edition
- The eyeon Fusion 6.3: A Tutorial Approach

Flash Textbooks
- Adobe Flash Professional CC 2015: A Tutorial Approach, 3rd Edition
- Adobe Flash Professional CC: A Tutorial Approach

Computer Programming Textbooks
- Introducing PHP 7/MySQL
- Introduction to C++ programming, 2nd Edition
- Learning Oracle 12c - A PL/SQL Approach
- Learning ASP.NET AJAX
- Introduction to Java Programming, 2nd Edition
- Learning Visual Basic.NET 2008

MAXON CINEMA 4D Textbooks
- MAXON CINEMA 4D R20 Studio: A Tutorial Approach, 7th Edition
- MAXON CINEMA 4D R19 Studio: A Tutorial Approach, 6th Edition
- MAXON CINEMA 4D R18 Studio: A Tutorial Approach, 5th Edition

Oracle Primavera Textbooks
- Exploring Oracle Primavera P6 Professional 18, 3rd Edition
- Exploring Oracle Primavera P6 v8.4

AutoCAD Textbooks Authored by Prof. Sham Tickoo and Published by Autodesk Press
- AutoCAD: A Problem-Solving Approach: 2013 and Beyond
- AutoCAD 2012: A Problem-Solving Approach
- AutoCAD 2011: A Problem-Solving Approach
- AutoCAD 2010: A Problem-Solving Approach
- Customizing AutoCAD 2010

Coming Soon from CADCIM Technologies
- Flow Simulation Using SOLIDWORKS 2022

Online Training Program Offered by CADCIM Technologies
CADCIM Technologies provides effective and affordable virtual online training on animation, architecture, and GIS softwares, computer programming languages, and Computer Aided Design, Manufacturing, and Engineering (CAD/CAM/CAE) software packages. The training will be delivered 'live' via Internet at any time, any place, and at any pace to individuals, students of colleges, universities, and CAD/CAM/CAE training centers. For more information, please visit the following link: *https://www.cadcim.com*.

www.ingramcontent.com/pod-product-compliance
Lightning Source LLC
Chambersburg PA
CBHW081758200326
41597CB00023B/4067